SECURITY in DISTRIBUTED, GRID, MOBILE, and PERVASIVE COMPUTING

SECURITY in DISTRIBUTED, GRID, MOBILE, and PERVASIVE COMPUTING

Yang Xiao

Auerbach Publications
Taylor & Francis Group
Boca Raton   New York

Auerbach Publications is an imprint of the
Taylor & Francis Group, an informa business

Auerbach Publications
Taylor & Francis Group
6000 Broken Sound Parkway NW, Suite 300
Boca Raton, FL 33487-2742

© 2007 by Taylor & Francis Group, LLC
Auerbach is an imprint of Taylor & Francis Group, an Informa business

---

Library of Congress Cataloging-in-Publication Data

Xiao, Yang.
    Security in distributed, grid, mobile, and pervasive computing / Yang Xiao.
        p. cm.
    Includes bibliographical references and index.
    ISBN-13: 978-0-8493-7921-5 (alk. paper)
    ISBN-10: 0-8493-7921-0 (alk. paper)
    1. Computer security. I. Title.

QA76.9.A25X53 2007
005.8--dc22
                                            2006033967

Visit the Taylor & Francis Web site at
http://www.taylorandfrancis.com

and the Auerbach Web site at
http://www.auerbach-publications.com

Contents

Preface

Distributed computing, grid computing, mobile computing, and pervasive computing have been dramatically advanced in recent years with a proliferation of services and applications. However, security issues are extremely important since attacks and threats are expected, and security is still a major impediment to the further deployment of these services. Security mechanisms are essential to protect data integrity and confidentiality, access control, authentication, quality of service, user privacy, and continuity of service. They are also critical to protect basic functionality in distributed computing, GRID computing, mobile computing, and pervasive computing.

This book covers the comprehensive research topics in security in distributed computing, grid computing, mobile computing, and pervasive computing, which include key management and agreement, authentication, intrusion detection, false data detection, secure data aggregation, anonymity, privacy, access control, applications, standardization, etc. It can serve as a useful reference for researchers, educators, graduate students, and practitioners in the field of security in distributed computing, grid computing, mobile computing, and pervasive computing.

The book contains 16 chapters from prominent researchers working in this area around the world. It is organized along four themes (parts) in security issues for distributed computing, grid computing, mobile computing, and pervasive computing.

- *Part I: Security in Distributed Computing*: Chapter 1 by Bertino and Koglin reviews security issues and challenges in content distribution networks and present enforcement of content security. Chapter 2 by Giruka, Chakrabarti, and Singhal reviews key agreement protocols based on the Diffie-Hellman key exchange, and key management protocols for complex distributed systems like the Internet. Chapter 3 by Fernandez and Larrondo-Petrie discusses securing design patterns for distributed systems including middleware security, its components, implementation issues, general methodology, etc.

- *Part II: Security in Mobile Computing and Wireless Networks*: Chapters 3–9 focus on security in mobile computing and wireless networks. Chapter 4 by Bradford, Grizzell, Jay, and Jenkins gives a survey of security issues for constrained wireless networks with a focus on a discussion of pragmatic issues. Chapter 5 by Chakrabarti,

Giruka, and Singhal discusses wireless authentication methods including GSM, IEEE 802.11, and ad hoc networks. Chapter 6 by Amini, Mišić, and Mišić reviews intrusion detection in wireless sensor networks, as well as the main differences between wireless sensor networks and *ad hoc* networks, and outlines main challenges. Chapter 7 by Çam and Ozdemir reviews false data detection, data aggregation, secure data aggregation, and key establishment schemes for wireless sensor networks. Chapter 8 by Hong and Kong studies privacy issues and anonymous routing protocol for mobile ad hoc networks. Chapter 9 by Xiao, Kay, Zhang, Li, and Ji provides a survey of security issues in the IEEE 802.15.1 Bluetooth wireless personal area network.

- *Part III: Security in Grid Computing*: Chapters 10–14 discuss security in grid computing. Chapter 10 by Kostopoulos, Sklavos, and Koufopavlou gives a comprehensive security overview in grid computing. Chapter 11 by Chadwick describes authentication and authorization security mechanisms that protect grid-enabled resources. Chapter 12 by Vivas, Lopez, and Montenegro provides an overview of grid security fundamentals, standards, requirements, models, architecture, and use patterns. Chapter 13 by Joshi, Du, and Joshi focuses on access control specification and enforcement for the protection of resources and shared information in a grid. Chapter 14 by Stephens, Nair, and Abraham focuses on safety and security challenges for distributed computing grids.

- *Part IV: Security in Pervasive Computing*: Chapters 15 and 16 study the security in pervasive computing. Chapter 15 by Venkatasubramanian and Gupta presents an overview of security solutions for pervasive healthcare systems. Chapter 16 by Walters, Liang, Shi, and Chaudhary surveys wireless sensor network security.

Although the covered topics may not be an exhaustive representation of all the security issues in distributed computing, grid computing, mobile computing, and pervasive computing, they do represent a rich and useful sample of the strategies and contents.

This book has been made possible by the great efforts and contributions of many people. First of all, we would like to thank all the contributors for putting together excellent comprehensive and informative chapters. Second, we would like to thank the staff members of CRC Press, for putting this book together.

Finally, I would like to dedicate this book to my family.

Yang Xiao

About the Editor

Yang Xiao is currently with the Department of Computer Science at the University of Alabama. He worked at Micro Linear as a MAC (Medium Access Control) architect involving the IEEE 802.11 standard enhancement work before he joined the Department of Computer Science at the University of Memphis in 2002. Dr. Xiao is the director of the W^4-Net Lab, and was with CEIA (Center for Information Assurance) at the University of Memphis and is an IEEE senior member. He was a voting member of the IEEE 802.11 Working Group from 2001 to 2004. He currently serves as editor-in-chief for the *International Journal of Security and Networks (IJSN)* and for the *International Journal of Sensor Networks (IJSNet)*. He is an associate editor or is on editorial boards for the following refereed journals: *(Wiley) International Journal of Communication Systems, (Wiley) Wireless Communications and Mobile Computing (WCMC), EURASIP Journal on Wireless Communications and Networking, International Journal of Wireless and Mobile Computing*, and *Recent Patents on Engineering*. He serves as a guest editor for the *IEEE Network*; special issue on "Advances on Broadband Access Networks" in 2007; a guest editor for the *IEEE Wireless Communications* special issue on "Radio Resource Management and Protocol Engineering in Future Broadband and Wireless Networks" in 2006; a (lead) guest editor for the *International Journal of Security in Networks (IJSN)* special issue on "Security Issues in Sensor Networks" in 2005; a (lead) guest editor for the *EURASIP Journal on Wireless Communications and Networking* special issue on "Wireless Network Security" in 2005; a (sole) guest editor for the *(Elsevier) Computer Communications Journal* special issue on "Energy-Efficient Scheduling and MAC for Sensor Networks, WPANs, WLANs, and WMANs" in 2005; a (lead) guest editor for the *(Wiley) Journal of Wireless Communications and Mobile Computing* special issue on "Mobility, Paging and Quality of Service Management for Future Wireless Networks" in 2004; a (lead) guest editor for the *International Journal of Wireless and Mobile Computing* special issue on "Medium Access Control for WLANs, WPANs, Ad Hoc Networks, and Sensor Networks" in 2004; and an associate guest editor for *International Journal of High Performance Computing and Networking*, special issue on "Parallel and Distributed Computing, Applications and Technologies" in 2003. He serves as editor/co-editor for ten edited books: *WiMAX/MobileFi: Advanced Research and Technology, Security in Distributed and Networking Systems, Security in Distributed, Grid, and Pervasive Computing, Security in Sensor Networks, Wireless Network Security, Adaptation Techniques in Wireless Multimedia Networks, Wireless LANs and Bluetooth, Security and Routing in Wireless Networks, Ad Hoc and*

Sensor Networks, and *Design and Analysis of Wireless Networks*. He serves as a referee/reviewer for many funding agencies, as well as a panelist for the U.S. NSF and a member of Canada Foundation for Innovation (CFI)'s telecommunications expert committee. He serves as TPC for more than 80 conferences such as INFOCOM, ICDCS, ICC, GLOBECOM, WCNC, etc. His research areas are wireless networks, mobile computing, and network security. He has published more than 180 papers in major journals and refereed conference proceedings related to these research areas.

Contributors

Jacob A. Abraham
Computer Engineering Research
  Center
Department of Electrical and
  Computer Engineering
The University of Texas at Austin
Austin, TX
E-mail: jaa@cerc.utexas.edu

Fereshteh Amini
Department of Computer Science
University of Manitoba
Winnipeg, Manitoba, Canada
E-mail: amini@cs.umanitoba.ca

Elisa Bertino
CERIAS and Computer Science
  Department
Purdue University
West Lafayette, IN
E-mail:
  bertino@cerias.purdue.edu

Phillip G. Bradford
Computer Science Department
The University of Alabama
Tuscaloosa, AL
E-mail: pgb@cs.ua.edu

Hasan Çam
Department of Computer Science
  and Engineering
Arizona State University
Tempe, AZ
E-mail: hasan.cam@asu.edu

David W. Chadwick
Computing Laboratory
University of Kent, Canterbury, U.K.
E-mail:
  d.w.chadwick@kent.ac.uk

Saikat Chakrabarti
Computer Science Department
University of Kentucky
Lexington, KY
E-mail: schak2@cs.uky.edu

Vipin Chaudhary
Department of Computer Science
Wayne State University
E-mail: vipin@wayne.edu

Siqing Du
School of Information Sciences
University of Pittsburgh
Pittsburgh, PA
E-mail: sdu@mail.sis.pitt.edu

Eduardo B. Fernandez
Department of Computer Science
  and Engineering
Florida Atlantic University
Boca Raton, FL
E-mail: ed@cse.fau.edu

Venkata C. Giruka
Computer Science Department
University of Kentucky
Lexington, KY
E-mail: venkata@cs.uky.edu

Benjamin M. Grizzell
Computer Science Department
The University of Alabama
Tuscaloosa, AL
E-mail: bgrizzell@cs.ua.edu

Sandeep K. S. Gupta
Department of Computer Science
    and Engineering
Ira A. Fulton School of Engineering
Arizona State University
Tempe, AZ
E-mail: sandeep.gupta@asu.edu

Xiaoyan Hong
Computer Science Department
The University of Alabama
Tuscaloosa, AL
E-mail: hxy@cs.ua.edu

Graylin T. Jay
Computer Science Department
The University of Alabama
Tuscaloosa, AL
E-mail: aka1@bpcc.ua.edu

Janet Truitt Jenkins
Computer Science Department
University of North Alabama
Florence, AL
E-mail: jltruitt@una.edu

James B. D. Joshi
School of Information Sciences
University of Pittsburgh
Pittsburgh, PA
E-mail:
    jjoshi@mail.sis.pitt.edu

Saubhagya R. Joshi
School of Information Sciences
University of Pittsburgh
Pittsburgh, PA
E-mail:
    srjoshi@mail.sis.pitt.edu

Ji Jun
Information Engineering Department
Changchun Institute of Technology
Changchun City, Jilin Province
P.R. China

Daniel Kay
Department of Computer Science
University of Memphis
Memphis, TN

Yunhua Koglin
Computer Science Department
Purdue University
West Lafayette, IN
E-mail: luy@cs.purdue.edu

Jiejun Kong
Scalable Network Technologies, Inc.
Los Angeles, CA
E-mail: jkong@cs.ucla.edu

Giorgos Kostopoulos
Electrical and Computer Engineering
    Department
University of Patras
Patras, Greece
E-mail:
    gkostop@ee.upatras.gr

Odysseas Koufopavlou
Electrical and Computer Engineering
    Department
University of Patras
Patras, Greece
E-mail:
    odysseas@ee.upatras.gr

Maria M. Larrondo-Petrie
Department of Computer Science
    and Engineering
Florida Atlantic University
Boca Raton, FL
E-mail: maria@cse.fau.edu

Tianji Li
Hamilton Institute
National University of Ireland
Maynooth, Ireland
E-mail: tianji.li@nuim.ie

Zhengqiang Liang
Department of Computer Science
Wayne State University
Detroit, MI
E-mail: sean@wayne.edu

Javier Lopez
Departamento de Lenguajes y
    Ciencias de la Comunicacion
University of Malaga
Malaga, Spain
E-mail: jlm@lcc.uma.es

Jelena Mišić
Department of Computer Science
University of Manitoba
Winnipeg, Manitoba, Canada
E-mail: vmisic@cs.umanitoba.ca

Vojislav B. Mišić
Department of Computer Science
University of Manitoba
Winnipeg, Manitoba, Canada
E-mail: vmisic@cs.umanitoba.ca

Jose A. Montenegro
Departamento de Lenguajes y
    Ciencias de la Comunicacion
University of Malaga
Malaga, Spain
E-mail: monte@lcc.uma.es

V. S. Sukumaran Nair
High Assurance Computing and
    Networking (HACNet) Lab
Department of Computer Science
    and Engineering
Southern Methodist University
Dallas, TX
E-mail: nair@engr.smu.edu

Suat Ozdemir
Department of Computer Science
    and Engineering
Arizona State University
Tempe, AZ
E-mail: suat@asu.edu

Weisong Shi
Department of Computer Science
Wayne State University
Detroit, MI
E-mail: weisong@wayne.edu

Mukesh Singhal
Computer Science Department
University of Kentucky
Lexington, KY
E-mail: singhal@cs.uky.edu

Nicolas Sklavos
Electrical and Computer
    Engineering Department
University of Patras
Patras, Greece
E-mail: NSklavos@ieee.org

Mark Stephens
High Assurance Computing and
    Networking (HACNet) Lab
Department of Computer Science
    and Engineering
Southern Methodist University
Dallas, TX
E-mail: mark.stephens
    @verizonbusiness.com

Krishna Venkatasubramanian
Department of Computer Science
    and Engineering
Ira A. Fulton School of Engineering
Arizona State University
Tempe, AZ
E-mail: kkv@asu.edu

Jose L. Vivas
Departamento de Lenguajes y
  Ciencias de la Comunicacion
University of Malaga
Malaga, Spain
E-mail: jlvivas@lcc.uma.es

John Paul Walters
Department of Computer Science
Wayne State University
Detroit, MI
E-mail: jwalters@wayne.edu

Yang Xiao
Department of Computer Science
The University of Alabama
Tuscaloosa, AL
E-mail: yangxiao@ieee.org

Yan Zhang
Simula Research Laboratory
Oslo, Norway
E-mail: yanzhang@ieee.org

Part I

Security in Distributed Computing

1

Security for Content Distribution Networks — Concepts, Systems and Research Issues

Elisa Bertino and Yunhua Koglin

CONTENTS

Abstract   Previous research on content distribution networks (CDNs) mainly focuses on improving system performance by deploying replication such that latency for data access could be reduced and bandwidth could be saved, especially when dealing with large amounts of data. Centrally-managed, trusted replicas are important characters in these traditional CDNs.

However, there is not enough attention given to the security of data in CDNs, even though data security is a crucial need for most Internet-based applications. Moreover, with the emergence of various network appliances and heterogeneous client environments, intermediaries are used for dynamic content delivery. Enforcing data security in such environments is more challenging than the traditional CDNs (client-server communication). Besides, new systems (such as publish/subscribe systems, peer-to-peer content distribution systems) are developed to meet different requirements of content distribution. Different mechanisms should be used in different systems to ensure content security.

In this chapter, we first review the security concepts related to CDNs and then present several systems, focusing on how they enforce content security. Finally, we discuss the other challenges in CDNs.

1.1 Introduction

Content distribution networks (CDNs) are all those applications that support data dissemination, searching, and retrieval. With the widespread use of Internet, CDNs have been studied extensively [1, 2, 3, 4, 5, 6, 7, 8, 9, 10, 11, 12, 13, 14, 15, 16, 17, 18, 19, 20, 21]. Most previous research focuses on enhancing performance of CDNs by replication. Different mechanisms (such as [22, 23, 24, 25, 26]) are used to deploy content replication on *trusted* cache proxies scattered around the Internet. When receiving a client request, instead of asking a content server for the requested contents, a proxy first checks if these contents are locally cached. Only when the requested contents are not cached or out of date are the contents transferred from the content server to the clients. If there is a cache hit, network bandwidth consumption can be reduced. A cache hit also reduces access latency for clients. System performance therefore improves, especially when large amounts of data are involved. Besides these improvements, caching makes the system robust by letting caching proxies provide content distribution services when the server is down or the network is congested.

Secure content distribution has received more attention from both academia and industry than before, due to the increasing emphasis on security in many applications. Ensuring content security in distributed environments is challenging. For example, content may be easily modified or accessed when it is transmitted across the Internet; a compromised replica may violate access control of content or damage integrity by maliciously modifying the content.

Different kinds of systems have been developed recently in order to meet the new requirements of content distribution. For example, with the emergence of various network appliances and heterogeneous client environments, content-aware systems are developed that involve intermediaries to transform content; publish/subscribe systems are developed to distribute content where publishers do not need to know the addresses of subscribers. These systems are different from the traditional client-server communication. They have different service requirements and different security challenges.

In this chapter, we first introduce the concepts of security related to CDNs, then present several systems, with focus on their security mechanisms. For each kind of these systems, we present its current research. Finally, we discuss some other research issues in CDNs.

1.2   Security Concepts

In this section, we briefly review some security concepts that are related to CDNs.

For any systems designed with security among its goals, a detailed security policy should exist.

Definition 1.1
A security policy is a statement of what is, and what is not, allowed ([27]).

Definition 1.2
An access control policy states the privileges of principals or users over content and services under certain conditions.

Security policies could be represented in high-level languages in which policy constraints are expressed abstractly, or low-level languages in which policy constraints are expressed in terms of program options, input, or specific characteristics of entities on system ([27]). Policies should be expressed precisely and unambiguously.

After specifying the security policies, a mechanism is chosen to enforce these policies.

Definition 1.3
A security mechanism is a method, tool, or procedure for enforcing a security policy ([27]).

In general, the security of content distribution systems is measured by how it supports data confidentiality, data integrity, and system availability.

Definition 1.4
Confidentiality is the assurance that content is shared only among authorized subjects.

Definition 1.5
Integrity is the assurance that the information is authentic and complete.

Definition 1.6
Availability is the assurance that the system which is responsible for dissemination, storing, and processing information is accessible when needed by those who need these services.

Both confidentiality and integrity are defined by *access control policies*.

In the next section, we will review some access control models that describe how the access policies for content are generated.

1.3   Access Control Models

In CDNs, an access control model specifies who is allowed to perform what kinds of operations on content under certain conditions. The following types of access control model are commonly used:

- *Discretionary access control (DAC)*: Access policy is completely determined by the owner of the content. The owner decides who is allowed to access the data and with what privileges (such as *read, write*, etc.).

  This type of access control has been widely used, even beyond CDNs. For example, Alice creates a file called temp.c. She can specify which subjects may access it and with what type of access (such as read or write). An access control list is normally used to make access decisions. Users usually present credentials (such as login and password) for authentication.

- *Mandatory access control (MAC)*: Access policy is determined by the system, not the owner of the content. In such a system, subjects receive a clearance label and objects (data) receive a classification label, also referred to as security level. A subject cannot read anything up, which means that a subject cannot read any objects that have labels higher than the subject's clearance. Moreover, a subject cannot write anything down, which means that a subject cannot write to objects or create new objects with lower security labels than the subject's clearance. This prevents subjects from sharing secrets with subjects with a lower security label, keeping information confidential.

  Note in MAC, only administrators can change the security labels of data. Data owners cannot make such a change.

  MAC is often used in systems that process highly sensitive data with confidentiality as the highest priority, such as classified government and military information. The original MAC model [28] (also called Bell-LaPadula model) was later expanded to Multi-Level Security (MLS), which handles multiple classification levels (i.e., "top secret," "secret," "confidential," and "unclassified") between subjects and objects.

- *Role-Based Access Control (RBAC)*: Access is dependent on functionality, not identity. In RBAC models ([29, 30, 31, 32, 33, 34, 35]), an administrator defines a series of roles that are created for various job functions. The permissions to perform certain operations are assigned to specific roles. An administrator assigns members of staff (or users) some roles, and through those roles members (or users) acquire the permissions to perform particular functions.

RBAC can save an administrator from the tedious job of defining permissions per user within an organization.

When defining an RBAC model, it normally includes the following relations:

- $UA \subseteq U \times R$ User-role assignment (a many-to-many mapping)
- $PA \subseteq P \times R$ Permission-role assignment (a many-to-many mapping)
- $RH \subseteq R \times R$ Partially ordered role hierarchy

where U = User, R = Role, P = Permissions

Moreover, a RBAC model normally includes a set of sessions (SESSIONS) where each session is a mapping between a user and an activated subset of roles that are assigned to the user. Such a model may also include function *session_roles* that returns the roles activated by the session and the function *user_sessions* that returns the set of sessions that are associated with a user.

A RBAC model may also have other features such as: 1) roles are granted permissions based on the principle of least privilege; 2) roles are determined with a separation of duties; 3) roles are activated statically or dynamically.

Some other access control models include:

- *Originator Controlled Access Control (ORCON)*: The originator (subjects or organizations who create data) controls data access. Note that the originator may not be the data owner. ORCON is a combination of MAC and DAC ([27]).
- *Rule-Based Access Control model*: This is sometimes referred to as Rule-Based Role-Based Access Control (RB-RBAC). It includes mechanisms that dynamically assign roles to subjects based on their attributes and a set of rules defined by a security policy ([36]).

1.4 Systems

In this section, we present several types of systems in CDNs, focusing on the current research in these systems.

1.4.1 Secure Distributed File Systems

One important application in CDNs is file distribution. Instead of storing files on the machines owned by the data owners, some owners put their data in a data server, which is responsible for distributing the data according to the access control policies related to the data. This approach not only removes the space requirement for the data owners, but also makes the data distribution scalable.

Most previous file distribution approaches are based on the assumptions that the data servers are trusted: They keep the confidentiality and integrity of the data, and they enforce the access control policies related to the data. However, these assumptions are hard to prove true. In the following text, we present some current research on distributed file systems that removes these assumptions.

Current Research

Current research on distributed file systems with untrusted data servers includes the following aspects:

- **Cryptographic access control**. Harrington and Jensen propose a cryptographic access control mechanism in [37]. Files are encrypted and stored on an untrusted server. Access control is enforced by distributing symmetric keys that are used for encrypt/decrypting files. Integrity of the files can be verified by the server with signature verification, even though the server may not access the file content. The files are maintained with modifications recorded in a log.

  The above approach provides a nice solution that gets rid of a centralized reference monitor, such that the server does not need to maintain an access control list for the file and enforces this access control policy. Users can read the log that is signed by the data owner with timestamps or version numbers.

- **Supporting operations on encrypted data:** Moving the computation to the data server that stores only encrypted data seems very difficult; the data server should perform the computation without decrypting the data. Song and others [38] propose a practical technique for searching on encrypted data. Their solution supports the following:

  - *Provable Secrecy:* The untrusted server cannot learn anything about the plaintext given only the ciphertext.
  - *Controlled Searching:* The untrusted server cannot search for a word without the user's authorization.
  - *Hidden Queries:* The user may ask the untrusted server to search for a secret word without revealing the word to the server.
  - *Query Isolation:* The untrusted server learns nothing more than the search result about the plaintext.

  Before presenting the protocol, we first introduce the notations it uses. If $f : \mathcal{K} \times \mathcal{X} \rightarrow \mathcal{Y}$ represents a pseudorandom function or permutation, then $f_k(x)$ is the result of applying f to input x with key $k \in \mathcal{K}$. $\langle x, y \rangle$ means concatenation of x and y.

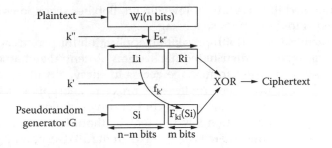

FIGURE 1.1
Encryption scheme (from [38]).

The protocol [38] has the following components:

1. *Storing data on the untrusted servers:* For each block W_i which has the fixed length of n, Alice gets the pseudorandom value S_i ($n-m$ bits long) from the pseudorandom generation G. Alice computes the ciphertext to be stored for W_i as $C_i = E_{k''}(W_i) \oplus \langle S_i, F_{k_i}(S_i) \rangle$ where $k_i = f_{k'}(L_i)$ and $E_{k''}(W_i) = \langle L_i, R_i \rangle$. L_i (respectively, R_i) denotes the first $n-m$ bits (respectively, the last m bits) of $E_{k''}(W_i)$. At the end, Alice keeps k', k'', and S_i, and sends the ciphertext to Bob (untrusted server) who stores the ciphertext. Figure 1.1 shows the encryption steps.

2. *Search operations:* To search the positions for word W_j, Alice sends Bob $X_j = E_{k''}(W_j) = \langle L_j, R_j \rangle$ and $k_j = f_{k'}(L_j)$. Bob performs a sequential scan on the encrypted data and returns $\langle p, C_p \rangle$ if $C_p \oplus X_j = \langle S_p, S'_p \rangle$ and $S'_p = F_{k_j}(S_p)$. In the returns, p denotes the position of the word. Note that there is small chance that some answers returned by Bob are garbage. This is due to the encryption collision.

3. *Retrieval operations:* To retrieve the data stored at position p, Alice sends Bob p. After Bob returns the ciphertext C_p at position p, Alice recalculates W_p by $C_p = \langle C_{p,l}, C_{p,r} \rangle$ where $C_{p,l}$ (respectively, $C_{p,r}$) denotes the first $n-m$ bits (respectively, the last m bits) of C_p, $X_{p,l} = C_{p,l} \oplus S_p$, $k_p = f_{k'}(X_{p,l})$, $T_p = \langle S_p, F_{k_p}(S_p) \rangle$, and finally, $W_p = D_{k''}(C_p \oplus T_p)$.

From the above description, we can see that each query takes one round of interaction and Bob performs one sequential scan on the ciphertext per query.

- *Proxy Re-encryption*
  In 1998, Blaze, Bleumer, and Strauss (BBS) [39] proposed an application called atomic proxy re-encryption, in which a semitrusted proxy converts a ciphertext for Alice into a ciphertext for Bob without seeing the underlying plaintext. This strategy is useful when Alice would like temporally to let Bob check the messages that are

addressed to her, without revealing to Bob her secret keys that are needed to decrypt these messages.

Ateniese et al. ([40]) present an application for proxy cryptography in securing distributed file systems. A centralized access control server is used to manage access to encrypted files stored on distributed, untrusted replicas. A proxy re-encryption scheme is proposed such that the access control server could re-encrypt the appropriate decryption key to clients without learning the key in the process. Thus, there is no need to grant full decryption rights to the access control server.

- *Byzantine fault tolerance*
  Besides using replication to increase content availability, other research focuses on byzantine fault tolerance. There are two types of system failure: fail-stop, which means data servers simply do not reply to clients' requests, and malicious failure, which means the data servers may behave arbitrarily; that is, they may reply with the wrong information to clients' requests.

  Castrol and Liskov ([41]) propose an approach that tolerates byzantine fault in *asynchronous* systems like the Internet. Their solution ensures that the system that includes a set of replicas performing deterministic services could survive byzantine faults. Moreover, their solution guarantees safety and liveness. In the system, a client sends the request for an operation to the primary of the replicas. The primary then multicasts the request to the other replicas, which then execute the request and send a reply to the client. After the client receives $f + 1$ replies from different replicas with the same conclusion, this is the result of the operation. The algorithm performed by replicas only requires five rounds of messages.

  The protocol in [41] has the following steps[1]:

  1. **Request:** Client c sends a request message $m = \langle REQUEST, o, t, c \rangle_{\sigma c}$ to the primary p, where o=operation, t=monotonic timestamp.

  2. **Preprepare:** Primary p assigns sequence number n to m and sends a message $\langle PRE\text{-}PREPARE, v, n, m \rangle_{\sigma p}$ to other replicas where v=current view.

  3. **Prepare:** If replica i accepts the message from p, it sends $\langle PREPARE, v, n, d, i \rangle_{\sigma i}$ to all other replicas, where d is the hash of the request m from client c. This indicates that i agrees to assign n to m in v.

  4. **Commit:** When replica i has a *PREPREPARE* and $2f + 1$ matching *PREPARE* messages, it sends $\langle COMMIT, v, n, d, i \rangle_{\sigma i}$ to all other replicas. At this point, correct replicas agree on an order of requests within a view.

---

[1] Message m signed by node i is denoted as $\langle m \rangle_{\sigma i}$.

5. **Reply:** Once replica i has $2f + 1$ matching $PREPARE$ and $COMMIT$ messages, it executes m, and sends to client c a message $\langle REPLY, v, t, c, i, r \rangle_{\sigma i}$ where r is the result of the operation.

The above approach requires at least $3f + 1$ replicas, where f is the max number of faulty replicas. It can tolerate malicious clients. Number of optimizations are described in [41] in order to have the proposed approach perform well in real systems.

1.4.2 Publish/Subscribe Systems

Publish/Subscribe (pub/sub) systems provide a new distributed paradigm for content dissemination. In such systems, a publisher publishes an event (or message) through a broker (also referred to as an event dispatcher). Subscribers specify their interests by registering with a broker. Brokers form a network in which they forward events to each other and, when needed, deliver events to subscribers that have registered with them. One major advantage of these systems is scalability: A publisher does not need to maintain subscription information, which may be changed dynamically, and a subscriber does not need to know which publishers may publish events of interest. Since there are no explicit destination addresses associated with an event, brokers are responsible for delivering each event to subscribers whose subscriptions are satisfied by the event, which is called event *matching*.

Figure 1.2 presents a general structure of pub/sub systems. Decoupling publishers from subscribers makes pub/sub systems scalable and powerful.

Basically, there are two types of pub/sub systems. The first, referred to as *subject-based* or *type-based* pub/sub, is a system in which events are labeled with predefined subjects (or types) to which subscribers may subscribe.

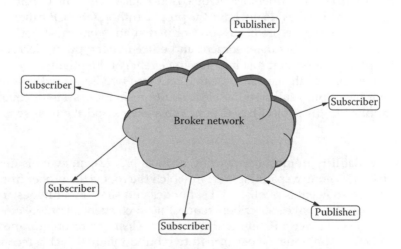

FIGURE 1.2
A general pub/sub system structure.

Since subscribers interested in a particular subject (or type) may be managed as a group, multicasting is an efficient method for event delivery in this kind of pub/sub system. The second one, referred to as *content-based pub/sub* system, is more flexible and powerful than the previous one. In this kind of system, both subscriptions and content are specified with respect to a set of *attributes*. Each attribute is an ordered pair of *name* and *type*. An attribute value is the type of the attribute. A subscriber subscribes to events by specifying *predicates* against attributes. For example, a classic schema used for a stock trade pub/sub system is (*company: string, price: integer, shares: integer*), a subscription could be: *(price < 20) AND company = "IBM."*

Current Research

Two important security issues for content-based pub/sub systems are *availability* and *confidentiality*. Availability in such a system means that after a subscriber registers with a broker by specifying subscription predicates, if the subscription request is accepted, the broker is responsible for delivering any event that satisfies such predicates to the subscriber in a timely manner. Furthermore, any broker failure along the event delivery path should not prevent the subscriber from receiving this event. This requirement means that if there is a broker failure, either the event delivery route should be reconfigured dynamically or multiple event delivery routes should be established at the beginning in such systems. For a large pub/sub system, it takes time to propagate subscription information into the whole network, therefore it may take a while for a subscriber to receive events. As a result, the subscriber may miss some matching events, even though these events are published after the registration. It is thus important that the *registration information propagating time (RIP time)* be minimized.

Confidentiality in pub/sub systems means that events should be available only to authorized subscribers. Malicious users must be prevented from reading events for which they do not have the proper authorization. Furthermore, even subscribers whose predicates do not match an event must not access the event. Therefore, key management and efficient encryption/decryption schemes play an important role in enforcing event confidentiality.

Next, we describe the current research on these two issues in the context of content-based pub/sub systems, since solutions in these systems could be easily applied to the subject-based pub/sub systems and the reverse is not true.

- **Availability** In most approaches, such as [42, 43], an event is distributed along a spanning tree, in which the root is the broker from which an event is published. Leaf nodes and some inner nodes are brokers that have subscribers requesting such events. If a broker at the root of a tree fails, either the events are lost or a reconfiguration (such as [44]) must be performed to rebuild the tree. Such a reconfiguration can be very expensive when pub/sub systems are large and a number of brokers are involved. Maintaining a tree structure for event forwarding also requires each broker replicate the whole network's subscription information. Therefore, the RIP time delay could be large and could further increase if brokers perform computations in order to minimize routing table information.

A simple approach to increase availability is to let each broker broadcast each event it receives. However, such an approach has the major disadvantage of resulting in system floods. Carzaniga et al. ([43]) propose an approach that broadcasts events only along the spanning tree, therefore, some unnecessary event broadcast could be avoided and event availability could be improved.

Srivatsa and Liu ([45]) propose a resilient network, which, instead of providing only a single path from each publisher to its subscribers, which is inherited from the spanning tree structure, several independent paths from a publisher to each of its subscribers are provided. These paths are built in a deterministic way. In their approach, building several independent paths from a publisher to every subscriber involves complex topology computations. In dynamic environments where subscriptions or unsubscriptions occur quite frequently, such computation is expensive.

Other approaches to improve availability such as multicasting an event by the broker that publishes the event also requires replicating subscription information at each broker. Besides the long time of RIP delay, broker space requirement is another challenge for the multicast approach in large scale pub/sub systems. This approach also causes the load unbalance, as some brokers where events are published frequently are overloaded, while other brokers that do not have an event published are idle.

- **Confidentiality** An event should be encrypted when it is delivered to subscribers, so that only authorized subscribers are able to decrypt it. Usually, a group key shared by the group members and the brokers is used to encrypt the event. However, since there could be many attributes and therefore a large number of complex predicates, for n subscribers, there are possibly 2^n subscription groups that may be interested in an event. Encrypting the event with a group key therefore could result in significant performance costs. Moreover, different events may be of interest to different sets of groups. In large-scale content-based pub/sub systems where the volumes of published events are huge, inefficiency may undermine availability.

  Opyrchal and Prakash [46] discuss how a broker can encrypt an event and deliver it to a possibly very large number of groups. As each group has a secret key shared by members and brokers, encrypting the event using a group key may involve performing many encryption operations, and there may be several groups to which this event should be delivered. Caching and clustering are therefore used to make fewer encryptions for an event.

Security issues in content-based pub/sub systems have not been so widely investigated. More detailed discussion on these issues can be found in [47].

1.4.3   Content-Aware Intermediary Transforming Systems

With the emergence of various network appliances and heterogeneous client environments, besides caching, there are other new requirements for content services by intermediaries [6, 7]. For example, content from the server needs to be transformed in order to adapt to the requirements of a client's security policy, device capabilities, preferences, and so forth. Several *content services* have been identified that include, but are not limited to, content transcoding [6, 7, 8, 13], in which data is transformed from one format into another, data filtering and value-added services, such as watermarking [10].

Intermediaries providing content services can be placed at the clients' end, at the servers' end or between them [12, 26]. Placing intermediaries at the client's end may not always be possible because of resource limitations. Because of these limitations, it is not possible to execute certain computation intensive transcoding functions at the clients' end. Placing intermediaries at the servers' end may result in reduced sharing. It is difficult to have one version of some content that satisfies diverse requests from clients. Placing intermediaries between clients and servers provides a better solution for content services.

Current Research

Though a lot of research on intermediary content service has been carried out [6, 7, 8, 13], there is not enough research on data security in this context. The approaches provided for securely transferring data from server to clients are not suitable when data are to be transformed by intermediaries. When a proxy mediates data transmission, if the data are completely enciphered during transmission, security is ensured; however, it is impossible for intermediaries to modify the data. It is difficult to enforce security when intermediaries are allowed to modify the data. Next, we list several research topics in this area:

- *SSL Splitting:* SSL splitting ([48]) is a technique that supports data integrity from untrusted caches. Upon receiving a request from clients, a proxy gets the data from caches and the Message Authentication Code (MAC) from the data server. Then, the proxy re-encrypts the merged data with the key shared by the server and the client and sends the encrypted data to the client. SSL splitting does not support data confidentiality, as the proxy has to access the key. The primary advantage of SSL splitting is that it reduces the bandwidth load of the data server.

- *Data Integrity Service Model:* Chi and Wu propose a Data Integrity Service Model (DISM) in [9]. In this model, integrity of intermediaries is enforced by using metadata expressing modification policies of content owners. However, in DISM everyone can access the data.

Thus confidentiality is violated. Another problem with DISM is the lack of efficiency. In several applications, such as multimedia content adaptation [6], efficiency is a vital factor.

- *JPSEC:* Wee and Apostolopoulos [19] present encryption methods and signaling syntax for JPEG-2000 images that allow an intermediary to transcode a JPEG-2000 codestream (JPSEC) without decryption. After unlocking the transcoded JPSEC, the transcoded JPEG2000 can be decoded to get the transcoded image. Moreover, an end user can verify that the transcoding operation was performed in a valid and permissible way.

1.4.4   Peer-to-Peer Content Distribution Systems

Peer-to-peer systems are characterized by the direct sharing of computer resources (such as content, storage, or CPU), rather than requiring the intermediation or support of a centralized authority.

Current Research

Many distributed file systems have been developed in peer-to-peer networks ([49, 50, 51, 52, 53, 54, 55]). These systems (such as Napster [49], Gnutella [50], and Freenet [51]) demonstrate a lot of benefit for content distribution. These benefits include node self-organization, load balance, fault tolerance, and scalability. Due to the lack of centralized administration and management, it is hard to ensure security in such environments. Androutsellis and Spinellis [56] have an extensive survey on the peer-to-peer content distribution technologies. Therefore, we will omit the discussion on the security issues in these systems. Interested readers are encouraged to read [56].

1.4.5   Collaborative Data Access and Updates Systems

The widespread use of the Internet for exchanging and managing data has pushed the need for techniques and mechanisms that secure information when it flows across the net. When several parties collaboratively perform certain transactions, each party needs to retrieve content and then perform certain authorized operations on it. Integrity and confidentiality have to be ensured for the data that flow among these parties.

Current Research

Several issues need to be addressed to support decentralized and cooperative document updates over the Web. A first requirement, which was investigated in [57], is the development of a high-level language for the specification of *flow policies*, that is, policies regulating the set of subjects that must receive a document during the update process, and *access control policies*. Starting from these policies, a server can determine the path that the document must follow and the privileges of each receiver. The second requirement is the development of an infrastructure and related algorithms to enforce confidentiality and integrity during the process of distributed and collaborative document updates.

- *Author-χ System:* This Java-based system ([58, 59, 60]) supports selective, secure, and distributed dissemination of XML documents. Specifically, Author-χ supports
  - the specification of security policies at varying granularity levels
  - the specification of user credentials
  - content-based access control
  - controlled release of XML documents according to the push-and-pull dissemination modes
  - document updates

  The system includes three Java server components: 1) X-Admin, 2) X-Access, and 3) X-Update.

  X-Admin component provides functions for administrative operations. Through this component, security administrators manage security policies, XML documents, subjects and credentials.

  X-Access component consists of two subcomponents: X-Push and X-Pull. X-Push supports document broadcast to clients at the server site. X-Pull supports the selective documents distribution upon clients' requests. All these kinds of distribution follow the policies stored in Policy Base ($\mathcal{PB}$).

  X-Update component manages the collaborative and distributed document update that we will describe later.

  Author-χ also includes X-bases repositories that consist of the following:
  - Policy Base ($\mathcal{PB}$) that contains the security policies for the XML documents and DTDs
  - Credential Base ($\mathcal{CB}$) that contains the user credentials and credential types
  - Encrypted Document Base ($\mathcal{EDB}$) that contains encrypted copy of portions of the documents in XML source
  - Authoring Certificate Base ($\mathcal{ACB}$) that contains generated certificates
  - Management Information Base ($\mathcal{MIB}$) that contains information required for updating process
  - XML Source
  - Push

  Next, we describe some protocols that are used to implement the X-Update component.

- *Self-Certifying Document Updates:* One important feature of these protocols that are used to implement the X-Update component is that the document integrity can be verified by each receiver. Before presenting these implementations, we first introduce the following notations for XML ([61, 62]) documents. These notations are used to enforce access control.

Definition 1.7
In an XML document, an atomic element (AE) is either an attribute or an element that includes the starting and ending tags of the element ([17]).

Definition 1.8
In an XML document, an atomic region (AR) includes a set of atomic elements. It is the smallest data portion to which the same access control policies apply ([17]).

Each atomic region is identified by an identifier. Therefore, an XML document could be divided into a set of atomic regions such that atomic elements of the same region are distinct and there is no atomic element that belongs to two different regions.

A region can be either *modifiable* or *nonmodifiable*. A region is nonmodifiable by a subject if this subject can only read it. A region is modifiable by a subject if this subject possesses the authorization to modify it, according to the access control policies.

Definition 1.9
In an XML document, a region object O is an instance of the information in a region R. A region object is associated with the region identifier R, the subject who authors it, and the time when the subject authors it.

Bertino et al. ([58, 63] propose a self-certifying document updating protocol in distributed systems. In their approach, the document is encrypted by the document server with the minimum number of keys such that different keys are used for encrypting different portions (a set of ARs) of the same document. Each participating subject receives only these keys for the portions that it is authorized to access from the document server. The encrypted document then circulates in sequence among the participating subjects.

When a subject receives the document, it could verify the correctness of the operations performed so far on the document, based on the control information the subject received from the document server. If there is no error, the subject can exercise its privileges on the document, sign these updates with its signature, then encrypt the portions it accessed, and send the encrypted document to the next subject. Only when the document fails the integrity check, a subject contacts the document server for document recovery.

A major limitation of this approach is that it does not exploit the possible parallelism that is inherent in data relationships and in the access control policies. Koglin et al. ([17]) propose an approach based on the use of a security region-object parallel flow (S-RPF) graph protocol. S-RPF graph protocol allows different users to simultaneously update different regions of the same document, according to the specified access control policies.

In an S-RPF graph, each node represents a subject in the flow path. An edge with label L from node i to node j denotes that there are region objects L sending from i to j. The S-RPF graph that the document server generated has the following properties:

- If no participating subject has access privilege to a region with the identifier of R, then no region object O associated with R will appear in the S-RPF graph.
- If a region object is modified by a subject *subj*, then this region object will not flow out from *subj* and a new region object with the same region identifier will start at *subj*.
- The same region object may be accessed by several subjects at the same time.
- The flow of each region object among the subjects is acyclic. This means that no region object flows back to the subject who authored it.
- If no subject has update rights on a region R, but there is at least one subject that has access privilege to this region, then a region object O associated with R starts its flow among the subjects from the document server and its author is the document server.

The S-RPF protocol is secure with respect to confidentiality and integrity. The proofs can be found in [17].

In all these mentioned approaches, the data server is not the bottleneck during the updating process. However, these approaches are not scalable. The data server has to perform some initial computation before the updating process starts. Furthermore, each participating subject is predefined. They cannot be changed once the updating starts. Also, these participants need to receive some control information from the data server in order to perform integrity checking of the document.

Further research in this area includes using roles to make the solution scalable. Moreover, the document server has too much control on the updated document, mechanisms should be proposed to enforce the principles such as separation of duty and least privilege.

1.5   Other Research Issues

Privacy preserving in content distribution networks is one important research area for study ([64]). Most research (such as [51]) in this area is on the techniques for supporting anonymity such that users could anonymously publish or retrieve various kinds of information; furthermore, the transaction between data servers and clients should be unlinkable.

Research on censorship-resistant document publishing (e.g., [51, 65, 66]) also demands further study. In these systems, the content stored on and distributed by servers should be free of censoring. Peer-to-peer systems are one promising area for such study, since they do not have a centralized administration.

Other research issues in CDNs include location-based access control ([67]). Different mechanisms are needed to ensure that content could be accessed only within certain locations. Therefore, precise location verification techniques are important to enforce this kind of access control.

Bibliography

1. M. Srivatsa and L. Liu. Securing publish-subscribe overlay services with eventguard. In *Proceedings of the 12th ACM Conference on Computer and Communications Security (CCS '05)*, 2005.
2. Yanlei Diao, Shariq Rizvi, and Michael J. Franklin. Towards an internet-scale XML dissemination service. In *VLDB Conference*, August 2004.
3. Fengyun Cao and Jaswinder Pal Singh. Efficient event routing in content-based publish-subscribe service networks. In *Proceedings of IEEE INFOCOM '04*, 2004.
4. Michael J. Freedman, Eric Freudenthal, and David Mazires. Democratizing content publication with coral. In *Proceedings of the USENIX/ACM Symposium on Networked Systems Design and Implementation (NSDI '04)*, March 2004.
5. Tieyan Li, Yongdong Wu, Di Ma, Huafei Zhu, and Robert H. Deng. Flexible verification of mpeg-4 stream in peer-to-peer cdn. In *Proceedings of the 6th International Conference on Information and Communications Security (ICICS)*, pages 79–91, 2004.
6. Girma Berhe, Lionel Brunie, and Jean-Marc Pierson. Modeling service-based multimedia content adaptation in pervasive computing. In *Conf. Computing Frontiers*, pages 60–69, 2004.
7. Armando Fox, Steven D. Gribble, Yatin Chawathe, and Eric A. Brewer. Adapting to network and client variation using active proxies: lessons and perspectives. *IEEE Personal Communications*, August 1998.
8. V. Cardellini, P. S. Yu, and Y. W. Huang. Collaborative proxy system for distributed web content transcoding. In *Proceedings of 9th ACM Intl Conf. on Information and Knowledge Management*, November 2000.
9. Chi-Hung Chi and Yin Wu. An XML-based data integrity service model for web intermediaries. In *Proceedings of the 7th International Workshop on Web Content Caching and Distribution*, August 2003.
10. Chi-Hung Chi, Yan-Hong Lin, Jing Deng, X. Li, and T.-S. Chua. Automatic proxy-based watermarking for www. *Computer Communications*, 24(2):144–154, 2001.
11. P. Thuraisingham, A. Gupta, E. Bertino, and E. Ferrari. Collaborative commerce and knowledge management. *Knowledge and Process Management*, 9(1):43–53, August 2002.

12. P. Maglio and R. Barrett. Intermediaries personalize information streams. *Communications of the ACM*, 43(8):99–101, August 2000.

13. J.-L. Huang, M.-S. Chen, and H.-P. Hung. A qos-aware transcoding proxy using on-demand data broadcasting. In *Proceedings of the IEEE INFOCOM Conference*, March 2004.

14. Yunhua Koglin and Elisa Bertino. Secure content services from cooperative internet intermediaries. Technical report, Purdue University, 2005.

15. S. Chandra and C. S. Ellis. Jpeg compression metric as a quality aware image transcoding. In *Proceedings of USENIX 2nd Symp. on Internet Technology and Systems*, October 1999.

16. R. Han, P. Bhagwat, R. LaMaire, T. Mummert, V. Perret, and J. Rubas. Dynamic adaptation in an image transcoding proxy for mobile web browsing. *IEEE Personal Communications*, 5(6):8–17, December 1998.

17. Yunhua Koglin, Giovanni Mella, Elisa Bertino, and Elena Ferrari. An update protocol for XML documents in distributed and cooperative systems. In *Proceedings of International Conference on Distributed Computing Systems*, June 2005.

18. John Apostolopoulos. *Secure media streaming and secure adaptation for non-scalable video*. Technical Report HPL-2004-186, Hewlett-Packard Laboratories, October 2004.

19. Susie Wee and John Apostolopoulos. *Secure transcoding with jpsec confidentiality and authentication*. Technical report HPL-2004-185, Hewlett-Packard Laboratories, October 2004.

20. Susie Wee and John Apostolopoulos. Secure scalable streaming enabling transcoding without decryption. In *IEEE International Conference on Image Processing*, 2001. Available as Hewlett-Packard Laboratories Technical Report HPL-2001-320.

21. Susie Wee and John Apostolopoulos. Secure scalable video streaming for wireless networks. In *IEEE International Conference on Acoustics, Speech, and Signal Processing*, May 2001.

22. Charu C. Aggarwal, Joel L. Wolf, and Philip S. Yu. Caching on the world wide web. *Knowledge and Data Engineering*, 11(1):95–107, 1999.

23. Bo Li, Xin Deng, Mordecai J. Golin, and Kazem Sohraby. On the optimal placement of web proxies in the internet. In *Proceedings of Infocom Conference*, March 1999.

24. S. Sivasubramanian, M. Szymaniak, G. Pierre, and M. V. Steen. Replication for web hosting systems. *ACM Computing Surveys*, 36(3):291–334, September 2004.

25. Lee Breslau, Pei Cao, Li Fan, Graham Phillips, and Scott Shenker. Web caching and zipf-like distributions: Evidence and implications. In *Proceedings of IEEE INFOCOM Conference*, March 1999.

26. S. Buchholz and A. Schill. Adaptation-aware web caching: caching in the future pervasive web. In *13th GI/ITG Conference Kommunikation in Verteilten Systemen (KiVS)*, 2003.

27. Matt Bishop. *Computer Security: Art and Science*. Addison Wesley Professional, 2002.

28. D. Elliott Bell and Leonard J. LaPadula. *Secure computer systems: unified exposition and multics interpretation*. Technical Report MTR-2997, MITRE Corporation, March 1976.

29. R. S. Sandhu. Lattice-based access control models. *IEEE Computer*, 26(11):9–19, November 1993.

30. J. Barkley, A.V. Cincotta, D.F. Ferraiolo, S. Gavrila, and D.R. Kuhn. Role-based access control for the world wide web. In *20th National Computer Security Conference*, 1997.

31. James Joshi, Elisa Bertino, Usman Latif, and Arif Ghafoor. A generalized temporal role-based access control model. *IEEE Trans. Knowl. Data Eng.*, 17(1):4–23, 2005.

32. Roberto Tamassia, Danfeng Yao, and W. H. Winsborough. Role-based cascaded delegation. In *Proceedings of the ACM Symposium on Access Control Models and Technologies (SACMAT '04)*, pages 146–155. ACM Press, June 2004.

33. E. Barka and R. Sandhu. Framework for role-based delegation models. In *Proceedings of the 16th Annual Computer Security Applications Conference (ACSAC'00)*, December 2000.

34. E. Freudenthal, T. Pesin, L. Port, E. Keenan, and V. Karamcheti. dRBAC: Distributed role-based access control for dynamic coalition environments. In *ICDCS 2002*, pages 411–420, 2002.

35. J. S. Park, R. Sandhu, and G.-J. Ahn. RBAC on the web. In *ACM Transactions on Information and Systems Security*, volume 4(1), 2001.

36. Mohammad A. Al-Kahtani and Ravi S. Sandhu. A model for attribute-based user-role assignment. In *ACSAC*, 353–364, 2002.

37. Anthony Harrington and Christian D. Jensen. *Cryptographic access control in a distributed file system*, 2003.

38. Dawn Xiaodong Song, David Wagner, and Adrian Perrig. Practical techniques for searches on encrypted data. In *IEEE Symposium on Security and Privacy*, pages 44–55, 2000.

39. Matt Blaze, Gerrit Bleumer, and Martin Strauss. Divertible protocols and atomic proxy cryptography. *Proceedings of Eurocrypt*, 1403:127–144, 1998.

40. Giuseppe Ateniese, Kevin Fu, Matthew Green, and Susan Hohenberger. Improved proxy re-encryption schemes with applications to secure distributed storage. In *Proceedings of the 12th Annual Network and Distributed System Security Symposium (NDSS)*, February 2005.

41. Miguel Castro and Barbara Liskov. Practical byzantine fault tolerance. In *OSDI: Symposium on Operating Systems Design and Implementation*. USENIX Association, Co-sponsored by IEEE TCOS and ACM SIGOPS, 1999.

42. Rongmei Zhang and Y. Charlie Hu. Hyper: A hybrid approach to efficient content-based publish/subscribe. In *Proceedings of International Conference on Distributed Computing Systems*, 2005.

43. A. Carzaniga, M. J. Rutherford, and A. L. Wolf. A routing scheme for content-based networking. In *IEEE INFOCOM*, 2004.

44. G. Cugola, D. Frey, A. L. Murphy, and G. P. Picco. Minimizing the reconfiguration overhead in content-based publish-subscribe. In *Proceedings of the 19th ACM Symposium on Applied Computing (SAC04)*, 2004.

45. M. Srivatsa and L. Liu. Securing publish-subscribe overlay services with event-guard. In *Proceedings of the 12th ACM Conference on Computer and Communication Security*, 2005.

46. Lukasz Opyrchal and Atul Prakash. Secure distribution of events in content-based publish-subscribe systems. In *Proc. of the 10th USENIX Security Symposium*, 281–295, 2001.

47. Chenxi Wang, Antonio Carzaniga, David Evans, and Alexander L. Wolf. Security issues and requirements for internet-scale publish-subscribe systems. In *Hawaii International Conference on System Sciences*, 2002.

48. Chris Lesniewski-Laas and M. Frans Kaashoek. Ssl splitting: Securely serving data from untrusted caches. In *Proceedings of 12th USENIX Security Symposium*, August 2003.

49. Napster. Available at: http://www.napster.com.

50. Gnutella. Available at: http://gnutella.wego.com.

51. Ian Clarke, Oskar Sandberg, Brandon Wiley, and Theodore W. Hong. Freenet: A distributed anonymous information storage and retrieval system. *Lecture Notes in Computer Science*, 2009:46–52, 2001.

52. Ion Stoica, Robert Morris, David Karger, Frans Kaashoek, and Hari Balakrishnan. Chord: A scalable peer-to-peer lookup service for internet applications. In *Proceedings of the 2001 ACM SIGCOMM Conference*, pages 149–160, 2001.

53. Ben Y. Zhao, Ling Huang, Jeremy Stribling, Sean C. Rhea, Anthony D. Joseph, and John Kubiatowicz. Tapestry: A resilient global-scale overlay for service deployment. *IEEE Journal on Selected Areas in Communications*, 22(1):41–53, January 2004.

54. John Kubiatowicz, David Bindel, Yan Chen, Patrick Eaton, Dennis Geels, Ramakrishna Gummadi, Sean Rhea, Hakim Weatherspoon, Westly Weimer, Christopher Wells, and Ben Zhao. Oceanstore: An architecture for global-scale persistent storage. In *Proceedings of ACM ASPLOS*. ACM, November 2000.

55. Frank Dabek, M. Frans Kaashoek, David Karger, Robert Morris, and Ion Stoica. Wide-area cooperative storage with cfs. In *Proceedings of the 18th ACM Symposium on Operating Systems Principles (SOSP)'01*, October 2001.

56. Stephanos Androutsellis-Theotokis and Diomidis Spinellis. A survey of peer-to-peer content distribution technologies. *ACM Comput. Surv.*, 36(4): 335–371, 2004.

57. Elisa Bertino, Silvana Castano, Elena Ferrari, and Marco Mesiti. Specifying and enforcing access control policies for xml document sources. *World Wide Web*, 3(3):139–151, 2000.

58. Elisa Bertino, Barbara Carminati, Elena Ferrari, and Giovanni Mella. Author-chi — A system for secure dissemination and update of xml documents. In *DNIS*, pages 66–85, 2003.

59. Elisa Bertino, Silvana Castano, and Elena Ferrari. Securing xml documents with author-x. *IEEE Internet Computing*, 5(3):21–26, 2001.

60. Elisa Bertino, Silvana Castano, and Elena Ferrari. Securing xml documents: The author-x project demonstration. In *SIGMOD Conference*, page 605, 2001.

61. Extensible markup language (XML). Available at: http://www.w3.org/XML/.

62. W3C XML schema. Available at: http://www.w3.org/XML/Schema.

63. Elisa Bertino, Elena Ferrari, and Giovanni Mella. An approach to cooperative updates of xml documents in distributed systems. *Journal of Computer Security*, 13(2):191–242, 2005.

64. Fatih Emekçi, Divyakant Agrawal, and Amr El Abbadi. Abacus: A distributed middleware for privacy preserving data sharing across private data warehouses. In *Middleware*, pages 21–41, 2005.

65. Marc Waldman. Censorship-resistant publishing systems-survey and thesis proposal.

66. Aviel D. Rubin Marc Waldman and Lorrie Faith Cranor. Publius: A robust, tamper-evident, censorship-resistant, web publishing system. In *Proc. 9th USENIX Security Symposium*, pages 59–72, August 2000.

67. Elisa Bertino, Barbara Catania, Maria Luisa Damiani, and Paolo Perlasca. Geo-rbac: A spatially aware rbac. In *SACMAT*, pages 29–37, 2005.

2

Key Management and Agreement in Distributed Systems

Venkata C. Giruka, Saikat Chakrabarti and Mukesh Singhal

CONTENTS

Abstract   Today's distributed systems typically support real-time dynamic groups, for instance, a secure video conferencing group, which requires security services like privacy, integrity, and nonrepudiation of data exchanged within the group. A *group secret key* or group key provides an efficient means for providing such services to the group members, and for any subsequent cryptographic use within the group. The challenge is to establish, distribute, and maintain a group key securely and efficiently while coping with the group dynamics. In this chapter, we present a walk-through of key agreement protocols based on the Diffie-Hellman key exchange, and key management protocols for complex distributed systems like the Internet.

2.1 Introduction

Distributed systems are an integral part of today's computing infrastructures. Loosely speaking, a distributed system is *a system that transparently connects geographically dispersed computers for resource sharing and information processing or exchange*. Complex systems like the World Wide Web (WWW), the Internet, load-balancing database servers, peer-to-peer systems like Napster, and mobile computing systems are some of the prominent examples of distributed systems that we encounter in our day-to-day life. These systems typically support applications that make it easy for users to communicate, access, share, and process information in a controlled manner. Such applications along with increased connectivity offered by the Internet have led to *group-oriented* applications.

In group-oriented applications, a group of users participate to achieve a common goal, like collaborative software development or video-conferencing. Several group-oriented applications require securing the data exchanged within the group to provide services like *authentication, access control, confidentiality, and nonrepudiation*, to name a few. A naive way to provide such services in a group is to have a secret key between every pair of nodes, which they use for pairwise encryption or decryption, or for any subsequent cryptographic use. This method becomes rather inefficient as the size of the group increases. Given the collaborative nature of the groups, most of the information is common to all the members of a group. Thus, a single *group key* is an efficient alternative to pairwise keys among group members.

The group key helps the group members in encrypting or decrypting group data, or for any subsequent cryptographic use. With such a group key, *only* those members that possess the key can access the group-specific data. However, distributing a group key securely to the group members is a nontrivial problem in itself. Furthermore, several practical systems involve dynamic groups where members join and leave the group at random. For each such event, the group key should be changed and distributed to the current group members securely to limit the access only to authorized members. For instance, a service provider of on-demand TV would certainly be interested in restricting the member-join *only* to the paid group members. Thus, when a member's subscription expires, the service provider should make sure that the member should not be able to access the on-demand TV while keeping this membership change transparent to other group members.

In the previous example, there is a controlling authority, typically referred to as a *group controller*, that is responsible for controlling the group activities like handling member-join, member-leave, and refreshing the key. When such a group controller is available, the processes of distributing a group key to the group members, and maintenance of keying relationships between authorized parties to cope with group-dynamism is called *key management*. However, there may be groups that may not have any group controller, or it may not be feasible to delegate the responsibility of the group controller to one or more nodes. Such groups are typically referred to as *peer groups*. Examples of peer groups include database servers, and ad-hoc networks. In peer groups, members establish and maintain a group key using a *key agreement* protocol.

In the rest of the chapter, we present a few known approaches for key management and key agreement in dynamic groups. Section 2.2 presents preliminaries of key management and key agreement protocols. Section 2.3 presents a few approaches for key management. Section 2.4 presents a few approaches for key agreement, and Section 2.5 summarizes the chapter.

2.2 Preliminaries

Key management is a set of techniques and procedures supporting the establishment and maintenance of keying relationships between authorized parties [10]. Key management is typically applicable to (large) multicast groups where a *group controller* is responsible for generating a group key and distributing it securely to the group members. Further, it is also responsible for changing the group key as and when necessary.

On the other hand, a (group or multiparty) key agreement is a mechanism in which a set of nodes agree on a group key for subsequent cryptographic use. A key agreement is *distributed* if every member in the group can generate the (same) group key based on information from the other members in the group. In a distributed multiparty key agreement, the responsibility of managing group membership events is distributed among the group, which offers high availability and avoids a central point of failure. A key agreement protocol is called *contributory* if the resulting group key is derived from contributions (a secret known to respective members but not known to other members) of all the members in the group. Since peer groups do not depend on any central server to generate, distribute, and manage a group key, a distributed and contributory key agreement protocol is well-suited to such groups.

2.2.1 Role of Key Management and Agreement Protocols

Both the key management and key agreement protocols should cope with the demands of various applications. Besides confidentiality, integrity, and authenticity requirements, the following features are desirable in a group key management, and a key agreement protocol.

1. Forward secrecy: Forward secrecy requires that a departed/expelled member of a group should not have access to future keys after it has left the group. This ensures that the departed/expelled members

cannot decrypt group data after it leaves the group. To provide forward secrecy, the protocol must change the group key after a current member leaves or is evicted from the current group.

2. Backward secrecy: Backward secrecy requires that a new member should not have access to the previous (old) group key. This ensures that the newly joined node cannot decrypt messages that are exchanged in the group before the node joined the group.

3. Rekeying: Rekeying refers to the processing of changing the group key securely upon a membership change. Rekeying should be triggered by the protocol after each membership change to ensure forward and backward secrecy. In addition to this, rekeying should also be triggered by the protocol at regular intervals to safeguard the secrecy of the group key.

4. Key independence: The group key generated/established by the protocol should be completely independent of all previous/future keys. This ensures that the compromise of one or more keys does not compromise other keys.

5. Resistance to attacks: Since the network infrastructure may be insecure, unauthorized members may eavesdrop on the group communication. For instance, a subset of departed members may collude to try to discover new group keys, or a subset of current members may collude to try to discover the keys of other members to impersonate the victims. Thus, the protocol should be resistant to attacks from both inside and outside the group.

6. Scalability: The size of a group may vary from tens to thousands of nodes. Furthermore, the rate of join/leave requests and the expected lifetime of a member (in the group) may vary largely in different applications. Thus, a protocol should not make arbitrary assumptions about group size. Further, membership changes should only affect a small subset of members so that the system can support large dynamic groups.

For a key management protocol, the major challenge is to efficiently/securely distribute a key. Thus, the core of the challenge concerns rekeying. Although rekeying is easy when a new member joins the group, it is nontrivial when a member leaves the group. After a member leaves the group, the old group key cannot be used to encrypt the new group key as the leaving member knows the current group key. For this reason, the group controller should use other mechanisms to distribute the new key.

In addition to challenges in key management, a key agreement protocol should establish a group key with only authentic members of the group. The situation is far more complicated than in a key management protocol where the group controller (alone) decides a key for the group. In case of a key agreement protocol, a group key is agreed by members of the group by exchanging messages. Thus, a key agreement protocol has to ensure that only authentic members can establish a key (using the messages), whereas others cannot.

2.3 Key Management Protocols

For the key management protocol that we describe in this section we use the following model. The communication group consists of N members M_1, $M_2, \ldots, M_N$. The group assumes a trusted server, called the group controller C, which is responsible for managing group memberships, as well as the services related with key distribution such as maintaining the key hierarchy, generating new keys, and initiating the rekeying process. Group members can join or leave (or be evicted by the controller) the group at will. The group controller can detect these membership changes and initiate the key distribution accordingly.

2.3.1 A Simple Approach

Simple key distribution center (SKDC) [1] is one of the simplest solutions for group key management. In SKDC approach, a group controller C shares a secret key $k_{C,i}$ with each group member M_i. The group controller is responsible for generating a group secret key k_G, as and when necessary, and distributing it to the group members. To distribute the group key, C encrypts the group key with $k_{C,i}$ and unicasts it to M_i.

When a new member M_{N+1} joins the group, the group controller generates a new group key k'_G, encrypts it with the old group key k_G, and multicasts it to $\{M_1, M_2, \cdots, M_n\}$. To distribute the group key to the new member, the group controller encrypts the new group key k'_G using $k_{C,N+1}$ and unicasts it to the M_{N+1}. Thus, the cost of rekeying is only two messages, one multicast and the other unicast. However, when a member leaves the group, the group controller cannot use the old group key k_G to encrypt the new key k'_G, since the departed member also knows k_G. Instead, k'_G has to be encrypted by each remaining member's individual key $k_{C,i}$ and unicast separately. Apparently, this approach does not scale well as the group size increases, as it requires N encryptions and N rekeying messages.

We can see that communicating the new group session key in a scalable and secure way, especially when members leave, is definitely a nontrivial task. Next, we present efficient and scalable group key management protocols and discuss corresponding encryption, messaging, and storage costs.

2.3.2 Key Graphs

One way to achieve scalability is to derive a hierarchy based on some group-specific information. Such a hierarchy helps nodes divide the whole communication group into several subgroups. Each subgroup can maintain a (subgroup) secret key, called the *auxiliary key* shared by members within the subgroup. The key corresponding to the whole group is the *group session key*, which is derived from the *auxiliary keys*. The hierarchy of these subgroups naturally leads to a tree-structure rooted at the group session key, with auxiliary keys as internal nodes and group members as leaves. Based on this approach, Wong et al. [15, 16] presented a formal notation of secure subgroups and key graphs. Further, they presented three different rekeying strategies, viz., *user-oriented*, *key-oriented*, and *group-oriented*, to cope with group dynamics like a member-join or a member-leave.

Key Graph Notations: The notion of a secure group is formalized as a three triple (U, K, R), where U is a finite and nonempty set of group members, K is a finite and nonempty set of keys, and $R \subset U \times K$ is a binary member-key relation. The group controller is assumed to know the member set U and the key set K, and is responsible for maintaining the member-key relation R. Two functions are associated with each secure group (U, K, R), viz., $keyset(M_i)$ and $userset(k)$. $keyset(M_i) = \{k|(M_i, k) \in R\}$ represents the set of all keys held by member M_i, and $keyset(\phi) = \phi$. $userset(k) = \{M_i|(M_i, k) \in R)\}$, represents the set of all members holding key k. For example, Figure 2.1 depicts a key graph where $keyset(M_2) = \{k_{12}, k_{234}, k_{1234}\}$, $keyset(M_3) = \{k_{23}, k_{1234}\}$, $userset(k_1) = \{M_1\}$ and $userset(k_{12}) = \{M_1, M_2\}$, where $k_{i...j}$ is the key shared by members $\{M_i, \ldots, M_j\}$.

Each member holds two kinds of keys: the group session key and one or more auxiliary keys. All group members are partitioned into several subgroups recursively. Members belonging to a subgroup share the same auxiliary key. A subgroup key is different from other subgroup keys, which ensures that members outside a subgroup cannot decrypt communication messages encrypted using the subgroup key.

Whenever a member of the group leaves or a new member joins the group, the group controller performs a rekeying operation to ensure forward and backward secrecy. The use of auxiliary subgroup keys helps reduce

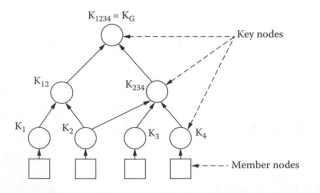

FIGURE 2.1
An example of a key graph.

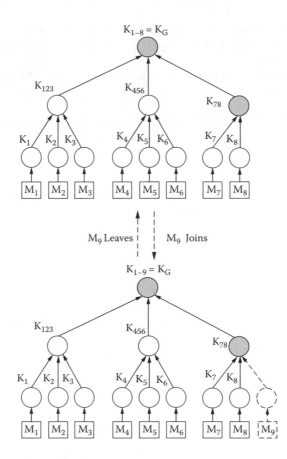

FIGURE 2.2
An example.

the rekeying overhead. When a member M_i leaves the group only those subgroups that M_i belongs to need to change the corresponding keys, i.e., $keyset(M_i)$. All other subgroups keep their subgroup keys intact and use them to encrypt new keys for future use. Thus, properly constructing subgroups and utilizing auxiliary keys to encrypt rekeying messages may substantially decrease the encryption cost and message overhead due to group dynamics. Next, we explain three rekeying strategies and illustrate them by using an example in Figure 2.2.

User-Oriented Strategy: During the rekeying process, each member M_i, $i = 1, \ldots, N$ may change a set of old auxiliary keys, which are affected due to group dynamics, with new keys denoted by the set K_i^{new}. In the user-oriented strategy, the controller C embeds M_i's new key set K_i^{new} in a single rekeying message, and then encrypts it by an appropriate auxiliary key. The auxiliary key is chosen by the controller in such a way that it is shared by the largest group (U_{max}) among all the existing subgroups. This strategy reduces the encryption cost by grouping the members into a small number of subgroups. Further, to reduce the traffic overhead, rekeying messages are multicast to the respective subgroups.

Let us consider the example shown in Figure 2.2. When M_9 joins the group, the following actions are performed by the group controller:

$$C \rightarrow \{M_1, \cdots, M_6\} : \{k_{1-9}\}_{k_{1-8}}$$
$$C \rightarrow \{M_7, M_8\} : \{k_{1-9}, k_{789}\}_{k_{78}}$$
$$C \rightarrow M_9 : \{k_{1-9}, k_{789}\}_{k_9}.$$

As a result of the member-join, the group $\{M_1, \cdots M_8\}$, changes to $\{M_1, \cdots, M_9\}$ and subgroup $\{M_7, M_8\}$ changes to $\{M_7, M_8, M_9\}$. $M_1, \cdots M_6$ belong to subgroups whose compositions are not affected by the new member-join, and therefore, they only receive the new session key k_{1-9} encrypted using the old group key k_{1-8}. Whereas the composition of the group containing M_7 and M_8 is affected by the new member-join. Thus, they receive the new group key k_{1-9} and a new subgroup key k_{789}, encrypted by the old subgroup key k_{78}. Finally, M_9 receives k_{1-9} and k_{789} encrypted using k_9.

When M_9 leaves the group, the group controller performs the following actions:

$$C \rightarrow \{M_1, M_2, M_3\} : \{k_{1-8}\}_{k_{123}}$$
$$C \rightarrow \{M_4, M_5, M_6\} : \{k_{1-8}\}_{k_{456}}$$
$$C \rightarrow M_7 : \{k_{1-8}, k_{78}\}_{k_7}$$
$$C \rightarrow M_8 : \{k_{1-8}, k_{78}\}_{k_8}.$$

When M_9 leaves the group, $M_1, \cdots M_6$ cannot use the old group session key $k_G = k_{1-9}$ to encrypt the new session key $k'_G = k_{1-8}$, because M_9 knows the old key k_G. The largest subgroups that share a common subgroup key are $M_1, \cdots M_3$ and $M_4, \cdots M_6$. Thus, k_{123} and k_{456} are used to encrypt the new group key, so that M_9 has no way to decrypt rekeying messages. Finally, M_7 and M_8 receive a new subgroup key k_{78} and the new group key encrypted using their respective keys, k_7 and k_8.

Key-Oriented Strategy: In the key-oriented approach, each rekeying message only contains a single new key. In order to minimize the encryption cost, the controller C selects an auxiliary key to encrypt a rekeying message such that the auxiliary is shared by the largest subgroup. For example, when M_9 joins the group, the group controller does the following:

$$C \rightarrow \{M_1, \cdots M_8\} : \{k_{1-9}\}_{k_{1-8}}$$
$$C \rightarrow \{M_9\} : \{k_{1-9}\}_{k_9}$$
$$C \rightarrow \{M_7, M_8\} : \{k_{789}\}_{k_{78}}$$
$$C \rightarrow \{M_9\} : \{k_{789}\}_{k_9}.$$

Note that every member needs the new group session key $k_G = k_{1-9}$. The largest subgroup that shares an auxiliary key (k_{1-8}) is $\{M_1, \cdots M_8\}$, so k_G is encrypted by k_{1-8}. The new subgroup key k_{789} is to be held only by $\{M_7, M_8, M_9\}$, and the largest group that shares an auxiliary key (k_{78}) is $\{M_7, M_8\}$. So old subgroup key k_{78} is used to encrypt new subgroup key k_{789}. The newly joined member M_9 gets the new session key k_G and the new subgroup key k_{789} in messages encrypted by its individual key k_9.

When M_9 leaves the group, the following actions are performed by the group controller:

$$C \rightarrow \{M_1, M_2, M_3\} : \{k_{1-8}\}_{k_{123}}$$
$$C \rightarrow \{M_4, M_5, M_6\} : \{k_{1-8}\}_{k_{456}}$$
$$C \rightarrow M_7 : \{k_{78}\}_{k_7}$$
$$C \rightarrow M_7 : \{k_{1-8}\}_{k_{78}}$$
$$C \rightarrow M_8 : \{k_{78}\}_{k_8}$$
$$C \rightarrow M_8 : \{k_{1-8}\}_{k_{78}}.$$

When M_9 leaves the group, the new group cannot use the old group key k_{1-9} to decrypt any rekeying messages. Thus the group controller encrypts the new group key k_{1-8} with individual subgroup keys k_{123}, k_{456}, and k_78 corresponding to the subgroups $\{M_1, M_2, M_3\}$, $\{M_4, M_5, M_6\}$, $\{M_7\}$ and $\{M_8\}$. First the new subgroup key k_{78} is delivered to M_7 and M_8 (encrypted using k_7 and k_8, respectively), and then the new subgroup key is used to deliver the group key.

Group-Oriented Strategy: In a group-oriented approach, a rekeying message contains as many new keys as possible. These new keys are encrypted by appropriate subgroup keys, which are chosen (by the group controller) to minimize the encryption cost of the rekeying messages. When a new member joins the group, the new group contains the old group and a new member; thus the controller constructs a rekeying message for each of these two subgroups. When a member leaves the group, the controller groups together all new keys encrypted by appropriate auxiliary keys and multicasts it to the new group. The main idea of adopting the group-oriented strategy is to take advantage of multicasting to reduce the control overhead.

For example, when M_9 joins the group, the controller executes the following actions:

$$C \rightarrow \{M_1, \cdots M_8\} : \{k_{1-9}\}_{k_{1-8}}, \{k_{789}\}_{k_{78}}$$
$$C \rightarrow \{M_9\} : \{k_{1-9, k_{789}}\}_{k_9}.$$

When a new member M_9 joins, the new group contains the (old) subgroup $\{M_1, \cdots, M8\}$ that shares a key k_{1-8}, and the subgroup $\{M_9\}$ that holds the key k_9. The controller C constructs and distributes two rekeying messages designated for these two groups, respectively.

TABLE 2.1

Storage Complexity of KG Protocol for Tree Topology

Total number of keys maintained in the whole group	$\frac{d*N}{d-1}$ to $\frac{1}{d-1}$
Number of keys held per user	$h = \log_d(N+1)$

When M_9 leaves the group, the controller executes the following actions:

$$C \rightarrow \{M_1, \cdots M_8\} : \{k_{1-8}\}_{k_{123}}, \{k_{1-8}\}_{k_{456}}, \{k_{1-8}\}_{k_{78}}, \{k_{78}\}_{k_7}, \{k_{78}\}_{k_8}.$$

The controller C constructs and multicasts rekeying message(s) to all remaining members, $\{M_1, \cdots M_8\}$, using appropriate auxiliary keys to encrypt different new keys. The goal is to minimize the encryption cost by choosing auxiliary keys shared by as many members as possible. When a member receives the rekeying message, it uses the keys in its *keyset* to extract the appropriate new keys from the message.

Remarks: Although we only discussed key management based on tree topology, key graphs are a generic structure that can be used in star, tree, and complete graph topologies; for more details interested readers are referred to [15, 16]. Compared to the SKDC approach, all the three strategies reduce the number of rekeying messages and the encryption costs are substantial. The numerical results for storage complexity are given in Table 2.1. The rekeying message complexity for the three strategies is presented in Table 2.2, where h is the height of the key tree with degree d, and N is the group size.

As we can see from Table 2.1 and Table 2.2, the controller needs to maintain $O(N)$ keys, and each user stores $O(\log(N))$ keys, and the encryption cost when a member joins is proportional to $O(1) \sim O(\log(N))$, and the encryption cost for a leave is $O(\log(N)) \sim O(\log^2 N)$. Note that the undesirable encryption cost $O(\log^2(N))$ is introduced by using user-oriented strategy when a member leaves, but we can easily avoid this relatively higher cost by choosing an alternative strategy, i.e., key-oriented or group-oriented. Hence, compared to the SKDC approaches, where the complexity of both encryption and rekeying messages is proportional to N, KG method substantially improves the scalability of the key distribution and management for group communications.

TABLE 2.2

Rekeying Complexity of KG Protocol for Tree Topology

	User-oriented Strategy		Key-oriented Strategy		Group-oriented Strategy	
	Join	Leave	Join	Leave	Join	Leave
Number of rekeying messages	$h = \log_d(N+1)$	$(d-1)(h-1)$	$2(h-1)$	$(d-1)(h-1)$	2	1
Encryption cost	$\frac{h(h+1)}{2} - 1$	$\frac{(d-1)h(h-1)}{2}$	$2(h-1)$	$d(h-1)$	$2(h-1)$	$d(h-1)$

However, the controller has to maintain $O(N)$ keys, which puts a heavy burden on both the controller's storage as well as computation. Next we describe another tree-based key management method that reduces the number of keys maintained by the controller to $O(\log(N))$, while providing a similar rekeying message complexity.

2.3.3 Boolean Function Minimization Techniques

Chang et al. [5] developed a method for group key management based on a novel idea of defining a user ID (UID), in the form of n-bit binary string, for each user. Under this representation, any two users differ with each other's UID in at least one bit. The set of keys held by a user M, denoted as *keyset(M)*, is the keys along the path from the root of the key tree to M. Further, *keyset(M)* is entirely determined by its UID. Since only $n = \log(N)$ bits are sufficient to represent a UID, the number of keys maintained by the controller C is reduced to $O(\log(N))$. Thus, this method achieves a substantial improvement over the key graph (KG) approach discussed previously, where the controller has to maintain $O(N)$ keys. Furthermore, the method intelligently handles *cumulative member removal*, which is frequent in large dynamic groups, using minimization techniques in Boolean algebra.

UID and Key Pair Notations: Each member in the group maintains n auxiliary key pairs, k_i and $\overline{k_i}$, where $0 \le i \le n-1$, $n = \log(N)$ and N is the size of the group. Each key pair corresponds to one bit in UID. The group controller distributes a group session(k_G) and n auxiliary keys to each member such that the member holds k_i if ith bit of its UID is 1, or $\overline{k_i}$ if ith bit of its UID is 0. For example, Figure 2.3 illustrates a UID and corresponding key assignment for a

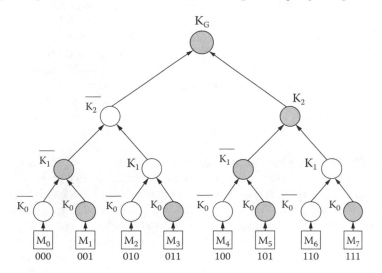

FIGURE 2.3
A key tree based on UIDs.

group of eight members. The root of the tree is the group session key k_G. The internal nodes, k_i and $\overline{k_i}$, $0 \leq i \leq n - 1$, represent the auxiliary keys, and the leaf nodes are the members of the group. Each member has a unique UID as illustrated in Figure 2.3. For example, the member M_5 has a UID 101, which implies that it possesses auxiliary keys k_2, $\overline{k_1}$, and k_0, plus k_G that is shared by all members. Note that auxiliary keys held by a member are the keys of nodes along its path to the root.

Rekeying Process: In general, both k_G and auxiliary keys update whenever a member leaves, so that the leaving member can no longer decrypt the future communications messages of the group. However, rekeying for each member leave can be very expensive especially in large dynamic groups where members join and leave frequently. Alternatively, the controller can use a batch processing approach where group changes are batched together and the rekeying is initiated periodically (after a predetermined timeout). Each such rekeying process is called a round. For the rth round, the group session key is denoted as $k_G(r)$, and the auxiliary keys as $k_i(r)$ and $\overline{k_i}(r)$.

Individual Member Removal: To update $k_G(r)$, the controller generates a new session key $k_G(r + 1)$. The new key is encrypted separately by the keys that are *complementary* to the auxiliary keys of the departing member. For example, in Figure 2.3, if M_5 leaves the group, the complementary keys are $\overline{keyset(M_5)} = \{\overline{k_2}, k_1, \overline{k_0}\}$. Thus, the controller encrypts $k_G(r + 1)$ using these three keys separately (i.e., $\{k_G(r + 1)\}_{\overline{k_1}}$, $\{k_G(r + 1)\}_{k_1}$, $\{k_G(r + 1)\}_{\overline{k_0}}$) and multicast to all group members. Since M_5 does not know any of these keys, it cannot decrypt the multicast rekeying message to get the new session key. On the other hand, any other member M_j's UID differs in at least one bit with the UID of M_5, therefore, possesses $keyset(M_j)$ such that $\overline{keyset(M_5)} \cap keyset(M_j) \neq \phi$, where $j \neq 5$. This ensures that any other member can decrypt at least one data chunk in the rekeying message. In addition, auxiliary keys are also updated using a one-way hash function f, such that $k_i(r + 1) = f(k_i(r), k_G(r + 1))$. This ensures that the departing member cannot use its auxiliary keys to decrypt future key update messages.

Removal of Multiple Members: As discussed earlier, multiple member removal is very common in large dynamic groups. In order to minimize the number of rekeying messages as well as encryption cost, a batch processing approach is more desirable. By carefully selecting the rekeying period, one can minimize both the information exposure (to expelled/departed members) and the computation and communication cost of rekeying. The cumulative member removal scheme is best understood by an example presented in Figure 2.3. Suppose that members M_0 and M_4 leave the group. Without batch rekeying, three messages encrypted by three auxiliary keys (corresponding to k_0, k_1, $\overline{k_2}$) are necessary for handling M_4's departure, and three messages encrypted by three auxiliary keys (corresponding to k_0, k_1, k_2) are necessary for handling M_0's departure. Thus, totally six messages are required.

In the cumulative member removal scheme, all the departed/expelled members are removed in a single round. This is achieved using the UIDs of the members, and a Boolean function $m()$. The Boolean function $m()$ helps in determining the group membership status of a UID. Let $X_{n-1} X_{n-2} \cdots X_0$ be the binary representation of a UID, if $m(X_0, X_1, \cdots, X_{n-1}) = 1$, then the UID belongs to the group, else the UID does not belong (i.e., departed/expelled) to the group. Thus the problem of rekeying reduces to finding a set of UIDs, for which $m(UID) = 1$ holds, and updating their auxiliary keys and the group key. This problem is equivalent to the minimization of the Boolean membership function $m()$. For a reasonable number of input variables, the Karnaugh map representation of Boolean function can be used to achieve the minimization. However, for a large number of input variables, the Quine-McCluskey algorithm can be used.

A detailed explanation of minimization technique in Boolean algebra is beyond the scope of this chapter; interested readers are referred to [5] for more details. However, we use the running example to explain the basic concepts involved in the Boolean function minimization. Continuing with our example, $\overline{keyset M_0} = \{k_2, k_1 k_0\}$ and $\overline{keyset(M_4)} = \{\overline{k_2}, k_1, k_0\}$, so $S = \{k_1, k_0\} = \overline{keyset(M_0)} \bigcap \overline{keyset(M_4)}$. Using keys in S to encrypt a new session key ensures that none of the departing members can figure out $k_G(r + 1)$, while all other remaining members can always determine it.

Table 2.3 illustrates the Boolean member function for the group of seven members. In Table 2.3 the input $X_2 X_1 X_0$ denotes the binary representation of the UID of a member, output "0" implies that the member does not belong to the groups, "1" implies that the member is in the group, and "X" implies that the UID is currently not assigned. Figure 2.4 illustrates the Karnaugh map minimization of membership function. The idea is to identify the minimum number of blocks in the Karnaugh table, so that all 1's are in a block, but none of the 0's. Such blocks correspond to minimum subgroups that share a common auxiliary key that is not shared by departing members. For each such subgroup, the controller sends a rekey message containing the group session key encrypted using the respective auxiliary keys.

TABLE 2.3

Boolean Member Function

Input X_2, X_1, X_0	Output
000	0
001	1
010	1
011	1
100	0
101	X
110	1
111	1

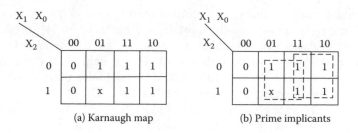

(a) Karnaugh map   (b) Prime implicants

FIGURE 2.4
Karnaugh map minimization of membership function.

Remarks: The method of using the Boolean function minimization technique (BFMT) for key management has some interesting features. First, it utilizes the most common computer science idea, i.e., using binary string to represent a member and designing the rekeying process accordingly. It simplifies the rekeying message generation and distribution algorithm. It just needs to compute the complementary key set S of the departing members and multicast the new session key encrypted by auxiliary keys in set S. Therefore the number of rekeying multicast messages is only 1. As for the encryption cost, it is at most $O(n) = O(\log N)$, where n is the number of auxiliary key pairs, since it is enough to have n bits to represent all N users. Second, it proposes the idea of batching rekeying messages and minimizes the number of encryptions by borrowing minimization technique from Boolean algebra. However, it does not present a satisfactory solution for reconstructing the key tree and reassigning the auxiliary keys, which incurs a cost proportional to the group size N.

2.3.4 One-Way Function Trees

Sherman et al. [12] proposed an algorithm based on one-way function trees (OWFTs) to establish a group session key for large dynamic groups. In this method, the controller maintains a binary key tree in which all group members are leaf nodes. However, all the leaf nodes need not be group members. Every node x is associated with two keys: an unblinded node key k_x and a blinded node key k'_x. The blinded key of a node is computed using a well-known one-way function g such that $k'_x = g(k_x)$. The internal nodes of the binary key are group keys of the respective subtrees rooted at them. The unblinded key of each internal node is computed as

$$k_x = f(g(k_{left(x)}), g(k_{right(x)}))$$
$$= f(k'_{left(x)}, k'_{right(x)}). \tag{2.1}$$

where $left(x)$ and $right(x)$ denote the left and the right children of an internal node x.

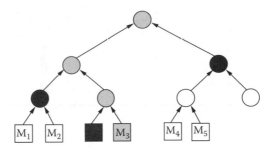

FIGURE 2.5
A one-way function tree.

The main idea of this approach is that the group session key is not delivered directly to each member, instead, members use the recursive definition in Equation 2.1 to compute a group session key (key of the root node) in a "bottom-up" fashion. The only requirement is that each member knows all blinded keys of sibling nodes along the path to the root. Based on the blinded keys of members from themselves to the root and the unblinded key itself, it can compute all the unblinded keys along the path from itself to the root in a bottom-up fashion. Figure 2.5 illustrates a one-way function tree, where member M_3 knows all the blinded keys on nodes in black and unblinded key of itself. Therefore the unblinded keys of all other gray nodes on the path to the root can be derived and known to M_3. The unblinded key associated with the root that is regarded as the group session key is finally computed independently by every member. With the "one-way" feature of the function g, even though a node's blinded key is exposed to nodes that are not its descendants (members), there is no way for them to compute its unblinded key, and therefore, it is used as the secure session key of the subgroup consisting of all its descendants.

Whenever a member joins/leaves, the controller changes the unblinded (as well as blinded) keys along the path from that member to the root. Thus, the number of new keys that are to be changed following a member-join/leave is equivalent to $O(\log(N))$ — the height of the key-tree.

Handling Group Dynamics: Figure 2.6 illustrates an example where a new member M_{new} joins the group. The controller selects a leaf node x that is close to the root and replaces it with a new internal node x' with two children, one of which is x itself and the other is M_{new}. The subgroups affected by the member-join are descendants of node x, as shown in gray. Therefore, the controller updates the unblinded keys of these gray nodes securely, to ensure backward secrecy.

In the example, node y should be given the updated blinded keys k'_z of node z. The set of nodes that needs k'_z is $S_z = \{u, v, y, M_0, M_1, M_2, M_3\}$, which exactly consists of z's sibling u and all descendants of u. This new blinded key k'_z is included in a rekeying message, encrypted by the unblinded

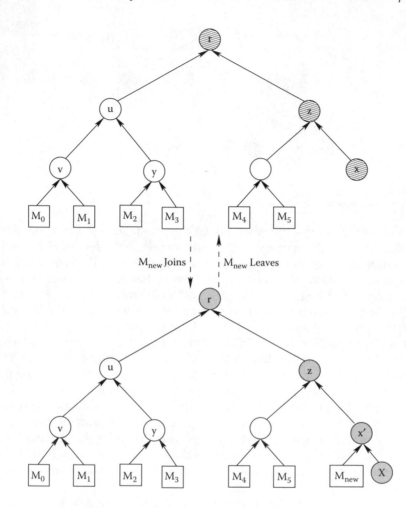

FIGURE 2.6
An example of member-join and member-leave.

key k_u of u, and then multicast to set S_z. Similarly, when the member M_{new} leaves, the controller removes the node x' and replaces it with x. All the keys along the path from x to root r are updated using the same method as when a join happens.

OWFTs also help in handling multiple member-joins and member-leaves efficiently using the following technique. When a set of members joins/leaves the group the controller identifies the nodes (in the key-tree) whose blinded keys are to be changed. All such nodes share a *common ancestor* and belong to a (sub)tree rooted at the common ancestor called *Common Ancestor Tree* (CAT). The size of the CAT determines the number of blinded keys that must be recomputed and broadcast to the group. Thus, the controller computes the blinded key that propagates the changes through the tree using a post-order traversal of CAT from the lowest level to the highest level of the tree.

Remarks: The novelty of OWFT is to improve the key management scalability by using a binary tree structure and getting around the direct delivery of the group session key to each member. Each member computes the group key using the key itself and blinded keys of sibling nodes along the path from the root to itself. The number of rekeying messages and encryption complexity are determined by the number of subgroups that need the updated blinded keys. Due to the binary tree structure of the tree-key, the number of blinded keys that need to be updated following a change in group membership is at most h (the height of the tree). Further, the number of the rekeying multicast message, as well as the encryption cost is $O(h) = O(log(N))$. The number of keys stored in the system is $O(N)$ and the number of keys stored at each member is $O(log(N))$.

However, in OWFT, the controller has to maintain the membership information of $2^N - 1$ subgroups, since each node in the tree corresponds to a subgroup consisting of itself and all its descendants.

2.3.5   Iolus

Unlike the approaches explained in the previous subsections, Iolus [11] is a framework for key management in secure multicasting groups. Iolus is a distributed hierarchical tree-based approach for key management in which a large group is decomposed into a number of smaller secure multicast subgroups. Key management in each subgroup can be independent of other subgroups, and any of the key management protocols described previously can be used for each subgroup.

Figure 2.7 depicts the Iolus architecture. In Iolus, each subgroup is managed by a subgroup controller called group security intermediary (GSI). GSIs form a hierarchy of subgroups and the top-level subgroup is managed by the group security controller (GSC). GSC is ultimately responsible for the security of the entire group. Each GSI joins the subgroup at the next higher level (or the subgroup of GSC) and acts as proxies of the GSC or its parent GSIs. In Iolus, there is no global group key; instead, each subgroup maintains its own subgroup key. When a member joins or leaves a subgroup, its effect is local to the subgroup. Therefore, only the subgroup key needs to be changed.

As a framework, Iolus specifies five basic operations, viz., the system startup, member-join, member-leave, key-refresh, and data transmissions. For starting the secure communication group, Iolus requires that at least the GSC of the group be started. After that, GSIs and other members join its subgroups.

Member-join and member-leave: To join a group, a member sends a JOIN request to its designated GSI (or GSC) using a secure unicast channel.[1]

---

[1] Any existing unicast security protocol that provides mutual authentication can be used for this purpose.

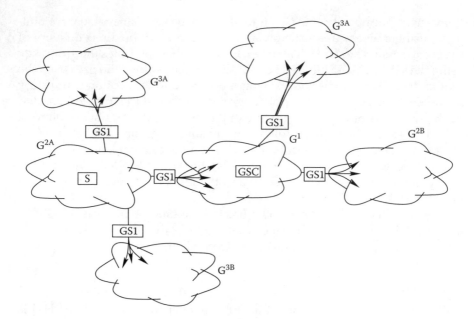

FIGURE 2.7
Iolus architecture.

Upon receiving the join request, the GSI (or GSC) decides whether to approve or deny the request. If the request is approved, the GSI (or GSC) performs the following actions:

1. It generates an individual key K_{MBR} with the new member.
2. Stores K_{MBR} along with any other relevant information concerning the new member in GSI's private database, and sends K_{MBR} to the new member securely.
3. Changes its current subgroup key K_{SGRP} to a new key K'_{SGRP}. Multicasts a GRP_KEY_UPDATE message containing K'_{SGRP} encrypted with K_{SGRP} to its subgroups and sends K_{SGRP} to the new member via the existing secure unicast channel.

A member can leave a group under two conditions: a member voluntarily leaves the subgroup, or GSI (or GSC) expels the member. In either case, the GSI (or GSC) needs to change the current subgroup key K_{SGRP} to a new key K'_{SGRP} to prevent the leaving member from participating in future communications. To distribute K'_{SGRP} to the subgroup members, it multicasts one message containing n copies of K_{SGRP} (n is the number of remaining members), each encrypted with a member's individual key K_{MBR}.

Further, to keep the subgroup key (K_{SGRP}) "fresh," each GSI changes the subgroup from time to time, and multicasts the new subgroup key to the subgroup encrypted with K_{SGRP}.

TABLE 2.4

A Performance Comparison

	KG (tree topology)	BFMT	OWFT
Keys maintained in the system	$O(N)$	$O(log(N))$	$O(N)$
Keys stored on each user	$O(log(N))$	$O(log(N))$	$O(log(N))$
Multicast rekeying messages	Group-Oriented: $O(1)$	$O(1)$	$O(log(N))$
Encryption cost	$O(log(N))$ to $O(log^2(N))$	O(log(N))	O(log(N))

Data Transmission: Due to the lack of a global group key, sending multicast data is not as simple as multicasting the data to the group encrypted with a group key. Instead, multicast data is relayed by GSIs. More specifically, the sender multicasts the data directly to its local subgroup encrypted with the subgroup key. The parent GSI (if this is not the top-level subgroup) receives multicast data, decrypts it, and remulticasts it to its parent subgroup encrypted with the subgroup key of its parent subgroup. Similarly, child GSIs get multicast data and remulticast it to their respective child subgroups.

The advantage of this approach is that there is no global group key. Thus both the frequency and computation/communication overhead of rekeying depends on the size of a subgroup instead of the size of the whole group. However, this approach requires full trust in the GSC and GSIs.

2.3.6 Performance Comparison and Summary

Table 2.4 presents a performance comparison among KG, BFMT, and OWFT. The table shows that BFMT performs better than both KG and OWFT in all aspects. Also, algorithms used in KG and OWFTs are relatively more complex than BFMT and require maintaining subgroup membership information. On the other hand, in BFMT no subgroup membership information is maintained, and the controller just needs to compute the complementary key set S of the departing members and multicast the new session key encrypted by keys in set S, which is much easier and straightforward.

Summary: Basically, key-tree-based approaches achieve scalability in key management by reducing encryption cost and the number of rekeying messages at the cost of larger storage space. The storage space is due to auxiliary keys, which are shared by members belonging to the same subgroup. The goal of adding auxiliary keys and organizing them as a tree architecture is to ensure that when new members join or old members leave the group, some rekeying messages can be encrypted aggregately using subgroup keys and multicast to all members in the subgroups rather than encrypted and unicast to each member separately. Several novel approaches have been explored and proved to succeed in achieving higher scalability, as mentioned above. However, the controller still remains the single point of failure.

2.4   Key Agreement Protocols

Key agreement protocols base their security on the computational complexity of an underlying mathematical problem. One such problem is the discrete logarithm problem (DLP), which is the basis of several key agreement protocols in the literature. The *discrete logarithm problem* is defined as follows: Let p be a prime number, Z_p denotes the set of integers modulo p. Let $\alpha \in Z_p$ be the generator such that each nonzero element in Z_p can be written as a power of α. Given a prime p, a generator α of Z_p, and a nonzero element $s \in Z_p$, find the unique integer $r, 0 \le r \le p - 2$, such that $s \equiv \alpha^r \bmod p$. Integer r is called the discrete logarithm of s to the base α. Based on the computational complexity of this problem Diffie and Hellman proposed the famous two-party key-agreement protocol [6], viz., Diffie-Hellman (DH) protocol.

2.4.1   Two-Party Diffie-Hellman Key Agreement

Diffie-Hellman (DH) key agreement protocol is a typical contributory key agreement protocol in which the session key is derived from the contributions of the two participating entities. Let A and B be two entities participating in the DH protocol. The goal of the two entities is to agree upon a shared secret that is computationally hard for others to compute, using the protocol messages. To achieve this, the two entities agree (a priori) on DH parameters, viz., p, α, and Z_p. The two-party DH protocol works as follows:

$$A \rightarrow B : \alpha^a \bmod p, \, a \in Z_p$$
$$B \rightarrow A : \alpha^b \bmod p, \, b \in Z_p$$
$$B \;\; : \;\; k_{ab} = (\alpha^a)^b \bmod p = \alpha^{ab} \bmod p$$
$$A \;\; : \;\; k_{ab} = (\alpha^b)^a \bmod p = \alpha^{ba} \bmod p.$$

A picks a secret number $a \in Z_p$, computes a public key $\alpha^a \bmod p$, and sends it to B. Similarly, B picks a secret number $b \in Z_p$, computes a public key $\alpha^b \bmod p$, and sends it to A. Once A receives the message from B, A computes a secret key $k_{ab} = (\alpha^a)^b \bmod p$. Similarly B computes the secret key $(\alpha^b)^a \bmod p$. Note that both A and B do not send their respective secrets to each other. Yet, they establish a secret that is derived from the secret of both. Further, given $\alpha^a \bmod p$ and $\alpha^b \bmod p$ it is exponentially hard to compute $\alpha^{ab} \bmod p$. But given $\alpha^a \bmod p, \alpha^b \bmod p, a$, and b, it is easy to compute $\alpha^{ab} \bmod p$. Thus, participating entities compute the session key easily, but others cannot (due to the computational complexity of discrete logarithms). Based on the two-party DH protocol, several key agreement protocols have been proposed. However, we restrict ourselves to a few well-known multiparty key agreement protocols.

2.4.2 Burmester and Desmedt's Protocol

Burmester and Desmedt [4] proposed a simple and efficient key agreement protocol for networks that support broadcasting. The protocol assumes that the group members, $M_i \cdots M_n$, are arranged in a logical ring topology such that $M_{n+1} = M_1$. This protocol includes three steps:

1. Each M_i selects its random exponent N_i and broadcasts $z_i = \alpha^{N_i} \bmod p$.
2. Each M_i computes and broadcasts $X_i = (z_{i+1}/z_{i-1})^{N_i} (\bmod \ p) = (\alpha^{N_{i+1}}/\alpha^{N_{i-1}})^{N_i} (\bmod \ p)$.
3. Each M_i computes the group key $K_i = z_{i-1}^{nN_i} \cdot X_i^{n-1} \cdot X_{i+1}^{n-2} \cdots X_{i-2}(\bmod \ p)$.

At the end of the three rounds, members compute the same group key $K = \alpha^{N_1 N_2 + N_2 N_3 \cdots + N_n N_1}$, which is computed as follows:

$$Define\, A_{i-1} = (Z_{i-1})^{N_i} = \alpha^{N_{i-1} N_i} (\bmod \ p)$$
$$A_i = (Z_i)^{N_i} = \alpha^{N_{i-1} N_i} \cdot (\alpha^{N_{i+1}/\alpha^{N_{i-1}}})^{N_i} = \alpha^{N_i, N_{i+1}} (\bmod \ p)$$
$$A_{i+1} = (Z_{i-1})^{N_1} \cdot X_i \cdot X_{i+1} = \alpha^{N_{i+1} N_{i+2}} (\bmod\, p)$$
$$K_i = A_{i-1} \cdot A_i \cdots A_{i-2} = \alpha^{N_{i-1} N_i} \cdot \alpha^{N_{i-1} N_{i+1}} = \alpha^{N_1 N_2 + N_2 N_3 + \cdots + N_n N_1}$$

The problem with this protocol is that most of the members need to change their secret (contribution) at every membership change event [7].

2.4.3 Group Diffie-Hellman Key Agreement

Group Diffie-Hellman (GDH) is a class of key agreement protocols developed by Steiner et al. [14]. These protocols are extensions of the two-party Diffie-Hellman key agreement to the multiparty scenarios. The idea is based on the following observation: If a member M_i knows the contribution of all other members in the form of $\alpha^{N_1 N_2 \cdots N_{i-1} N_{i+1} \cdots N_n}$, then using its own share it can compute the group key as $\alpha^{N_1 N_2 \cdots N_i \cdots N_n}$. However, using $\alpha^{N_1 N_2 \cdots N_{i-1} N_{i+1} \cdots N_n}$ it cannot obtain the secret of any individual member. Based on this observation Steiner et al. proposed a class of multiparty key agreement protocols, which include GDH.2 and GDH.3 protocols. Both these protocols assume a logical ring topology of the group and result in a contributory group key.

Suppose N_i is the secret exponent of member M_i and α is a generator in the algebraic group (say Z_p). The GDH.2 works as follows:

$$(1)\, M_i \rightarrow M_{i+1} : \left\{ \alpha^{\frac{N_1 \cdots N_i}{N_j}} \mid j \in [1, i] \right\}, \alpha^{N_1 \cdots N_i}, i \in [1, n-1]$$

$$(2)\, M_n \rightarrow ALL : \left\{ \alpha^{\frac{N_1 \cdots N_n}{N_j}} \mid j \in [1, n-1] \right\}.$$

The protocol has two stages. Stage 1, also called up-flow stage, requires $n-1$ rounds of message exchanges, which is used to collect contributions from all group members. In each round i, an M_i unicasts a collection of i values to M_{i+1}. Of these, $i-1$ items values are intermediate values, $\alpha^{N_2 N_3 \cdots N_i}$, $\alpha^{N_1 N_3 \cdots N_i}$, $\cdots$, $\alpha^{N_1 N_2 \cdots N_{i-1}}$, and one is a cardinal value, $\alpha^{N_1 \cdots N_i}$. When up-flow

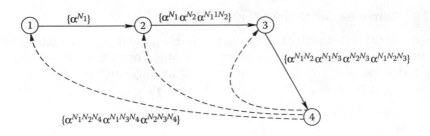

FIGURE 2.8
GDH.2 with n = 4.

reaches M_n, M_n can compute the group key as $\alpha^{N_1 N_2 \cdots N_n}$. Also, M_n computes $(n-1)$ intermediate values $\alpha^{N_2 N_3 \cdots N_n}$, $\alpha^{N_1 N_3 \cdots N_n}$, $\alpha^{N_1 N_2 \cdots N_n}$. In stage 2, M_n broadcasts these $n-1$ intermediate values to all the group members. When member M_i receives the broadcast message, it computes the group key as $\alpha^{N_1 N_2 \cdots N_n}$ by using its own share N_i to the corresponding intermediate. Figure 2.8 illustrates the working of the GDH.2 protocol for a case where $N = 4$.

In the GDH.2 protocol, every member M_i performs a total of $(i + 1)$ exponentiations and thus the computational burden on each member increases with an increase in the group size. The GDH.3 protocol addresses this issue by reducing the computational burden on each member. In the GDH.3 protocol, during the up-flow stage, each member M_i computes only the cardinal value (instead of i values as in GDH.2) and sends it to the member M_{i+1}. This avoids the overhead of computing $i - 1$ intermediate values. The GDH.3 protocol has four stages.

$$(1)\ M_i \rightarrow M_{i+1} : \alpha^{\prod N_p | p \in [1, i]}, \quad i \in [1, n-2]$$

$$(2)\ M_i \rightarrow ALL : \alpha^{\prod N_p | p \in [1, n-1]}$$

$$(3)\ M_{n-1} \rightarrow M_n : \alpha^{\frac{\prod N_p | p \in [1, n-1]}{N_i}}$$

$$(4)\ M_n \rightarrow ALL : \left\{ \alpha^{\frac{\prod N_p | p \in [1, n-1]}{N_i}} \mid i \in [1, n-1] \right\}.$$

The first stage collects the contributions of $n - 1$ members. After first stage is complete, M_{n-1} obtains $\alpha^{\prod N_p | p \in [1, n-1]}$. In stage 2, M_{n-1} broadcasts $\alpha^{\prod N_p | p \in [1, n-1]}$ to every member. In stage 3, every member factors out its own exponent from $\alpha^{\prod N_p | p \in [1, n-1]}$ and sends the result to M_n. In stage 4, M_n raises every message received in stage 3 with its own secret and unicasts the result back to the respective member. Upon receiving the message from M_n each member computes the groups key as $\alpha^{\prod N_p | p \in [1, n]}$. The problem with GDH.3 is that $n - 1$ unicast messages are sent to M_n in stage 3, which may congest M_n.

Asokan and Ginzboorg [2] proposed a similar group key agreement protocol for *ad-hoc* networks. In their method, all members share a password P, which helps avoid the man-in-the-middle attack on the two-party DH protocol. Each member M_i generates a secret share S_i. The protocol works as follows:

$$(1) \ M_i \quad \rightarrow M_{i+1} : g^{s_1 s_2 \cdots s_i}, \ i = \cdots n - 2$$
$$(2) \ M_{n-1} \rightarrow ALL : g^{s_1 s_2 \cdots s_{n-1}}$$
$$(3) \ M_i \quad \rightarrow M_n : \{g^{s_1 s_2 \cdots s_{n-1} \hat{s}_i / s_i}\}_P$$
$$(4) \ M_n \quad \rightarrow M_i : g^{s_1 s_2 \cdots s_n \hat{s}_i / s_i}$$
$$(5) \ M_i \quad \rightarrow ALL : M_i, \{M_i, H(M_1, M_2, \cdots, M_n)\}_K.$$

The first four stages are the same as those in GDH.3, except stage 3. In stage 3, every member encrypts the revised intermediate key using the share password P and sends it to M_n. In stage 4, instead of using multicast as in GDH.3, M_n unicasts the result to every member. Stage 5 is used for key confirmation, in which few (random) group members broadcast the key message to make sure that other members compute the same group key.

2.4.4 Tree-Based Group Diffie-Hellman Protocol

The tree-based Group Diffie-Hellman (TGDH) protocol [9] is a contributory group key agreement protocol that unifies two important trends in group key management: 1) TGDH uses key trees to efficiently compute and update group keys; and 2) It uses the Diffie-Hellman key exchange to achieve probably secure and fully distributed multiparty key agreement. Figure 2.9 depicts an example of TGDH key-tree model. The root is at level 0 and the lowest leaves are at level h. The key tree is a binary tree, thus every node has exactly two or zero children. Every leaf node $< h, v >$ ($0 \leq v \leq 2^l - 1$, and each level l hosts at most 2^l nodes) is associated with a member M_i of the group. Each node $< h, v >$ in the tree has a key $K_{<l,v>}$ and a blinded key $BK_{<l,v>} = \alpha^{K_{<l,v>}}$. Every member M_i at node $< h, v >$ knows every key along the path from $< h, v >$ to $< 0, 0 >$. This path is called the member's key-path, denoted

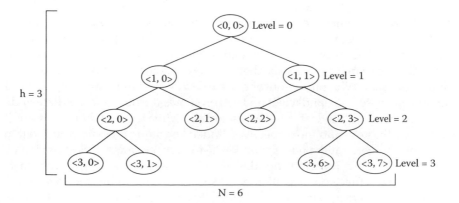

FIGURE 2.9
Notations for TGDH protocol.

as KEY_i^*. In Figure 2.9 member M_2's key-path is $KEY_2^* = \{< 3, 1 >, < 2, 0 >, < 1, 0 >, < 0, 0 >\}$. M_2 knows all the keys along its key-path, $\{K_{<3,1>}, K_{<2,0>}, K_{<1,0>}, K_{<0,0>}\}$, and knows the blinded keys of all the nodes in the tree, $BK_2^* = \{ BK_{<0,0>}, BK_{<1,0>}, \cdots, BK_{<3,7>}\}$.

Every key $K_{<l,v>}$ is of the form $\alpha^{K_{<l+1, 2v>} K_{<l+1, 2v+1>}}$, where $K_{<l+1, 2v>}$ and $K_{<l+1, 2v>}$ are the keys of the left and right child of node $< l, v >$, respectively. So in order to compute $K_{<l, v>}$, a member needs to know a key at level l and a blind key sibling at the same level. Using this recursive definition all the members compute, $K_{<0,0>}$, the group key, $K_{<0,0>}$. For example, the group key $K_{<0,0>}$ for the group presented in Figure 2.9 is computed as:

$$
\begin{aligned}
K_{<0, 0>} &= \alpha^{K_{<1,0>} K_{<1,1>}} \\
&= \alpha^{\alpha^{K_{<2,0>} K_{<2,1>}} \alpha^{K_{<2,2>} K_{<2,3>}}} \\
&= \alpha^{\alpha^{\alpha^{K_{<3,0>} K_{<3,1>}} K_{<2,1>}} \alpha^{K_{<2,2>} \alpha^{K_{<3,6>} K_{<3,7>}}}}
\end{aligned}
$$

As an example, M_2 can compute $K_{<2,0>}$, $K_{<1,0>}$, $K_{<0,0>}$ using its own key and the blinded keys $BK_{<3,0>}$, $BK_{<2,1>}$ and $BK_{<1,1>}$. It computes $K_{<2,0>}$ using $BK_{<3,0>}$ and its own key $K_{<3,1>}$, $K_{<1,0>}$ using $BK_{<2,1>}$ and $K_{<2,0>}$ and computes $K_{<0,0>}$ using $BK_{<1,1>}$ and $K_{<1,0>}$.

In TGDH protocol, any time one of the group members needs to assume a special role as *sponsor*. The criterion for assuming the role of a sponsor member depends on the membership events (join, leave, etc.). The sponsor is responsible for broadcasting the blinded keys to the group during a member-join or member-leave. However, the sponsor node is not a privileged entity, like the group controller or group leader in previous protocols. TGDH protocol includes support for the following operations: join, leave, merge, partition, and key refresh. However, we only discuss join and leave protocols as the rest of the operations are special cases or almost similar to either member-join or member-leave operation.

Join Protocol: When a new member M_{n+1} wishes to join the group, it broadcasts a join request message that contains its own blinded key $BK_{<0,0>}$. For a member-join operation a sponsor node is the one that is close to the root (if more than one such node exists, then the node with least ID assumes the role of the sponsor). When the sponsor node receives the join request, it generates an intermediate node and a leaf node. It promotes the new intermediate node as its parent, and the leaf node as its sibling. After the member-addition, it computes the keys and the respective blinded key along the path from itself to the root, and generates a new group key using the recursive definition: $K_{<l,v>} = \alpha^{K_{<l+1, 2v>} K_{<l+1, 2v+1>}}$. After computing the group key, the sponsor broadcasts the new tree, which contains all blinded keys. Upon receiving such a message all other members update their tree using this message, and compute the new group key using the recursive definition.

Figure 2.10 shows an example of member M_4 joining the group. The sponsor M_3 performs the following actions: It renames node $< 1, 1 >$ to $< 2, 2 >$,

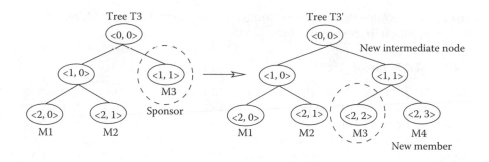

FIGURE 2.10
Tree update in a join operation.

generates a new intermediate node $< 1, 1 >$ and a new member node $< 2, 3 >$, promotes $< 1, 1 >$ as the parent node of $< 2, 2 >$ and $< 2, 3 >$, computes the new group key $K_{<0,0>}$, and broadcasts the new tree containing the blinded keys of all the group members. Upon receiving the broadcast message, every member computes the new group key.

Leave Protocol: When an existing member M_d leaves the group, the sponsor is the sibling node SN of M_d. If SN is not a leaf node, then the sponsor is the right-most leaf node of the subtree rooted at SN. In the leave protocol, the sponsor updates its key tree by deleting the node of M_d and its parent node. The sponsor picks a new secret share, computes all keys on its key path up to the root, and broadcasts the new blinded keys of its key path to the group. This information allows all members to recompute the new group key.

For example, consider Figure 2.11, if member M_3 leaves the group, the sponsor M_5 deletes node $< 1, 1 >$ and $< 2, 2 >$, promotes its parent node to $< 1, 1 >$, its sibling M_4 to node $< 2, 2 >$, and itself to $< 2, 3 >$. After updating the tree, the sponsor M_5 picks a new key $K_{<2,3>}$, recomputes $K_{<1,1>}$, $K_{<0,0>}$, $BK_{<2,3>}$ and $BK_{<1,1>}$, and broadcasts the updated tree with $BK_5{}^*$. Upon receiving the broadcast message, the rest of the group members compute the group key. Note that the member that left the group, M_3, cannot compute the

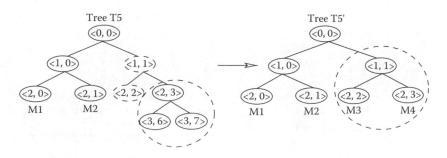

FIGURE 2.11
Tree update in a leave operation.

group key because its share is no longer in the group key, and M_5 refreshed its contribution in the group key.

2.4.5　STR Protocol

Kim et al. proposed STR [8], a tree-based key agreement protocol, which is an extension of a protocol proposed by Steer et al. [13]. STR [8] protocol is based on the observation that the rapid advances in computing are shifting the bottleneck (overhead) requirements of a protocol from computation to communication. Thus, communication cost has a more pronounced effect on the system performance than the computation cost. With this philosophy, STR protocol allows more liberal use of cryptographic operations while attempting to reduce the communication overhead, which dominates in a WAN environment. STR is basically an "extreme" version of TGDH, where the key-tree structure is completely unbalanced or stretched out.

Like TGDH, the STR protocol uses a tree structure that associates the leaves with individual random session contributions of the group members. Each leaf node (LN), M_i, has a random secret r_i and a corresponding blinded secret $br_i = \alpha^{r_i} \bmod p$. Similarly, every internal node (IN) M_j has an associated secret key k_j and a public blinded key $bk_j \, \alpha^{k_j} \bmod p$. The secret key of an internal node is the result of a Diffie-Hellman key agreement between the node's two children. Thus, k_i (i > 1) can be computed recursively as follows:

$$k_i = (bk_{i-1})^r \bmod p \qquad\qquad (2.2)$$
$$= (br_i)^{k_{i-1}} \bmod p$$
$$= \alpha^{r_i k_{i-1}} \bmod p \qquad if \ i > 1.$$

Generally, the group key for a group of N members is computed as follows:

$$k_2 = (br_2)^{r_1} \bmod p = \alpha^{r_1 r_2} \bmod p, \ bk_2 = \alpha^{k_2} \bmod p \qquad (2.3)$$
$$k_N = (br_N)^{k_{N-1}} \bmod p \qquad\qquad (2.4)$$

The group key is the key associated with the root node. As an example, the group key in Figure 2.12 is: $k_4 = \alpha^{r_4 \alpha^{r_3 \alpha^{r_2 r_1}}} \bmod p$. Similar to the TGDH

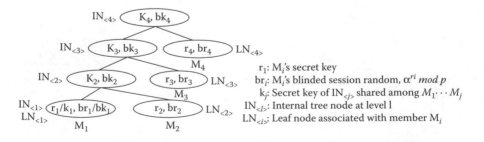

FIGURE 2.12
Tree notation for STR protocol.

TABLE 2.5

Comparison of TGDH and STR

	TGDH		STR	
	Join	Leave	Join	Leave
Rounds	2	1	1	1
Messages	3	1	2	1
Unicast	0	0	1	0
Broadcast	3	1	1	1
Exponentiation	$2log(n)$	$log(n)$	2	$3n/2+2$

protocol, STR also needs a sponsor member, which is responsible for handling member join/leave operations.

Join Protocol: When a new member M_{n+1} wishes to join the group, it broadcasts a join request message that contains its own blinded session secret br_{n+1}. Upon receiving the request, the current group's sponsor (M_n) creates a new root node and adds the old key tree as its left node and the new member as its right node. The sponsor computes a blinded version of the old group key (bk_n) and sends the old tree $BT_{<n>}$ to M_{n+1} with all blinded keys and blinded session secrets. Upon receiving the message, each member increases the size of the group by one and generates a new tree structure, and a group key using the same procedure as that of the sponsor.

Leave Protocol: When a member M_d ($d \leq n$) leaves the group, if $d > 1$, the sponsor M_s is the member M_{d-1}, otherwise the sponsor is M_2. The sponsor updates its key tree by deleting the nodes $LN_{<d>}$ corresponding to M_d and its parent node $IN_{<d>}$. Then the sponsor renumbers the nodes that are at a higher level than $LN_{<d>}$ to accommodate the member-leave. It promotes the former sibling $IN_{<d-1>}$ of M_d to (former) M_d's parent. After updating the tree structure, the sponsor selects a new secret session random, computes all keys and blinded keys up to the root, and broadcasts the $BT_{<s>}$ to the group. This $BT_{<s>}$ allows all members to recompute the new group key. Table 2.5 shows a comparison of TGDH protocol and STR protocol. As seen from the table, STR costs less in communication on every membership event.

2.4.6 Hypercube Protocol

Becker and Wille proposed the 2^d-cube protocol [3]. The basic idea of the protocol is to divide a large group into four (2^2) subgroups recursively. Each subgroup agrees on a key using a 4-party (DH-based) key agreement protocol. The group key is constructed bottom-up fashion from the subgroup key. Thus, by grouping four members (or subgroups) in a 4-party key agreement, the protocol minimizes the number of rounds for a generic n-party key agreement. The 2^d-cube protocol is best understood by an example. Consider the four nodes as depicted in Figure 2.13. Let four nodes A, B, C, D be arranged in a square, and let a, b, c, d be their respective secrets. In the first round, A and B

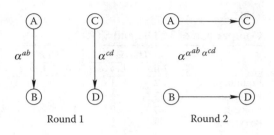

Round 1 ⟍⟋ Round 2

FIGURE 2.13
An example of 2^d-cube agreement exchange.

agree on a DH key $K_{AB} = \alpha^{ab}$, and parallel on a C and D agree on a DH key $K_{CD} = \alpha^{cd}$. In the next round, A and C establish a DH key using secrets K_{AB} and K_{CD}. Similarly, B and D establish a DH key using secrets K_{AB} and K_{CD}. After two rounds, all four members will have the same key $K = \alpha^{\alpha^{ab}.\alpha^{cd}}$.

This process can be extended to work in any group that has number N, where $N = 2^d$, which is arranged on the vertices of a d-dimensional cube. In such a group, in jth round, each participant executes a two-party DH key agreement with its peer on the jth-dimension using the key of the $(j - 1)th$ round as its secret exponent. Thus, 2^d participants agree on a key after d rounds of DH key exchange on the edges of a d-dimensional cube.

For the case where the number of group members, N, is not a power of 2, Becker and Wille proposed the 2^d-Octopus protocol. In the Octopus protocol, if $2^d < N < 2^{d+1}$, the first 2^d participants play the role of central controllers. The rest of the participants form wards that are attached to one of the central nodes. First, the controllers execute a Diffie-Hellman key agreement with the wards. Then, the controllers perform 2^d-cube exchange using the keys gathered in the first stage. Finally the key derived in the second stage is distributed to the wards. The Octopus protocol is efficient (in number of rounds) for a member-join operation. However, it is very inefficient for member-leave operation.

Table 2.6 summarizes the performance of the four key agreement protocols for establishing a key. As we see, each of the protocols has a niche of

TABLE 2.6

A Comparison of the Performance of Key Agreement Protocols

Protocol	Messages	Message-size	Total Exponentiations	Exchanges[a]	Rounds
Burmester et al's protocol	2n (broadcast)	2n	n^2	$2n$	2
GDH.2	n	$(n-1)(n+1)/2 - 1$	$(n+3)n/2 - 1$	n	n
GDH.3	$2n - 1$	$3(n-1)$	$5n - 6$	$2n - 1$	$n+1$
Hypercube	$n\log_2(n)$	$n\log_2(n)$	$n(1 + \log_2(n))$	$0.5n\log_2(n)$	$\log_2(n)$
Octopus[b]	$3n + 2^d(d-3)$	$3n + 2^d(d-3)$	$4n + 2^d(d-3)$	$2n + 2^{d-1}(d-4)$	$d+2$

[a] An exchange means a DH key exchange by two parties simultaneously or a key exchange from one party to another at one time.

[b] For any d $2^d < n < 2^{d+1}$.

applicability. For the system that supports broadcasting Burmester et al.'s protocol is the best among the four protocols. But as discussed earlier, it is inefficient in handling group dynamics. The GDH.2 and GDH.3 protocols require a number of rounds that are linear in number of members in the group, but handle group dynamics efficiently compared to Burmester et al.'s protocol. The hypercube achieves a logarithmic number of message exchanges and rounds, but it is inefficient in handling member-leave operations.

2.5 Summary

With a rapid growth in dynamic group-oriented systems and applications, security in such dynamic groups has become an increasingly important issue. Security in dynamic groups is necessary to provide services like authentication, access control, confidentiality, nonrepudiation, and privacy to name a few. A group key is an efficient means to provide such services to a group. Further, a group key can be used by group members for any subsequent cryptographic operations.

In groups where it is feasible to have a controlling authority, like a group controller, the process of generating and distributing a group key, and managing keying relationships among group members is called key management. For peer-to-peer groups, where no single member can be assigned a special role like the group controller, it is not feasible to have a controlling authority. Thus, members have to establish and manage keys in the group by themselves, which is typically achieved via key agreement.

In this chapter, we described a few well-known protocols for key management and key agreement in dynamic groups. Each of the protocols described has its own niche of applicability, and shows an improvement in one or more performance metrics (say, like the number of rounds) by trading off some other performance metric(s). From a rich volume of literature on these topics, it is evident that there is no single solution that satisfies all the requirements of all possible dynamic groups. Furthermore, it is a well-known fact that designing a secure protocol is a notoriously hard task. Thus, a standard practice is to analyze security requirements of a specific group, and design or select an efficient key management or a key agreement protocol for that group.

Bibliography

1. Group key management protocol (GKMP) specification. RFC 2094, IETF Internet Draft, 1997.
2. N. Asokan and Philip Ginzboorg. Key agreement in ad hoc networks. *Computer Communications*, 23(17):1627–1637, 2000.

3. Klaus Becker and Uta Wille. Communication complexity of group key distribution. In *CCS '98: Proceedings of the 5th ACM Conference on Computer and Communications Security*, pages 1–6, 1998.
4. Mike Burmester and Yvo Desmedt. A secure and efficient conference key distribution system (extended abstract). In *EUROCRYPT*, pages 275–286, 1994.
5. Isabella Chang, Robert Engel, Dilip D. Kandlur, Dimitrios E. Pendarakis, and Debanjan Saha. Key management for secure internet multicast using boolean function minimization techniques. In *INFOCOM*, pages 689–698, 1999.
6. Whitfield Diffie and Martin E. Hellman. New directions in cryptography. *IEEE Transactions on Information Theory*, IT-22(6):644–654, 1976.
7. Yongdae Kim, Adrian Perrig, and Gene Tsudik. Simple and fault-tolerant key agreement for dynamic collaborative groups. In *CCS '00: Proceedings of the 7th ACM conference on Computer and Communications Security*, pages 235–244, 2000.
8. Yongdae Kim, Adrian Perrig, and Gene Tsudik. Communication-efficient group key agreement. In *Sec '01: Proceedings of the 16th International Conference on Information Security: Trusted Information*, pages 229–244, 2001.
9. Yongdae Kim, Adrian Perrig, and Gene Tsudik. Tree-based group key agreement. *ACM Trans. Inf. Syst. Secur.*, 7(1):60–96, 2004.
10. Alfred J. Menezes, Scott A. Vanstone, and Paul C. Van Oorschot. *Handbook of Applied Cryptography*. CRC Press, Inc., Boca Raton, FL, 1996.
11. Suvo Mittra. Iolus: a framework for scalable secure multicasting. In *SIGCOMM '97: Proceedings of the ACM SIGCOMM '97 Conference on Applications, Technologies, Architectures, and Protocols for Computer Communication*, pages 277–288, 1997.
12. Alan T. Sherman and David A. McGrew. Key establishment in large dynamic groups using one-way function trees. *IEEE Trans. Softw. Eng.*, 29(5):444–458, 2003.
13. D. G. Steer, L. Strawczynski, Whitfield Diffie, and Michael J. Wiener. A secure audio teleconference system. In *CRYPTO*, pages 520–528, 1988.
14. Michael Steiner, Gene Tsudik, and Michael Waidner. Diffie-hellman key distribution extended to group communication. In *CCS '96: Proceedings of the 3rd ACM Conference on Computer and Communications Security*, pages 31–37, 1996.
15. Chung Kei Wong, Mohamed Gouda, and Simon S. Lam. Secure group communications using key graphs. In *SIGCOMM '98: Proceedings of the ACM SIGCOMM '98 Conference on Applications, Technologies, Architectures, and Protocols for Computer Communication*, pages 68–79, 1998.
16. Chung Kei Wong, Mohamed Gouda, and Simon S. Lam. Secure group communications using key graphs. *IEEE/ACM Trans. Netw.*, 8(1):16–30, 2000.

3

Securing Design Patterns for Distributed Systems

Eduardo B. Fernandez and Maria M. Larrondo-Petrie

CONTENTS

3.1 Introduction

Complex applications such as medical, financial, military, and legal, have become very important in the last few years. These applications are typically distributed and implemented in systems that have additional nonfunctional requirements such as reliability, fault tolerance, or real-time constraints. They are composed of a variety of units, some built ad hoc and some bought or outsourced. Another typical aspect of these systems is that they may need to follow regulatory standards, e.g., HIPAA [23], Sarbanes/Oxley [32], Graham-Leach-Bliley [22], or military standards. Their architectures may include databases of different types and typically require Internet and wireless access. The applications in these systems are usually integrated using a Web Application Server (WAS), a type of middleware that has a global enterprise model, implemented with object-oriented components such as J2EE or .NET. These applications are of fundamental value to enterprises and their security is extremely important. Security is complicated by the need to support distribution, heterogeneity, and different types of policies and mechanisms.

A systematic approach is required to build these applications so they can reach the appropriate level of security. We look here at some security aspects of the middleware structure needed to support such applications.

Embedding security into middleware systems requires a secure development methodology. We have proposed a methodology that helps developers build secure systems without being security experts [12, 15]. This methodology accomplishes its purpose through the use of patterns. Our discussion here is independent of this methodology although this work resulted as a consequence of that work. We concentrate on system architecture aspects; network security relies heavily on cryptography and is not considered in our discussion. Agents are also of interest in this context but are not considered either.

Patterns provide solutions to recurrent problems and many of them have been catalogued. We see the use of patterns as a fundamental way to incorporate security principles in the design process even by people having little experience with security practices. In our work we have found many security patterns, e.g., [6, 10]. We also developed a type of pattern called a Semantic Analysis Pattern (SAP), which implements a set of basic use cases [9]. We have shown that we can combine SAPs and security patterns to create authorized SAPs, which can be converted into conceptual models for secure designs [12]. These models could be used to define combined application/security models for WAS applications. We have also addressed how to carry over the security model of the analysis stage into the design stage [15]. In this stage the middleware functionality is used to support the applications. We show here how patterns allow us to define a secure architecture for middleware systems able to accommodate strict requirements. Our approach to middleware security is to consider the architecture as a composition of functional (unsecured) patterns with patterns that provide specific security functions. We show in some detail how we can start from general distribution and component patterns and add security patterns to build a secure middleware architecture.

There has been a good amount of work on software architectures for middleware systems [35]. However, there is much less work about their security. Schmidt studies the use of patterns to build extensible brokers [34] and to build telecommunications systems [33], but he does not consider security aspects, although his more recent papers consider security [39]. Crane et al. [5] consider patterns for distribution but again they do not include security aspects. Keller et al. [24] discuss patterns for network management but they do not include security. Security aspects are considered in [18], which analyzes how to combine security policies in heterogeneous middleware, and [26] that defines how to find identities for clients and servers. Reference [8] applies aspect-oriented programming to separate middleware services, including security. These papers consider very specific security problems. We are interested in the global security architecture. Global security aspects are discussed in [7], which focuses on the combination of RBAC and multilevel models but does not consider architectures using patterns.

Section 2 presents an overview of our approach to middleware security. Section 3 discusses security aspects of components. Section 4 considers security in the distribution architecture. Section 5 discusses implementation issues. Section 6 summarizes our general methodology to develop secure systems, while the last section presents some conclusions.

3.2 Middleware and Security

Middleware is the software layer between the operating system and the user-distributed applications. Middleware hides details of the communication layers usually controlled by the operating system and provides convenient application programming and management. The Internet has brought new challenges to the basic middleware architectures [20], one of the most important being security. We start from the functional patterns that describe the architecture of a middleware system [3, 19, 35]. Figure 3.1 shows some of the patterns that could affect the security of a middleware system. These patterns are secure versions of the corresponding functional patterns.

The isolated patterns, secure layers, secure facade, and secure reflection, are orthogonal to the others and can be combined with any of the other patterns. Secure layers [10, 41] is a specialization of the layers pattern of [3]. It describes the use of architectural layers to provide security to the functional mechanisms allocated to these layers. We discuss secure facade later in this chapter. Secure reflection emphasizes one of the uses of reflection (a general reflection pattern appears in [3]).

Starting from the conceptual model of the application (maybe composed from a set of analysis patterns) we define security constraints (rules) at that level. These rules are stored in a Policy Administration Point (PAP) [42] and are enforced when a request is sent to the Policy Enforcement Point (PEP) that consults the Policy Decision Point (PDP). The PDP uses the information in the PAP and in the Policy Information Point (PIP) to decide if the request is valid. The PIP includes additional information to make the decision.

The next architectural level includes architectural patterns to implement the application into a specific platform. The standard functional patterns for this level have been complemented with security functions: secure model view controller, secure adapter, secure broker, security enterprise component framework, and secure Web services (these include a variety of patterns for security standards, e.g., [6].

The following level corresponds to databases and high-level communications. Here we have a secure relational database mapping, a secure proxy, and a secure client/dispatcher/server. Part of this level is the lower boundary of middleware. A typical middleware system may also include legacy systems and COTS (Commercial Off-the-Shelf) components.

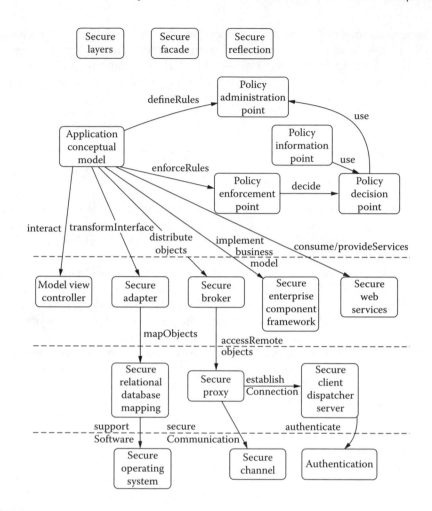

FIGURE 3.1
Pattern diagram for middleware architectural patterns.

The bottom level includes a secure operating system [36], a secure channel [2], and an authenticator [36].

We discuss in the next section how to add security to functional patterns to obtain their secure versions. To illustrate the approach we consider two aspects of basic importance for the security of a middleware architecture:

- How to store and execute a business enterprise model. Business models are handled through component frameworks, typically using an object-oriented model. Part of this model may consume or provide Web services.

- Its distributed systems architecture. Distribution is handled through distributed objects or Web services protocols.

3.3 Components and Security

Several patterns solve specific problems of components:

- The Enterprise Component Framework pattern [25] describes the container structure of components (Figure 3.2). This representation of components can describe J2EE and .NET components by proper specialization.
- The Component Configurator [35] lets an application dynamically attach and detach components or processes.
- The Interceptor [35] allows the transparent addition of services to an application or framework. These services are automatically invoked when certain events occur.
- The Extension Interface [35] defines multiple interfaces for a component. The Component Interface pattern focuses on the composition of interfaces [29].
- The Home pattern separates the management of components from their use by defining an interface for creating instances of components.

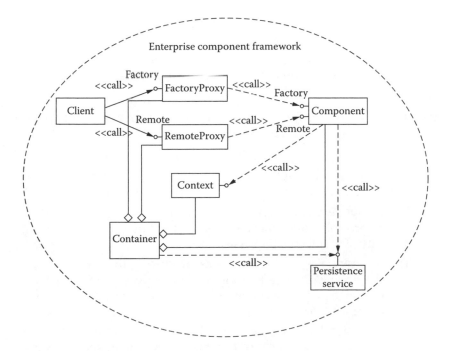

FIGURE 3.2
Class diagram of the Enterprise Component Framework pattern.

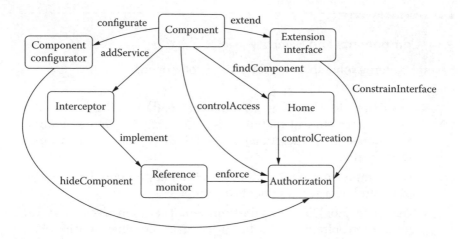

FIGURE 3.3
Pattern diagram for securing component patterns.

Three patterns are used to handle persistent data and to hide low-level details:

- The Facade [19] provides a unified, higher-level interface to a set of interfaces in a subsystem.
- The Adapter [19] converts the interface of an existing class into a more convenient interface.
- The Wrapper Facade [35] encapsulates the functions and data provided by existing subsystems or levels and defines a higher-level interface.

We can add security to these functional patterns to define a set of patterns that can implement a secure component-based architecture (Figure 3.3):

- The Enterprise Component Framework can include security descriptors that define authorization rules [30]. These rules can then be enforced by a concrete version of the Reference Monitor pattern [11].
- The Component Configurator can be used to reduce the time when critical processes are exposed to attacks by hiding them from the visibility of suspicious processes.
- The Interceptor is useful to add security to a framework, e.g., a CORBA-based system, if the original implementation did not have it. The intercepted requests can be checked by a concrete version of the Reference Monitor.
- The Extension Interface can be used to define views that let a user or role access only some parts of the information in specific ways, according to their authorizations. This is similar to the use of views for database security [38].

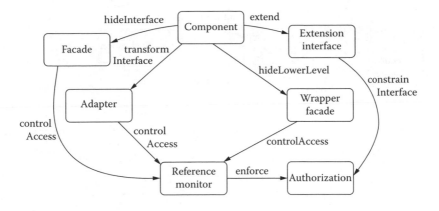

FIGURE 3.4
Pattern diagram for securing hiding patterns.

- The Home pattern can be used to apply authorization rules to control the creation of objects in components as it has been done in operating systems [11]. There is a pattern in operating systems, the Controlled Object Creator, which can be used to assign rights to a newly created object.

Similarly, we can add security to the three other patterns (Figure 3.4):

- The Facade can hide implementation details that could be exploited by hackers and can apply security checks in the operations of the Facade according to authorization rules.
- The Adapter can be used to define a new interface with fewer operations for some uses according to their security restrictions, to map database security constraints to application constraints, or to just enforce access restrictions in the operations of the interface.
- The Wrapper Facade can be used to hide the implementation of the lower levels. This prevents attackers from taking advantage of implementation flaws. A higher-level interface restricts the possibilities of a hacker. Access control is also possible in the operations of the interface.

The combined use of these patterns would be able to provide a good level of security to the component aspect of a middleware system.

3.4 Distribution Architecture and Security

A Secure Broker provides transparent and secure interactions between distributed components. Figure 3.5 shows how we can add security to the broker pattern by composing it with several security patterns [16, 27].

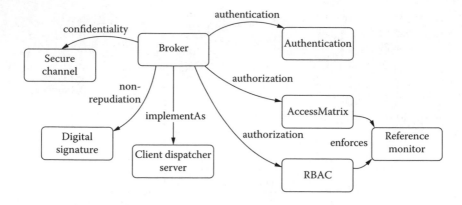

FIGURE 3.5
Pattern diagram for secure broker.

To interpret the meaning of the broker pattern [3] we need to apply access control at the local proxy and the object adapter, and we need to define a secure channel between the client and the server. We also need mutual authentication between the local proxy and the object adapter. We show two possibilities for authorization (Access matrix and RBAC); we did not do this in the earlier figures for the sake of simplicity.

The Client Dispatcher Server pattern is, in turn, implemented using Lookahead, Connector/Acceptor, and other lower-level patterns. These may apply some of the required enforcement; for example, the Connector when establishing a new connection would apply authentication.

There are also many standards for Web services security, e.g., XACML [42] and SAML [31]. They must be considered when producing or consuming Web services in the middleware. Their combination with the remote object security architecture makes middleware security quite complex. For example, a pattern for XACML [6] can be combined with functional Web service patterns.

An important direction is the use of Web services in wireless devices. A cell phone can be a consumer and provider of Web services. Web services standards can be transported to wireless devices by appropriately reducing them [4, 13]. The architectures used for remote objects do not appear very appealing for wireless devices and we know of no commercial system using this distribution approach. A problem may be their lack of convenient interoperability. However, distributed objects may be of interest for systems that require a higher level of security or performance.

3.5 Implementation Possibilities

Patterns provide an abstract view of architecture including security aspects. There are different ways to implement security in middleware that can make use of patterns. We consider a few below.

Eichberg and Mezini [8] modularize middleware security services using Aspect-Oriented programming. They perform this modularization by working from the code and adding pointcuts using Java annotations. In principle, the separated aspects could correspond to patterns as in [21].

Another way to add security to middleware is the use of microkernels [37]. A microkernel itself has been represented as a pattern [14], and it just corresponds to a lower-level mechanism. The security services could be external servers of this architecture.

Demurjian et al. [7] compare the security models of CORBA, J2EE, and .NET and propose a secure architecture to realize the models. Their proposed Unified Security Resource, comprising Security Policy Service, Security Authorization Service, Security Registration Service, and Security Analysis and Tracking, can be implemented using patterns.

D. Schmidt's group has done significant work on middleware for several years. In [39] they consider the use of metaprogramming mechanisms to enforce security as well as other properties. Smart proxies are application-oriented stubs that replace the stubs created by the ORB compiler to customize client behavior. Interceptors are objects in the path of the ORB to monitor or modify the behavior of the invocation. These mechanisms can be used to add or replace functionality into distributed object applications.

3.6   A Secure Software Development Methodology

These ideas are part of a secure software methodology [15]. A basic idea in the proposed methodology is that security principles must be applied at every development stage and that each stage can be tested for compliance with those principles. Figure 3.6 shows a secure software lifecycle. The white

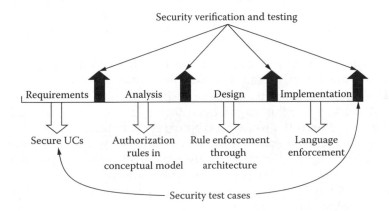

FIGURE 3.6
Secure software lifecycle.

arrows show where security can be applied and the black arrows where we can audit compliance with security policies. We describe each stage in the text that follows.

Requirements stage: When building object-oriented software, use cases define the required interactions with the system. We study each action within a use case and see which attacks are possible [17]. We then determine which policies would stop these attacks. From the use cases we can also determine the needed rights for each actor and thus apply a need-to-know policy. Note that the set of all use cases defines all the uses of the system and from all the use cases we can determine all the rights for each actor. The security test cases for the complete system can also be defined at this stage. Risk analysis should be applied at this stage.

Analysis stage: Analysis patterns, and in particular, semantic analysis patterns, can be used to build the conceptual model in a more reliable and efficient way [9]. Security patterns are used to describe security models or mechanisms. We can build a conceptual model where repeated applications of a security model pattern realize the rights determined from use cases. In fact, analysis patterns can be built with predefined authorizations according to the roles in their use cases. Then we only need to additionally specify the rights for those parts not covered by patterns. We can start defining what mechanisms (countermeasures) are needed to prevent attacks.

Design stage: Design mechanisms are selected to implement the policies that can stop the attacks identified in the requirements stage. User interfaces should correspond to use cases and may be used to enforce the authorizations defined in the analysis stage. Secure interfaces enforce authorizations when users interact with the system. Components can be secured by using the security structure of Java or .NET components. Distribution provides another dimension where security restrictions can be applied, and we have shown some possibilities in this chapter. Deployment diagrams can define secure configurations to be used by security administrators. A multilayer architecture is needed to enforce the security constraints defined at the application level. In each level we use patterns to represent appropriate security mechanisms. Security constraints must be mapped between levels.

Implementation stage: This stage requires reflecting in the code the security rules defined in the design stage. Because these rules are expressed as classes, associations, and constraints, they can be implemented as classes in object-oriented languages. In this stage we can also select specific security packages or COTS, e.g., a firewall product, a cryptographic package.

Deployment and maintenance stages: Our methodology does not yet address issues in these stages. When the software is in use other security problems may be discovered by users. These problems can be handled by patching, although the amount of patching after applying our approach should be significantly smaller compared to current systems.

If necessary, the security constraints can be made more precise by using the Object Constraint Language (OCL) [40], instead of textual constraints. Patterns for security models define the highest level of the architecture. At each lower level we apply the model patterns to specific mechanisms that enforce these models. In this way we can define patterns for file systems, Web documents, J2EE components, etc. We can also evaluate new or existing systems using patterns. If a system does not contain an embodiment of a correct pattern then it cannot support the corresponding secure model or mechanism.

3.7 Conclusions

We have analyzed how the security of a typical middleware, a Web Application Server, can be decomposed into the security of several patterns used in building the middleware. In particular, we considered a system with distributed objects. Systems using Web services can be analyzed similarly except that they use a larger variety of security standards. Web services have a strong affinity to remote objects [1]. They need lifecycle support, dynamic object creation/deletion, state management, transaction support. While most current Web services platforms do not provide this, it is clear that these functions will be in them in the future. This implies an evolution of middleware systems where combinations of patterns like these or new patterns will be needed.

We indicated that these patterns should be used through a specific methodology and we have proposed an approach of this type [15]. That approach and the patterns presented here could be the basis of a specific approach to build secure middleware.

Patterns like these have value for approaches such as Model-Driven Architecture (MDA). Patterns are a fundamental aspect of MDA. MDA is a conceptual framework that separates business-oriented aspects from platform aspects together with some related standards (UML, MOF, CWM) [28].

Future work includes the integration of the approaches discussed here with our methodology, one should be able to map classes to components and secure these components at the same time. We also need to write or complete some of the patterns mentioned here.

Acknowledgments

This work was supported through a federal earmark grant from DISA, administered by Pragmatics, Inc.

Bibliography

1. K.P. Birman. Like it or not, web services are distributed objects. *Comm. of the ACM*, vol. 47, No. 12, December 2004, 60–62.
2. A. Braga, C. Rubira, and R. Dahab. Tropyc: A pattern language for cryptographic object-oriented software. Chapter 16 in *Pattern Languages of Program Design 4* (N. Harrison, B. Foote, and H. Rohnert, Eds.).
3. F. Buschmann, R. Meunier, H. Rohnert, P. Sommerland, and M. Stal. *Pattern-Oriented Software Architecture*. Wiley, 1996.
4. H. Chu, C. You, and C. Teng. Challenges: Wireless web services. *Proc of the 10th International Conference on Parallel and Distributed Systems (ICPADS '04)*.
5. S. Crane, J. Magee, and N. Pryce. Design patterns for binding in distributed systems. *Proc. of the OOPSLA-95 Workshop on Design Patterns for Concurrent, Parallel and Distributed Object-Oriented Systems*. October 1995. Austin, Texas.
6. N. Delessy, E.B. Fernandez, and T. Sorgente. Patterns for the eXtensible Access Control Markup Language. *Proc. of the Pattern Languages of Programs Conference (PLoP 2005)*. Allerton Park, IL.
7. S. Demurjian, K. Bessette, T. Doan, and C. Phillips. Concepts and Capabilities of Middleware Security. Chapter 9 in *Middleware for Communications*, Q. H. Mahmoud (Ed.). John Wiley and Sons, 2004.
8. M. Eichberg and M. Mezini. Alice: Modularization of middlewar using aspect-oriented programming. *Proc. of the Conf. on Software Eng. and Middleware (SEM) 2004*. Linz, Austria, Sept. 2004.
9. E.B. Fernandez and X. Yuan. Semantic analysis patterns. *Proc. of 19th Int. Conf. on Conceptual Modeling*. ER2000, 183-195. Also available from: http://www.cse.fau.edu/ed/SAPpaper2.pdf.
10. E.B. Fernandez and R. Pan. A Pattern Language for security models. *Proc. of PLoP 2001*. http://jerry.cs.uiuc.edu/plop/plop2001/accepted_submissions.
11. E.B. Fernandez and J.C. Sinibaldi. More patterns for operating system access control, *Proc. of the 8th European Conference on Pattern Languages of Programs, EuroPLoP 2003*. http://hillside.net/europlop, 381–398.
12. E.B. Fernandez. A methodology for secure software design. *2004 Intl. Symposium on Web Services and Applications (ISWS'04)*. Las Vegas, NV, June 21–24, 2004.
13. E.B. Fernandez, S.A. Rajput, M. VanHilst, and M.M. Larrondo-Petrie. Some security issues of wireless systems, in *Advanced Distributed Systems*, F.F. Ramos, V. Larios R. and H. Unger (Eds.). *Proc. of 5th International School and Symposium on Advanced Distributed Systems, ISSADS 2005*, IEEE, Guadalajara, Mexico, January 24–28, 2005. Revised Selected Papers, LNCS 3563, Springer, 2005, 388–396.
14. E.B. Fernandez and T. Sorgente. A pattern language for secure operating system architectures. *Proc. of the 5th Latin American Conference on Pattern Languages of Programs*. Campos do Jordao, Brazil, August 16–19, 2005.
15. E.B. Fernandez, M.M. Larrondo-Petrie, T. Sorgente, and M. VanHilst. A methodology to develop secure systems using patterns, Chapter 5 in *Integrating security and software engineering: Advances and Future Vision*. H. Mouratidis and P. Giorgini (Eds.), IDEA Press, 2006. 107–126.
16. E.B. Fernandez, S. Huang, and M.M. Larrondo-Petrie. Building secure middleware using patterns. *Proc. of the IEEE Int. Symposium and School on Advanced Distributed Systems (ISSADS 2006)*.

17. E.B. Fernandez, M. VanHilst, M.M. Larrondo-Petrie, and S. Huang. Defining security requirements through misuse actions, *Advanced Software Engineering*: Expanding the Frontiers of Software Technology. S.F. Ochoa and G.-C. Roman (Eds.), International Federation for Information Processing. Springer, 2006, 123–137.

18. S.N. Foley, T.B. Quillinan, M. O'Connor, B.P. Mulcahy, and J.P. Morrison. A framework for heterogeneous middleware security. *Proc. of the 13th Heterogeneous Computing Workshop (HCW'04)*, 1–11.

19. E. Gamma, R. Helm, R. Johnson, J. Vlissides. *Design Patterns: Elements of Reusable Object-Oriented Software*, Addison-Wesley, Boston, 1994.

20. D. Geibs, Middleware challenges ahead. *Computer*. June 2001, IEEE, 24–31.

21. G. Georg, Indrakshi Ray, and R. France. Using aspects to design a secure system, *Proceedings of the Eighth IEEE International Conference on Engineering of Complex Computer Systems*. Greenbelt, MD, December 2002.

22. Gramm-Leach-Bliley Act, Senate Banking Committee, Monday, November 1, 1999 http://www.senate.gov/banking/conf/fincon.pdf.

23. http://www.hipaa.org/.

24. R.K. Keller, J. Tessier, and G. von Bochmann. A pattern system for network management interfaces. *Comm. of the ACM*, vol. 41, No. 9, September 1998, 86–93.

25. C. Kobryn. Modeling components and frameworks with UML. *Comm of the ACM*, October 2000, 31–38.

26. U. Lang, D. Gollmann, R. Schreiner. Verifiable identifiers in middleware security. *Proceedings 17th Annual Computer Security Applications Conference, ACSAC, 2001*, 10–14 Dec. 2001.

27. P. Morrison and E.B. Fernandez. Securing the broker pattern, *Proc. of the 11th European Conf. on Pattern Languages of Programs (EuroPLoP 2006)*.

28. Object Management Group, http://www.omg.com.

29. R. Pereira e Silva and R.T. Price. Component interface pattern. *Proc. of the Pattern Languages of Programs Conference, PLoP 2002*.

30. M. Pistoia, N. Nagaratnam, L. Koved, and A. Nadalin. *Enterprise Java Security: Building Secure J2EE Applications*. Addison-Wesley, 2004.

31. SAML, http://www.oasis-open.org/committees/tc_home.php?wg_abbrev=security.

32. One Hundred Seventh Congress of the United States of America, Sarbanes Oxley Act of 2002, January 23, 2002, http://news.findlaw.com/hdocs/docs/gwbush/sarbanesoxley072302.pdf.

33. D.C. Schmidt. Using design patterns to develop reusable object-oriented communication software. *Comm. of the ACM*, vol. 38, No. 10, October 1995, 65–74.

34. D.C. Schmidt and C. Cleeland. Applying patterns to develop extensible ORB middleware. *IEEE Comm. Mag.*, April 1999, 54–63.

35. D.C. Schmidt, M. Stal, H. Rohnert, and F. Buschmann. *Pattern-Oriented Software Architecture*. Wiley, 2000.

36. M. Schumacher, E.B. Fernandez, D. Hybertson, F. Buschmann, and P. Sommerlad. *Security Patterns: Integrating Security and Systems Engineering*. J. Wiley and Sons, 2006.

37. A. Silberschatz, P. Galvin, G. Gagne. *Operating System Concepts 7th Ed.*. John Wiley & Sons, 2005.

38. R.C. Summers. *Secure Computing: Threats and Safeguards*. McGraw-Hill, 1997.

39. N. Wang, K. Parameswaran, and D.C. Schmidt. The design and performance of meta-programming mechanisms for object request broker middleware. *Proc. of the 6th Usenix Conf. on Object Oriented Technology and Systems (COOTS)*, 2001.

40. J. Warmer and A. Kleppe. *The Object Constraint Language, 2nd Ed.* Addison-Wesley, 2003.

41. J. Yoder and J. Barcalow. Architectural patterns for enabling application security. *Proc. of the 4th Conference on Patterns Language of Programming (PLoP'97)*, http://jerry, cs.uiuc.edu/ plop/plop97. Also Chapter 15 in *Pattern Languages of Program Design*, Vol. 4 (N. Harrison, B. Foote, and H. Rohnert, Eds.), Addison-Wesley, 2000.

42. XACML, http://www.oasis-open.org/committees/tc_home.php?wg_abbrev =xacml.

Part II

Security in Mobile Computing

4

Pragmatic Security for Constrained Wireless Networks

Phillip G. Bradford, Benjamin M. Grizzell, Graylin T. Jay and Janet Truitt Jenkins

CONTENTS

Abstract   This chapter gives a select survey of security issues for constrained wireless networks. This chapter starts with security in the design of constrained devices; it goes on through security issues for classical attacks on constrained wireless devices, and finally it ends with security and the regulatory environment. This high-level view aims to give a different perspective on securing constrained wireless devices. Its focus is on the discussion of pragmatic issues.

The main objective of this chapter is to impart an unusually broad understanding of security issues for constrained devices. This chapter does not exhaustively cover security topics; rather it focuses on select issues to allow a substantial breadth of perspective. It is hoped this breadth of perspective will shed light on securing constrained wireless devices.

4.1   Introduction

This chapter follows the *weakest link principle* in the following way: to have secure constrained wireless systems, then every aspect of these systems must be secure.

High-level perspectives on security in general and for constrained devices in particular are not new. McGraw [22], Viega and McGraw [32] argue that security should be built in from the start as a first-class design criteria. Kocher, Lee, McGraw, Raghunathan, and Ravi [17] argue that security is another design dimension for embedded systems. Moreover, Perrig, Stankovic, and Wagner [28] emphasize that security must pervade every aspect of system design.

Perrig et al. [28] argue that security should not be viewed as a stand-alone component. They note that security must be incorporated into every component or module for a system to be secure. Components lacking integrated security are usually the first point of attack.

This chapter is divided into the following sections.

Section 4.2: Background and Foundations

Section 4.3: Engineering Issues in Building Constrained Network Nodes

Section 4.4: Classic Attacks and Countermeasures on Constrained Wireless Networks

Section 4.5: Regulations, Policies, Procedures, and Security

Section 4.6: Conclusions

Current trends promise a plethora of small wireless devices in a variety of venues. These small devices are often power and size constrained. In many cases, these devices also lack physical security.

Resource-constrained devices must overcome significant hurdles to establish solid security. These hurdles include: computational capacity, memory, power, sensing, and transmission capabilities.

In addition to the constraints of the devices themselves, the deployment environment further constrains security systems. Small devices may be physically compromised. Their broadcasts may be monitored, altered, etc. The deployment environment also provides a determined adversary with the potential to reverse-engineer security systems.

4.2 Background and Foundations

Constrained wireless networks consist of small devices or even very basic sensors being linked together via wireless connections. These devices gather, process, aggregate, and transfer data or information. In many cases they also make decisions based on the data and information they capture or integrate. Such networks have been used for monitoring in numerous situations and contexts. On the high end these devices may be PDAs or cell phones.

In many cases, constrained wireless networks are made up of devices that are physically constrained and therefore have little room for memory, batteries, and auxiliary chips. These constraints introduce significant challenges that have to be addressed in order to maintain a secure network.

These constrained wireless networks have numerous applications. A selection of these applications and some of their security concerns follows.

Constrained wireless networks may make life more luxurious. Such networks may be routinely used to make medical care more efficient and better delivered. They may even directly save lives. Constrained wireless networks can also be used in [9] "smart spaces, . . ., medical systems and robotic exploration." Ideally, these devices will secure their subjects' personal information.

Networks of constrained devices can be applied in remote places. For example, networks of small devices have been used in ocean, wildlife, and earthquake monitoring [28]. Devices may be deployed in natural disaster relief situations [19]. In these cases, authenticity and accuracy may be the largest concerns.

Small wireless devices may be deployed in hostile or dangerous environments. For example, in the military these networks may be put to use in scoping out a hostile environment and helping with reconnaissance missions for target spotting [19]. They may even be used to detect nuclear weaponry. In hostile environments there are numerous direct attacks in addition to a good number of solid countermeasures.

On the high end of constrained devices, handheld computing devices are intermediaries between people and their environments. The widespread adoption of these devices leads to a plethora of security concerns. These concerns range from individual privacy to institutional property issues. Regardless of the precise security concern, handheld systems will continue to be built and evolve.

4.3 Engineering Issues in Building Constrained Network Nodes

This section explores the following idea. *Refitting and refactoring code for new systems can cause unintended consequences for security.* Refitting code encompasses updating systems, thus ensuring the appropriate functionality. Refactoring is a code refinement paradigm that is particularly applicable to code refitting. The issues covered here highlight the importance of designing security into systems from the start.

Refactoring is intended to improve code design. An opportune time to refactor is when software systems are refit for new devices. Refitting systems for smaller constrained devices can require paring or enhancing functionality. This gives way to ideal circumstances for refactoring. For instance, targeting established systems for small wireless computers provides many dimensions on which to refactor. In this case, some areas to refit and refactor include: (1) user interfaces, (2) communications functionality, (3) algorithm and system efficiency issues, and (4) power and energy management. However, refittings and refactorings in each of these areas can indirectly lead to obscure security failures. Such insidious security failures may be extremely hard to protect against.

4.3.1   Refactoring and Refitting Constrained Devices

Refitting code is the updating of existing code functionality to adapt to a new environment or constraint. *Refactorings* are structural software transformations that keep observable system behavior invariant [11]. The goal of refactoring is to improve code design. Top-down design and refitting with iterative refinement followed by refactoring to get the right fit for a handheld device is a very attractive approach to code development and improvement.

Often refactorings are implemented with strong testing regimes [3]. This ensures that the system's behavior remains invariant. However, a great deal of this (automated) testing is based on expected behaviors using likely scenarios. It was argued by Simmons [26] that a great number of computer security failures are essentially from "bad implementations" of "good algorithms." He gives important examples where not adhering to implicit assumptions leads to security failures. Implicit assumptions from another segment of the code tend to be forgotten. In fact, this problem can extend to "good refittings and refactorings" that lead to unintended consequences. The distinction is refitting and refactoring may change the level of security unintentionally. One would not intentionally have refitting change security where refactoring incrementally drives improvements of code structure while keeping observable functionality invariant. However, a great number of security issues are not readily observable. This is of particular concern in handheld devices.

A significant vulnerability for constrained wireless networks is battery power. As a hypothetical example, consider a refactoring that replaces a block of code by more efficient power-managed code [14, 23]. Eventually, this may be done directly by a compiler switch. This new code could change the performance of thermal-based (pseudo) random number generation, thereby affecting many security systems. For example, Intel's 82802 "random number generator" chip uses thermal noise to generate (pseudo) random numbers. If the temperature goes below a certain threshold, then the chip outputs a constant value.

An example of refactoring *communication functionality* comes from "tightening the communications." Suppose software for a handheld device is refit to allow more synchronous communication, then it may become vulnerable to Brumley-Boneh-like timing attacks [5]. In such an attack, the adversary determines likely prime factors for the RSA key in open SSL by understanding and exploiting its well-known prime generation algorithms [25]. The refitting establishes the change in functionality, where the refactoring restructures the code design, keeping the observable functionality intact and perhaps operating more synchronously. It is the change in functionality that enables the timing attack. The refactoring may contribute to the attack by allowing more efficient functionality. Moreover, refactoring may give a false sense of security by keeping the observable behavior invariant.

In general, design with refitting and refactoring can give way to unintended consequences for security. The security semantics of a device are often hard to observe and are based on all of its systems and all of their implicit or explicit interactions. This is critical when porting systems to handheld wireless devices.

4.4 Classic Attacks and Countermeasures

The following subsections discuss the next constrained wireless network attacks:

- denial-of-service attacks,
- path-based denial-of-service attacks,
- selfishness-based denial-of-service attacks,
- network authentication related attacks, and
- physical node attacks.

In the next subsections, each attack is introduced and some known countermeasures are given in the following subsubsections.

4.4.1   Denial-of-Service Attacks

A temptation for an attacker of any wireless network is to launch a Denial-of-Service (DoS) attack. Such an attack may be realized in many different ways.

Some DoS attacks are brought on by one or more malicious nodes. Alternatively, a damaged node may lead to a DoS situation. For example, cleverly damaging an opponent's nodes may lead them to cause DoS attacks on an opponent's network.

For any type of wireless network, a jammer can cause radio interference. A jammer is defined by Xu, Trappe, Zhang, and Wood as [35]:

> ... an entity who is purposefully trying to interfere with the physical transmission and reception of wireless communications.

Jamming attacks come in many forms. For example, there are constant, deceptive, random, and reactive jamming attacks. A constant jammer simply sends a continuous radio signal on a channel that would normally be used to send information on a wireless network. Valid nodes make note of congestion and are in a continual wait formation until the channel clears. Since the interference is continuous, that channel never clears for valid transmission to take place. A deceptive jammer sends packets that seem legitimate. This forces the receiving node to stay in a continual receive mode, regardless of whether it has packets to send or receive. A random jammer conserves its own energy by randomly alternating between transmitting a malicious signal for an apparently random time and sleeping for another random time. A reactive jammer responds to the level of communication on a channel. If little communication takes place, then the jammer sleeps. If action begins on the network, then so does the jammer. This allows the jammer to both conserve energy and it also helps the jammer to be less noticeable [35].

Countermeasures for Denial-of-Service Attacks

The strength of a constrained wireless network to defend against DoS attacks is critical in the security of the networks. Jammers themselves are a constant threat to any open wireless signal-based network. Jammers can greatly constrict the potential of constrained wireless networks by disturbing the normal flow of information. Jammers can have an adverse effect on a given network depending on the strength and nature of attack. Defending against these malicious acts can be a challenging process [35]:

> Detecting jamming attacks is important because it is the first step toward building a secure and dependable wireless network. It is challenging because jammers can employ different models, and it is often difficult to differentiate a jamming scenario from legitimate scenarios.

Spread spectrum or frequency hopping, see, for example, Engelberg [10], pseudorandomly distributes a transmission signal across many frequencies to evade detection. More recently, these protocols have been used to prevent transmission collisions. For example, Code Division Multiple Access (CDMA) systems used for cell phones use spread spectrum techniques. Likewise spread spectrum methods may be used to avoid jamming attacks. In fact, Perrig et al. [28] suggest using spread spectrum communications as a defense against jamming attacks. A challenge for this countermeasure is that cryptographically secure-spread spectrum radios are not currently readily available for very small constrained devices. A second issue with this approach is that it does not provide a shield from physical capture in which the cryptographic keys driving the spread-spectrum transmissions may be extracted.

4.4.2 Path-Based Denial-of-Service Attacks

One type of DoS attack on constrained wireless networks is a Path-Based Denial-of-Service attacks (PDoS). A PDoS attack takes place on multihop end-to-end networks. These networks send packets through several nodes before they finally arrive at a destination node. Nodes on a particular path can be flooded with packets (repeated or falsified). Certain nodes may be compromised and used to send such spurious data. This bogs down the resources of those particular nodes and prohibits them from transferring data down the path rendering those nodes ineffective on that particular pathway. This problem can potentially bring down the effectiveness of the entire network [9].

As with any wireless network, a constrained wireless network also faces the possibility of being flooded with packets from an attacker. An overwhelming amount of data may quickly degrade the system and render it of little use. Many large constrained wireless networks have three different types of nodes in the network: member nodes, aggregator nodes, and base stations [9]. The basic design of a network using collection and aggregator nodes employs member nodes to collect data and transport it to aggregator nodes. Aggregator nodes collect and summarize data and send the results to one or more base stations. These node types are often organized in tree structures. An attacker may cleverly choose which nodes of the trees to attack in order to have the highest impact.

Countermeasures for Path-Based Denial-of-Service Attacks

The capability to implement secure routing protocols in a constrained wireless network provides the means to pass data from one node to another without the inference of an attacker inserting altered data or altering the routing path in place. The lack of secure routing in a network could be potentially dangerous given the possibility of using altered data. In critical situations, it is imperative to secure the information during transit from end to end without the possibility of the data being compromised by an attacker.

The next definitions are used in a classic countermeasure for PDoS attacks.

A *hash chain* is a sequence of values $v_0, v_1, \cdots, v_n$, each $v_i \in \{0, 1\}^\ell$ for $i : n \geq i \geq 0$ and some integer ℓ. Furthermore, v_n is the hash chain *seed* and assumed to be randomly and uniformly chosen from $\{0, 1\}^\ell$. Subsequent values v_i, for $i : n > i \geq 0$, are computed as $v_i = f(v_{i+1})$ where f is a hash function (Table 4.1).

For analysis, it is convenient to assume the hash function f is one way [25]. This means, given x, it is "easy" to compute $y \leftarrow f(x)$. On the other hand, given y, it is "hard" or intractable on average to find x. It is not known if one-way functions exist [8].

TABLE 4.1

A Hash Chain Element for Each Time Interval [4]

Time Interval	0	1	$\cdots$	$n-1$
Hash Chain Elements	$v_0 = f(v_1)$	$v_1 = f(v_2)$	$\cdots$	$v_{n-1} = f(v_n)$

The important fact about hash chains is their elements are used in the reverse order that they are generated in.

If the hash functions f have ideal hash properties, then they may be expensive to put on a small constrained device, see Bradford and Gavrylyako [4].

Deng, Han, and Mishra [9] use hash chains to defend against PDoS attacks. One of their basic schemes works as follows. All nodes in the network know the common hash function f and an initial hash value v_0. Each aggregator or root node stores a hash chain $v_0, v_1, \cdots, v_n$. When this originating node sends packets through the network, then v_i is sent in the ith packet. An intermediate or final node verifies that there is some integer $w \geq 1$ so that

$$v_i = f^k(v_0), \quad \forall k : w \geq k \geq 1.$$

Consider some $k : w \geq k \geq 1$, then the value v_i becomes the new initial hash value for this intermediate node. Moreover, if there is no $k : w \geq k \geq 1$ so that $v_i = f^k(v_0)$, then the packet including this hash value is simply dropped from the network. A critical issue is choosing a suitable value w.

Due to the one-way nature of f, if a node is physically compromised, then it will be intractable on average for it to find v_{i+1} given v_i. Therefore, PDoS attacks will be very challenging to launch given this protocol.

4.4.3 Selfishness-Based Denial-of-Service Attacks

Another interesting possibility is when a node joins a constrained wireless network that has goals that are not aligned to the goals of the network. This node may not be malicious, but could be selfish by placing a higher priority on individual goals over goals of the entire network. This type of node might acquire more bandwidth than the protocol allows and consequently reduce the efficiency of the entire network, see Kyasanur and Vaidya [18]. Koutsoupias and Papadimitriou [16] give a formalization of the unregulated or "anarchist" nature of resource sharing on the Internet. Several results have improved their bound and extended the model. For an overview of work in this area for the Internet, see Czumaj [7]. The Internet results are often applicable to small networks of constrained devices.

4.4.4 Countermeasures for Selfishness-Based Denial-of-Service Attacks

Refaei, Srivastava, DaSilva, and Eltoweissy [24] give reputation-based countermeasures against selfish nodes. Basically, each node keeps a reputation measurement for each of its neighboring nodes. The reputation scores are based on completions of service requests. If a node is determined to be selfish, then it is isolated thus preventing it from degrading other nodes. This scheme depends on reporting successful transfers from the destination nodes.

4.4.5 Network Authentication-Based Attacks

Secrecy and authenticity are critical issues for constrained wireless networks. Constrained wireless networks depend on a good deal of personal or proprietary information. This information must be accurate. Constrained wireless networks need the functionality to defend against malicious agents that may attempt to alter network packets. In order to allow communication on a wireless network it is imperative that wireless networks are able to identify other legitimate users on the network.

In some very constrained networks, public key systems and public key exchanges are too expensive. In these cases, preseeding the processors is necessary for authentication. Preseeding these devices allows symmetric key systems (such as AES or TEA) to be run on even very constrained processors. The Tiny Encryption Algorithm (TEA) was developed by Wheeler and Needham [34] specifically for small constrained devices. It was updated to XTEA in Needham and Wheeler [15].

Countermeasures for Network Authentication-Based Attacks

Symmetric key systems give authentication provided the random seeds used to generate the keys are not compromised.

Andem [1] gives a thorough analysis of TEA with an eye toward implementing it on constrained devices. See also Liu, Gavrylyako, Bradford [21] for a discussion of implementing TEA on very constrained devices. TEA may be used to generate hash chains, but Hernández, Sierra, Ribagorda, Ramos, and Mex-Perera [12] show that TEA does not adhere to the strict avalanche criterion when it does not run through all 64 of its rounds. It may be a temptation to have TEA run through fewer rounds to conserve power. To show what this means, let T_k be the TEA algorithm using only six rounds [12] with some key k. Let m be a 64-bit plaintext input to T_k and let e be a 64-bit vector with a single bit set to 1. The strict avalanche criterion [33] requires that

$$\mathbf{P}[T_k[m] \oplus T_k[m \oplus e] = 1] = \frac{1}{2}.$$

Since such a round-restricted implementation of TEA does not have the strict avalanche criterion it may not be a good choice for certain types of encryption.

Public key systems appear to require more resources than symmetric key systems. There is ongoing research to make highly secure public key systems run as efficiently as symmetric key systems. See, for example, Lenstra and Verheul [20].

Perrig's Biba protocol [27, 29] gives authentication by exploiting the challenge of finding hash collisions in a small number of "bins." These bins may be computed by moding hash outputs with small values. The hash function is dynamically chosen for each message, thus it depends on the message being sent. A signature is made of two inputs that form a collision given a particular message. Thus, given a plaintext message m and its associated hash function f_m, find two different inputs s_1 and s_2 so that $f_m(s_1) = f_m(s_2)$. It is assumed that given m then the generation of the function f_m is secret. Colliding inputs may be found with high probability. The signature of m is $< m, s_1, s_2 >$, which will be transmitted. Thus, an adversary may see $< m, s_1, s_2 >$, but since it does not know the hash function f_m it will be hard pressed to find another pair of inputs s_i and s_j where $f_m(s_i) = f_m(s_j)$.

The BiBa signature protocols allow fast verification while having a small signature size. These characteristics make BiBa suitable for constrained devices. There have been several improvements to the BiBa, as an example see Reyzin and Reyzin [30].

The SPINS protocol (Security Protocols for Sensor Networks) is another protocol that is for resource-constrained environments and wireless communications, see [13] or [29]. SPINS uses Rivest's RC5 algorithm [31] for encryption. SPINS is built on two parts: SNEP and μTESLA. SNEP is a data confidentiality protocol implementing two-part data authentication, data integrity, and data freshness. SNEP works using message authentication code-based packets, see [29]. μTESLA is a subcase of the TESLA protocol and it provides efficient authentication broadcasts for constrained wireless sensor networks.

μTESLA works by periodically transmitting hash chain elements from the transmitting base station. Each node has a well-known hash function f on it. Assume f is one-way and thus it is intractable to invert, see subsection 4.4.2. If a node receives v_i in time period i and it receives v_{i+1} in time $i+1$ then the node may verify that $v_i = f(v_{i+1})$. This verification indicates that the transmitter who sent v_i is the same one who transmitted v_{i+1} assuming the hash function f is one-way. Since f is easy to compute, then we can efficiently use delayed disclosure to verify subsequent elements. Thus, delayed disclosure allows verification of periodic broadcast that incorporate hash chains.

4.4.6 Physical Node Attacks

A unique issue that makes a constrained network particularly vulnerable is the environments in which the nodes are placed. Given large numbers of nodes or sensors in a network, an attacker could easily seize one of them and reverse-engineer it to extract the cryptographic key or routing information. At other points, determining the level of security of a node's information depends on several factors including the type of information the nodes hold, along with how easily the information may be read. In any case, unique strategies combined with special designs must both be addressed.

The creation of a secured key that is shared by all nodes within a network may be a direct solution to this issue. However, the physical capture of even a single node on the network may lead to the extraction of such a key, allowing for decryption of all transmissions.

Countermeasures to Physical Node Attacks

It is difficult to determine the single weakest point in a constrained wireless network. However, physically capturing or compromising nodes or sensors may expose numerous weak points. The large-scale deployment of nodes makes it likely that some will be vulnerable to physical capture.

It seems for every new tamper-resistant technology, then there are eventually effective countermeasures. Moreover, in large networks of constrained devices there are many nodes to physically attack. In some cases it has been argued that tamper resistance is not enough [2].

Perrig et al. [28] discusses an idea of a single key as well as establishing a set of linked keys in which there is one per pair of nodes. This would eliminate the need for a network-wide key after the establishment of the session keys. Perrig et al. [28] recognize that this theory does not allow for deployment of addition nodes in the network.

The LEAP (Localized Encryption and Authentication Protocol) identifies compromised nodes and rejects their data. LEAP was given by Zhu, Setia, and Jajodia [36]. LEAP allows different types of communications to require different security measures. LEAP has devised a set of four keys to be implemented at different stages of the network. These keys include an individual key, a pairwise key, a cluster key, and a group key. The individual key is exchanged between an individual node and the base station, a pairwise key is shared between nodes, the cluster key is shared between clusters of nodes, and the group key is shared by all nodes. The use of multiple keys strengthens wireless networks against physical attacks.

4.5   Regulations, Policies, Procedures and Security

As pervasive wireless devices proliferate industries such as the healthcare industry, regulatory issues will become a critical focus. Moreover, in numerous industries many of these wireless devices will be constrained wireless devices. This section focuses on where policy justifications are made for technical decisions with an eye toward constrained wireless networks.

Almost every organization's technical decisions must conform to some sort of standard or guideline. Such guidelines include:

- HIPPA (the Health Insurance Portability and Accountability Act), or
- SOX (the Sarbanes-Oxley Act), or
- FIPS (Federal Information Processing Standards).

Security is critical to all of these guidelines. If a system is not secure, then guidelines may be irrelevant.

The primary author of this section has worked as the sole IT worker at a small residential mental-health treatment center for children. They have been successfully audited for HIPAA conformance twice. Given these experiences, this section presents a sampling of the practices that best helped us overcome stumbling blocks for conforming to HIPAA. These best-practice heuristics are readily applicable to any standards conformance process.

This presentation assumes this is your first exposure to what is essentially a nontechnical standard or technical guideline.

System developers learn to adhere to numerous technical standards. Now, due to the pervasive and significant roles IT and computer science have assumed, computer specialists must learn and adhere to numerous (legal) regulations revolving around computer systems. This is particularly important for constrained wireless systems.

4.5.1   Experiences Conforming to HIPAA

The bulk of conformance issues often reduce to processes or quality measurements. Conformance issues are inevitably enforced bureaucratically. These elements of conformance result in structures such as chains of responsibility.

There may be some culture shock among IT professionals when faced with this particular type of challenge. The important thing is to recognize that this is a bureaucratic procedure and act accordingly. When non-engineers refer to standards they mean something very different from the well-defined specifications with which an engineer may be familiar. Many in IT or computer science are tripped up by the ambiguities and "play" inherent in such regulatory processes. The key to success may be realizing that you have been given such freedom for a reason. The system often assumes that you are qualified to decide what level of protection is needed in a given situation.

Standards or guidelines often define procedures and not specific choices. From a technological point of view this may be good since technology changes fast. Technological standards are obsoleted, updated, or changed quickly. However, the regulatory environments driving regulations such as HIPAA generally change more slowly if at all.

Once IT professionals realize that standards may be viewed as a way of framing (and justifying) their choices, they are a long way toward working productively within a standard.

4.5.2   The Web of Indemnification

There is a very specific mindset that is useful when making technical decisions with an eye toward conforming to a set of guidelines. A large portion of that mindset is understanding the bureaucratic context in which a decision is being made. We call this context "the web of indemnification."

Here is a real-world example of a typical web. Let us examine a relatively small issue that might come up under HIPAA. Should health information transmitted wirelessly be encrypted? There is a great deal of (nonlegally binding) discussion in HIPAA regulations, but about the most concrete statement you will find in HIPAA may be: "Implement a mechanism to encrypt electronic protected health information whenever deemed appropriate."

Some engineers may want to leave this discussion as by now it should be clear that the HIPAA specification is fairly free of detailed technical specifications. This may seem strange to anyone who has worked in a healthcare environment and had to move their desk away from a window "to comply with HIPAA." How do vague phrases such as "deemed appropriate" become hard and fast rules? Recall, computer scientists sometimes go as far as using formal semantics to very tightly define their specifications. Thus, the worlds of regulatory rules defined in English seem a good deal more ambiguous.

One way to deal with such statements is to define a process to rectify what is "appropriate" and what is "not appropriate." Providing process documentation is critical. This challenge may be solved by executing a well-defined process and storing a trace of how the process worked. Computer science may also be used to enhance these processes. For example, documentation of the processes and their traces may be saved using version control systems such as CVS or Microsoft Sourcesafe. It is always good to reframe your challenges in terms of the technology at hand.

4.5.3 Working with the Hierarchy

For the purposes of this discussion let us assume the top of the bureaucracy is the state. Considering federal regulations such as HIPAA, a state may choose a strategy to show it attempted to satisfy HIPAA "in good faith." To maintain its independence and to garner experience in meeting HIPAA requirements a state may hire a HIPAA consulting firm that helps the state create slightly more concrete versions of HIPAA's recommendations. These may even become state laws. One step below the state is (in our case) the state's department of mental health. Faced with nontechnical state laws the department of mental health realizes a tenable legal strategy is to show it attempted to comply "in good faith." The department of mental health thus hires a HIPAA consulting firm and writes some slightly more concrete versions of the state's regulations or laws. These often become department of mental health policy. By the time this process gets to the person making decisions about a wireless network he or she is usually answering to an auditor from a consulting firm who has very specific recommendations that may not be technical. Examples include: policies on desktop background colors.

While this is certainly a simplification, how the system works should be clear. Practically everyone's decisions have been at least somewhat vetted by someone further up the chain of command. Everyone has ready-made arguments for their proverbial day in court. Everyone is vouching for everyone else. This is extremely good news for the people making actual technical decisions because if they work within the context of the system it seems hard for these individuals to be left out to dry should something go wrong. This is a good thing because the penalties for violating most standards are one of the few things in them that are well defined. In the case of HIPAA, intentionally revealing someone's healthcare information is punishable by up to a quarter-of-a million-dollar fine and a decade in prison.

It is useful to make clear the technical decision makers' exact responsibilities. They need to have absolutely everything possible "signed off" by someone more senior. Moreover, they need to provide their superiors with a well-documented written justification for every decision.

4.5.4 General Tips on Conformance

Compliance is expensive. It is important to get everyone involved to understand the expenses in making a system comply.

To mitigate the cost of compliance and to sharpen the standards conformance process it is very useful to appeal to more rigorous standards. This should be done whenever possible.

Next we outline some heuristics that may be helpful to someone trying to comply with HIPAA regulations. These heuristics are slanted toward constrained wireless devices when possible.

Remember Reality. This is not a practice so much as something not to lose sight of. As an IT professional you have the job of following both the letter and spirit of regulation, policy, and law. In the case of HIPAA we had to follow the letter of the law in the form of whatever our auditor told us to do, but we also had to follow the spirit of the law and provide actual security. These two requirements clash more often than you might think. The web of indemnification exists, but you are still going to be the person closest to any actual incidents.

Know What Is Not Covered. Several things fall under any given guideline but most things do not. If the first thing you do is separate what is under the purview of an auditor or certification committee, then you have half of the process covered. It may be necessary to physically split these concerns, but usually a conceptual split is fine. Compliance is expensive. Do not comply any more than you must!

Make Policy—Not Decisions. This guideline is sometimes called the "shall rule" in administrative circles. Imagine that you have decided that for a particular case of simple wireless Web surfing WEP or WPA filtering is enough. Do not be content to have this be a decision for this specific case. Write up your decision in a format similar to the following: For networks that do not transmit sensitive data WEP shall be considered adequate protection. If you honestly stick by this rule your policies will practically write themselves.

Program with Policy. This is something systems analysts have known for years: people are programmable. Take the above example with nonsensitive data. How are we to prevent users from sending sensitive data over unsecured networks? We make that a violation of policy! There is a limit on how much responsibility you should offload onto others. Prudently make this judgment following written policy. Always keep in mind that you are working with people who can be shaped by policy. This may help you easily resolve what might, if approached technically, seem like impossible issues.

Do It Their Way—Do It Your Way. While there are exceptions worth fighting for, in the general case you must accept anything specific given to you from an auditor or other "higher" authority. It is good to keep on top of your costs, keeping a spreadsheet of time and resources you spend addressing policy issues and related matters. This information may drive discussions on how to accomplish future objectives. Documentation is always critical. Also, work to deliver exactly what your customer needs while understanding the technology and costs yourself. A part of your job will be educating the hierarchy as to the technical tradeoffs and their costs.

As an example, when we were being audited our auditors wanted our machines to have a password-protected screensaver. We wanted our machines to automatically log out, which is much more secure. With a little bit of cajoling we were able to rig Windows so that when logging back in from a logged-out session you "logged in" to a screensaver. The auditors were satisfied with this. We did it their way, and we did it our way. Many people run redundant systems or programs. One set of systems complies with guidelines and one we actually trust.

When All Else Fails Change the Spec. This programmer's favorite is not as impossible as you might think. No, you probably are not going to get a section of HIPAA itself changed. However, acts like HIPAA are usually vague enough that what you really want changed is a state or institutional policy. Sometimes such a change is as simple as a few phone calls. A corollary to this practice is to be plugged in to your state or institution's policymaking procedures or committees. As constrained wireless technology becomes more pervasive, then the tradeoffs you must discuss with your hierarchy will become more critical.

4.5.5   Building Your House on a Rock

As suggested earlier, one of the best resources in conforming to regulations and guidelines is other standards. Consider NIST (the National Institute of Standards and Technology). For example, in the world of cryptography, NIST develops and publishes cryptography standards such as the AES (Advanced Encryption Standard). NIST is part of the U.S. Department of Commerce and it has ties to government agencies as diverse as NASA and the NSA.

What makes NIST useful for technologists implementing or conforming to regulations? A number of factors make NIST an important resource:

1. NIST standards are written by technologists with commercial or government use in mind. NIST standards and recommendations are usually concrete and are developed with substantial technical feedback from stakeholders.

2. NIST has technical labs and provides certification for some technologies. In some cases, NIST actually certifies products! Unlike laws or acts NIST standards are meant to be followed. They include materials such as compliance checklists and approved vendors.

How does all of this help you? The strategy goes something like this: First, find a NIST standard that is more rigorous than the standard you actually have to meet. Second, meet the NIST standard. Third, use compliance with the NIST standard as proof of your compliance with the standard. To give a concrete HIPAA example: My encryption is good enough because I am using a FIPS-approved version of OpenSSL, which is approved for all but "Eyes Only" levels of secret documents. Similarly: We make use of BlackBerrys in accordance with the same recommendations as the Department of Homeland Security.

Here is the best part. NIST standards are probably better than what you would have done on your own. NIST's special publication 800-48 (Wireless Network Security) includes the following as recommendation 46:

> Use a local serial port interface for AP configuration to minimize the exposure of sensitive management information.

This is a clear, security-conscience recommendation we honestly would not have come up with on our own. Recommendation 19 is also a solid well-thought-out recommendation:

> Validate that the SSID character string does not reflect the agency's name (division, department, street, etc.) or products.

The NIST special publication has 56 total recommendations for wireless networks. You could do worse than simply blindly following each recommendation regardless of what standard you might actually be trying to meet.

NIST FIPS standards can in most cases settle any doubts that wireless networks can comply with standards requiring security. AES (the replacement for DES) is certified for satellite and many other inherently "open channel" uses. As long as it is using proper appropriately NIST-certified encryption, wireless networks can be authorized to carry eyes-only data. We recommend you get to know the NIST Web sites and read all the appropriate standards. Acquiring certified products and services can also certainly ease how much you have to do on your own.

4.6 Conclusions

Securing constrained systems requires numerous levels and types of security. To develop a solid security picture of constrained wireless devices we must understand issues ranging from the design constraints to the political environment in which these devices are deployed. The deployment environments for constrained wireless devices vary from hostile military environments to leisure home environments. As technology advances, many constrained wireless devices will fall behind technologically and become more vulnerable.

Bibliography

1. V. R. Andem. *A Cryptanalysis of the Tiny Encryption Algorithm, Masters thesis, The University of Alabama.* Tuscaloosa, 2003.
2. R. Anderson, M. Kuhn. "Tamper resistance—a cautionary note," The Second USENIX Workshop on Electronic Commerce Proceedings, November 18–21, 1–11, 1996.
3. D. Astels, G. Miller, and M. Novak. *A Practical Guide to eXtreme Programming.* Prentice Hall, 2002.
4. P. G. Bradford, O. V. Gavrylyako. "Foundations of security for hash chains in ad hoc networks." *Cluster Computing* 8(2–3): 189–195, 2005.
5. D. Brumley, D. Boneh. "Remote timing attacks are practical," *Computer Networks* 48(5): 701–716, 2005.
6. L. Buttyán, J.-P. Hubaux (with P. Kyasanur and N. H. Vaidya). "Report on a working session on security in wireless ad hoc networks," *ACM SIGMOBILE Mobile Computing and Communications Review,* 7(1), 74–94, 2003.
7. A. Czumaj. "Selfish routing on the internet," Chapter 42 in *Handbook of Scheduling: Algorithms, Models, and Performance Analysis,* edited by J. Leung, CRC Press, Boca Raton, FL, 2004.
8. H. Delfs and H. Knebl. *Introduction to Cryptography, Principles and Applications,* Springer-Verlag, 2002.
9. J. Deng, R. Han, and S. Mishra. "Defending against path-based DoS attacks in wireless sensor networks," *Proceedings of the 3rd ACM Workshop on Security of Ad Hoc and Sensor Networks,* 89–96, 2005.
10. S. Engelberg. "Spread spectrum from two perspectives," *SIAM Review,* 45(3), 574–587, 2003.
11. M. Fowler. *Refactoring: Improving the Design of Existing Code,* Addison-Wesley, 1999.
12. J. C. Hernández, J. M. Sierra, A. Ribagorda, B. Ramos, and J. C. Mex-Perera. "Distinguishing TEA from a random permutation: Reduced round versions of TEA do not have the SAC or do not generate random numbers," In *Proceedings of the IMA Int. Conf. on Cryptography and Coding 2001,* pages 374–377, 2001.
13. N. Hu, R. K. Smith, P. G. Bradford. "Security for fixed sensor networks," *ACM Southeast Regional Conference,* 212, 213, 2004.
14. F. Li, G. Chen, M. Kandemir, M. J. Irwin. "Compilation: Compiler-directed proactive power management for networks," *Proceedings of the 2005 International Conference on Compilers, Architectures and Synthesis for Embedded Systems* (CASES '05), 137–146, 2005.
15. R. M. Needham and D. J. Wheeler. "Tea extensions," Technical Report, Computer Laboratory, University of Cambridge, October 1997.
16. E. Koutsoupias and C. Papadimitriou. "Worst case equilbria," in *STACS 1999,* C. Meinel and S. Tison (Eds.), Springer-Verlag, LNCS #1563, 404–413, 1999.
17. P. Kocher, R. Lee, G. McGraw, A. Raghunathan, and S. Ravi. "Security as a new dimension in embedded system design," *ACM Symposium on Applied Computing (SAC 2004),* 2004.
18. P. Kyasanur and N. H. Vaidya. "Selfish MAC layer misbehavior in wireless networks," *IEEE Transactions on Mobile Computing,* 4(5), 502–516, September/October 2005.

19. L. Lazos, R. Poovendran. "SeRLoc: Robust localization for wireless sensor networks," *ACM Transactions on Sensor Networks,* 73–100, 2005.
20. A. K. Lenstra and E. R. Verheul. "Selecting cryptographic key sizes," *Journal of Cryptology,* Vol. 14, No. 4, 255–293, 2001.
21. S. Liu, O. V. Gavrylyako, and P. G. Bradford. "Implementing the TEA algorithm on sensors," In *Proceedings of the 42nd Annual Southeast Regional Conference,* 64–69, 2004.
22. G. McGraw. *Software Security: Building Security In,* Addison-Wesley Professional, 2006.
23. R. A. Ravindran, P. D. Nagarkar, G. S. Dasika, E. D. Marsman, R. M. Senger, S. A. Mahlke, R. B. Brown. "Compiler managed dynamic instruction placement in a low-power code cache," *Proceedings of the International Symposium on Code Generation and Optimization (CGO '05),* 179–190, 2005.
24. M. T. Refaei, V. Srivastava, L. DaSilva, M. Eltoweissy. "A reputation-based mechanism for isolating selfish nodes in ad hoc networks," *The Second Annual International Conference on Mobile and Ubiquitous Systems: Networking and Services,* 3–11, 2005.
25. B. Schneier. *Applied Cryptography,* Second Edition, John Wiley & Sons, 1996.
26. G. J. Simmons. "Cryptanalysis and protocol failures," *Communications of the ACM,* Vol. 37, No. 11, 56–65, November 1994.
27. A. Perrig. "The BiBa one-time signature and broadcast authentication protocol," *ACM 8th Computer and Communication Security Conference (CCS '01),* 28–37, 2001.
28. A. Perrig, J. Stankovic, and D. Wagner. "Security in wireless sensor networks," *Communications of the ACM,* 47(6), 54–57, June 2004.
29. A. Perrig, J. D. Tygar. *Secure Broadcast Communication in Wired and Wireless Networks,* Kluwer Academic Publishers, 2003.
30. L. Reyzin and N. Reyzin. "Better than *BiBa*: Short one-time signatures with fast signing and verifying," In the *Seventh Austrailasian Conference on Information Security and Privacy (ACISP),* 2002.
31. R. L. Rivest. "The RC5 encryption algorithm," *Proceedings of the 1994 Leuven Workshop on Fast Software Encryption,* 86–96, 1995. (See Rivest's Web pages for an update.)
32. J. Viega and G. McGraw. *Building Secure Software: How to Avoid Security Problems the Right Way,* Addison-Wesley Professional Computing Series, 2002.
33. A. F. Webster, S. E. Tavares. "On the design of S-boxes," *Advances in Cryptology— Crypto '85,* Lecture Notes in Computer Science # 218, Springer-Verlag, 523–534, 1986.
34. D. J. Wheeler and R. M. Needham. "TEA, a tiny encryption algorithm," In Bart Preneel, editor, *Fast Software Encryption: Second International Workshop,* #1008 of Lecture Notes in Computer Science, 363–366, Leuven, Belgium, 14–16, 1994.
35. W. Xu, W. Trappe, Y. Zhang, and T. Wood. "The feasibility of launching and detecting Jamming attacks in wireless networks," *Proceedings of the 6th ACM International Symposium on Mobile Ad Hoc Networking and Computing,* 46–57, 2005.
36. S. Zhu, S. Setia, and S. Jajodia. "LEAP: Efficient security mechanisms for large-scale distributed sensor networks," *Proceedings of the 10th ACM Conference on Computer and Communications Security (CCS '03),* 62–72, 2003.

5

Authentication in Wireless Networks

Saikat Chakrabarti, Venkata C. Giruka and Mukesh Singhal

CONTENTS

5.1 Introduction

Authentication (more precisely, entity authentication) is the process of identifying an entity in a reliable manner. It provides a means to verify that an entity is indeed who it claims to be. The most common technique for verification is to check whether the claimant possesses a "secret" that a genuine entity is supposed to. In an everyday situation of making a telephone call, we authenticate the person answering the phone by his/her voice. The "secret" here (voice) is a quality inherent to the person. We can identify people already known to us by their visual appearance. But authentication gets complicated when we do not have voice and/or sight of appearance to help identify the person we are trying to communicate with. For example, a computer trying to authenticate a human user or two computers (communicating on behalf of two human users) trying to authenticate each other.

An essential goal of the process of authentication is to allow both authenticated entities to engage in a secure communication. This necessitates the generation of a cryptographically strong secret key (called the session key) to be shared by the entities after a successful run of the authentication protocol. The entities would use the session key to encrypt and decrypt subsequent messages. It has been long recognized that there should be some mechanism by which two entities, who do not share a secret key and do not have any knowledge of each other beyond an identifier, may establish a shared key for engaging in a secure communication. An authentication protocol provides such a mechanism.

Authentication in wireless networks is a challenging task. The very absence of a secure wired medium in a wireless network creates new threats in building authentication protocols in wireless networks. In a wireless environment, the radio medium can be accessed by anyone who has the proper wireless equipment. A mobile adversary can receive and send messages in the wireless network at will, and even if such adversaries are detected, it is difficult to remove them from the network because of their mobile nature. Moreover, the mobile wireless devices are typically resource constrained, in terms of computational capability, memory, bandwidth and power availability. The authentication protocol should be carefully designed, keeping in mind the resource-constrained nature of the wireless devices.

5.1.1　Outline

Different kinds of wireless networks pose unique security issues in the design of authentication protocols. In this chapter, we classify wireless networks into three distinct categories: (1) the Global System for Mobile Communications (GSM) system, which provides an architecture for digital cellular communications, (2) the IEEE 802.11 standard for wireless networks, which provides specifications for Wireless Local Area Networks (WLANs), and (3) Wireless ad hoc networks which are self-organized networks, rapidly deployed for a special purpose in situations where no infrastructure exists. In this chapter, we discuss various approaches of designing authentication protocols in the three types of wireless networks mentioned above.

The rest of the chapter is organized as follows. In Section 5.2, we present the definition and basis of authentication, describe the phases of an authentication protocol and discuss the general design goals of a wireless authentication protocol. In Section 5.3, we discuss the GSM architecture and authentication in the GSM system. In Section 5.4, we discuss the IEEE 802.11 standard for wireless networks and authentication in IEEE 802.11 WLANs. Section 5.5 deals with wireless ad hoc networks and authentication in ad hoc networks. We conclude the chapter with a summary in Section 5.6.

5.2   Authentication

Entity authentication is the process by which an entity (the verifier) is assured of the identity of another entity (the claimant) involved in a protocol and that the claimant has actually participated in the protocol. The outcome of the process of authentication can be either (1) the verifier acquires corroborative evidence and is assured of the identity of the claimant, in which case the verifier accepts; or (2) the verifier does not receive sufficient evidence and cannot identify the claimant, in which case the verifier rejects. An entity authentication protocol involves actual communication between the verifier and the claimant during the execution of the protocol, creating a feeling of "real-time," while "the verifier awaits" [20]. This is in contrast to message authentication, which provides no timeliness guarantees as to when the message was created. In this chapter we use the term "authentication" to denote entity authentication. Suppose entities A and B are engaged in an authentication protocol. The primary objective of the authentication protocol is to defeat impersonation, which implies the following: an adversary C cannot play the role of A, run the authentication protocol and cause B to complete the protocol and accept A's identity with nonnegligible probability.

Basis of Authentication: Entity authentication can be classified on the basis of three main categories [20]:

1. Something the claimant knows: In this category of authentication protocols, a claimant demonstrates its knowledge of a secret to re-assure the verifier that it is who it claims to be. Examples include passwords, Personal Identification Numbers (PINs) and shared secret keys whose knowledge is demonstrated in challenge-response protocols (described in Section 5.5.1).

2. Something the claimant possesses: Sometimes, the claimant has to prove the knowledge of something it possesses to reassure the verifier that it is who it claims to be. Magnetic-striped cards, physical keys, and ATM cards fit into this category.

3. Something inherent to the claimant: This category includes authentication techniques using human physical characteristics and involuntary actions (biometrics), such as fingerprints, voice, and retinal patters which the claimant uses to prove its identity to the verifier.

In this chapter, we will describe techniques and models for authentication in wireless networks that use the first category, i.e., "Something the claimant knows." The second category, "Something the claimant possesses," involving physical keys and ATM cards, and third category, "Something inherent to the claimant," involving noncryptographic measures, are beyond the scope of this chapter.

5.2.1 Phases of an Authentication Protocol

An authentication protocol is composed of two distinct phases: the bootstrapping phase and the authentication phase. In the bootstrapping phase, the claimant is securely provided with "something the claimant should know," such as a password or a PIN, by the verifier. We call this "bootstrapping material." The verifier would need the claimant to demonstrate the knowledge of the "bootstrapping material" as a proof of the claimant's eligibility to access protected resources or use paid services. Since in the bootstrapping phase, the claimant is yet to be authenticated, the "bootstrapping material" needs to be sent to the claimant over a secure channel. The second phase (the authentication phase) commences after the first phase (the bootstrapping phase) has been successfully completed. In the authentication phase, the claimant provides evidence to the verifier that the claimant has the "bootstrapping material" that was provided by the verifier. In the case of mutual authentication, both entities have to identify each other by using the "bootstrapping material" that was exchanged in the first phase.

If the "bootstrapping material" consists of symmetric key data, the secure channel needs to be authentic and confidential. Stajano et al. [26] call this bootstrapping phase "imprinting" in their resurrecting duckling model. The "imprinting" involves the duckling and the mother to exchange some "secret information" by physical contact (over a location-limited channel [8]). In the realm of public-key cryptography, the notion of the "bootstrapping material" is different. In the first phase, both entities need to use an authentic channel to exchange their public keys, which constitute the "bootstrapping material." In the second phase, the entities will authenticate each other by proving the possession of their corresponding private keys. Here, the "bootstrapping material" and the associated private keys are needed for successful completion of the authentication phase. When using public-key cryptography, the channel used to exchange the "bootstrapping material" need not be confidential as in the symmetric case, because the adversary is allowed to passively eavesdrop on the channel and learn the public key.

Remarks: An essential goal of a mutual authentication protocol is to allow both authenticated entities to engage in a secure communication. This necessitates the generation of a cryptographically strong secret key (called the session key) shared by the entities after a successful run of the authentication protocol. The entities would use the session key to encrypt and decrypt subsequent messages. This can be considered as the third phase (the session key generation phase) of authentication.

5.2.2 General Design Goals of a Wireless Authentication Protocol

The very absence of a secure wired medium in a wireless network creates new threats in building security protocols in wireless networks. In a wired network, eavesdropping involves physically "tapping" into a wire on the network [18]. Standard security measures like restricting building access and creating tamper-resistant lockers containing computing equipment can be used to prevent illegal access to the wired network. If a "tap" into the wire is detected and located, it can be easily removed. In a wireless environment, the radio medium can be accessed by anyone who has the proper wireless equipment. This means a mobile adversary can receive and send messages in the wireless network at will, and even if such adversaries are detected, it is difficult to remove them from the network because of their mobile nature. In such an "open" environment, like wireless networks, authentication of communicating entities becomes crucial for secure exchange of sensitive data.

The mobile wireless devices are resource constrained, in terms of computational capability, memory, bandwidth, and power availability. The authentication protocol should be carefully designed, keeping in mind the resource-constrained nature of the wireless devices. Viable implementations of the authentication protocol demand the following:

- Computational efficiency: This denotes the total number of operations required to execute the authentication protocol. Encryption and decryption following the symmetric approach (using secret keys) requires less computations than following the asymmetric approach (using public-key cryptography). However, the advantage of using public-key cryptography is that the channel used for distributing the "bootstrapping material" need not be confidential. Furthermore, when the authentication protocol is run in unbalanced conditions (meaning one entity is a recource-constrained, mobile wireless device, and the other entity belongs to a fixed network and has sufficiently large resources), the computational load for executing the authentication protocol should be directed more toward the entity that is resource rich [29]. In balanced conditions such as ad hoc networks (assuming that all devices have similar constraints) the authentication protocol should be carefully designed so that all devices are required to perform the same number of equally resource-consuming operations.

- Storage of secrets: The "bootstrapping material" needed to initiate the authentication phase should require memory space small enough to fit the available memory of the wireless devices.

- Communication efficiency: Wireless devices spend most of their energy (battery power) for transmitting messages. The authentication protocol should be carefully designed so as to minimize the number of message exchanges and the total number of bits transmitted.

- Involvement of a trusted third party: The "bootstrapping material" is often distributed by a trusted third party. In the symmetric approach, a trusted third party distributes (and initializes the wireless devices with) symmetric keys. In the asymmetric approach, the Certification authority (CA) acts as the third party: The CA helps with distribution of public keys and certificates and verification of the validity of certificates via certificate revocation lists. Design of the authentication protocol should take into consideration the availability of such trusted third parties, like online or offline availability and availability in certain network phases (e.g., the trusted third party might be only available in the bootstrapping phase but not in later phases when an entity wants to join a group of already authenticated entities) [31].

- Security guarantees: Before implementation, the authentication protocol should undergo thorough cryptanalysis to verify the security that the protocol guarantees. Wireless authentication protocols need more than heuristic arguments to provide guarantees of security, and the use of formal methods to analyze and validate security issues is very important in building "acceptable" authentication protocols.

Remark: A mutual authentication protocol generates a cryptographically strong secret key (called the session key) shared by the entities after a successful run of the protocol. The entities would use the session key to encrypt and decrypt subsequent messages. The session key should possess certain desired attributes, such as known-key security, (perfect) forward secrecy, unknown key-share resilience, key compromise impersonation resilience and key control. The details of such attributes are beyond the scope of this chapter. Interested readers are refered to [14] for further reading.

5.3   Wireless Networks: The GSM System

The GSM architecture is broadly composed of three entities, viz., the mobile stations, the base station subsystem, and the network subsystem [23]. Figure 5.1 shows the system architecture of GSM.

The Mobile Station: The mobile station consists of a mobile wireless equipment (a piece of hardware like a cell phone) and the subscriber information. The subscriber information includes a unique identifier of the subscriber, called the International Mobile Subscriber Identity (IMSI). The IMSI is stored in a smart card, called the Subscriber Identity Module (SIM). The subscriber uses the SIM card in his mobile device to make and receive calls and use other subscribed services (like the Short Message Service, which allows sending and receiving messages containing alphanumeric characters and/or images).

Base Station Subsystem (BSS):   The Base Station (BS) subsystem consists of the Base Transceiver Station (BTS) and the Base Station Controller (BSC).

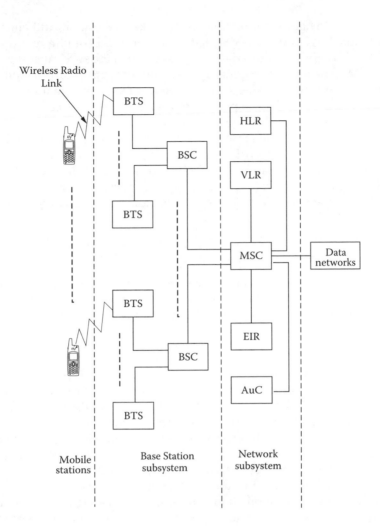

FIGURE 5.1
The GSM architecture.

The BTS contains radio transceivers and engages in radio-link protocols with the mobile station. BSC controls and manages the radio resources of several BTSs. The BSC is responsible for radio channel setup, frequency hopping and handovers between two BTSs that the BSC controls. There can be several BSCs in a BSS, as shown in Figure 5.1.

Network Subsystem (NSS): The Network Subsystem (NSS) consists of five entities, viz., the Mobile Switching Center (MSC), the Home Location Register (HLR), the Visitor Location Register (VLR), the Equipment Identity Register (EIR), and the Authentication Center (AuC). The MSC provides all functionalities to handle a mobile subscriber like registration, authentication, location updating, and inter-MSC handovers and call routing to a roaming subscriber. The HLR contains the current location of the subscriber registered in the GSM network and also manages all administrative information of the subscribers. The VLR obtains selected administrative information from the HLR that is necessary for the provision of subscribed services to the mobile stations currently located in the geographical area under the supervision of the VLR. The EIR is a database that stores the list of identities of all valid mobile stations in the network. The AuC is a protected database containing the shared secret keys stored in the mobile subscriber's SIM card. The secret key is used for authentication and encrypting/decrypting messages on the radio channel between the mobile subscriber and the base station.

5.3.1　Authentication in GSM Systems

In the GSM architecture, the authentication procedure involves the SIM card stored in the mobile station of the subscriber, and the AuC in the NSS. In this section, we describe various approaches toward GSM authentication.

Symmetric Approach:　In this approach, the authentication procedure is carried out using symmetric-key cryptography (also called secret-key cryptography). Classic examples of symmetric-key encryption mechanisms include the Data Encryption Standard (DES) and its recent successor, the Advanced Encryption Standard (AES), the mathematical details of which are beyond the scope of this chapter. Interested readers are referred to [27] for further reading. In the GSM system, during the initial registration (the bootstrapping phase) of a mobile subscriber, the AuC generates a 128-bit shared secret key K_i (called Subscriber Authentication key), which is stored in the SIM card of the mobile station. The mobile station uses the shared secret key K_i to authenticate itself to the AuC. The shared secret key K_i, which is known only to the SIM card and the AuC, is highly protected and never leaves the SIM card. The SIM itself is protected by an optional Personal Identity Number (PIN), which the subscriber enters on the cell phone's keypad. If the PIN is correct, the SIM proceeds with the user authentication phase with the BS. Otherwise (if an incorrect PIN is entered), the SIM generates an error message and displays it on the cell phone.

Authentication in GSM uses two basic functions, viz., the Authentication Function A_3 and the Ciphering Key Generating Function A_8 (both A_3 and A_8 are implemented as keyed-hash functions). The function A_3 is depicted in Figure 5.2 [21], and is defined as follows:

$$A_3 : r \times K_i \mapsto SR; \quad A_3 : \{0, 1\}^{128} \times \{0, 1\}^{128} \mapsto \{0, 1\}^{32}$$

where r denotes a 128-bit random challenge, K_i denotes the 128-bit shared secret key (Subscriber Authentication key) and SR denotes a 32-bit signed

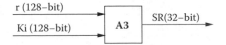

FIGURE 5.2
The authentication function $A3$.

response. In the symmetric approach, authentication in GSM is carried out as follows:

$$MS \rightarrow NSS : IMSI/TMSI$$
$$NSS \rightarrow MS : r$$
$$MS \rightarrow NSS : SR = A_3(r, K_i)$$

Here for simplicity, the details of generation, storing, and transmitting of secrets internally within the NSS is not shown. (For example, the AuC provides the 128-bit random challenge r, the 32-bit signed response SR and the 64-bit session key K_{sess}, the HLR provides the MSC with the 3-tuple $\langle r, SR, K_{sess} \rangle$, etc.). The authentication protocol takes place over the wireless (radio) interface between the BTS and the MS. Since the BTS does not compute the secrets used in the authentication phase but acts as a broker between the MS and the AuC, we use the NSS (Network Subsystem) to denote the entity with which the MS is carrying out the authentication protocol. Initially, the MS sends the IMSI (International Mobile Subscriber Identity) to the NSS. The NSS generates a 128-bit random challenge r and sends r to the MS. The MS computes a 32-bit signed response SR using the A_3 algorithm (described in Equation 5.3.1) and sends the SR to the NSS. The NSS performs a similar computation to produce SR and matches the SR it has calculated with the SR it received from the MS. If both SRs match, the NSS is guaranteed (to a very high probability) that the MS used the same shared secret key K_i as itself, and the MS is successfully authenticated by the NSS. Otherwise, the NSS decides to either send an "authentication reject" message to the MS or restarts the authentication protocol.

Remark: Here A_3 does not refer to a particular algorithm but is a fucntion specification that any algorithm has to implement for use in GSM authentication. The most common implementations for A_3 are in the form of keyed-hash functions, viz., *COMP128V1* and *COMP128V2* [21].

The goal of an authentication protocol is to produce a cryptographically strong session key at the end of a successful run of the authentication protocol. Both the MS and the NSS compute the session key K_{sess} using a function A_8, shown in Figure 5.3 [21]. The function A_8 is defined as follows:

$$A_8 : r \times K_i \mapsto K_{sess}; \quad A_3 : \{0, 1\}^{128} \times \{0, 1\}^{128} \mapsto \{0, 1\}^{64}$$

where K_{sess} is a 64-bit session key. This session key K_{sess} is used to encrypt and decrypt messages between the MS and the NSS. It is worth noting that

FIGURE 5.3
Ciphering key generating function A_8.

the random challenge r is unique for every run of the authentication protocol. Given the algorithm A_8 (implemented as a keyed hash function), it is ensured (with a very high probability) that every run of the authentication protocol generates a unique 64-bit session key, K_{sess}. This is necessary to provide the known key security. Known key security implies that the compromise of the session key from one instance of the protocol should not compromise session keys of other instances.

A primary design goal for the GSM authentication protocol is to provide subscriber anonymity. This implies that IMSI (which uniquely identifies the mobile subscriber) should not be sent in plaintext every time the authentication protocol is initiated by the MS. When the SIM card is used in the authentication protocol for the first time, the MS does not have a choice but to send its unique identifier, IMSI, in plaintext over the wireless radio link to the NSS. However, for subsequent runs of the authentication protocol a temporary identity, known as Temporary Mobile Subscriber Identity (TMSI) is used in place of IMSI. The TMSI is sent to the MS by the NSS after a successful run of the authentication protocol, encrypted by the session key K_{sess}. The mapping between the TMSI and the IMSI is typically handled by the VLR. The TMSI prevents an adversary from (1) gaining resource information of the subscriber, (2) tracing the location of the subscriber, and (3) matching the subscriber and the transmitted signal.

The GSM authentication protocol described above has several security flaws, which are as follows [21]:

- The GSM authentication is not mutual: The MS authenticates itself to the network by proving its knowledge of the Subscriber Authentication key, K_i, but the network does not authenticate itself to the MS. An adversary can set up a false base station with the mobile subscriber's network code. The network initiates the GSM authentication protocol and an adversary wanting to impersonate the network can choose to do the following: (1) start the authentication procedure by sending the random challenge r and then ignore the challenge or (2) not start the authentication protocol at all. The MS would unknowingly make calls with (and/or use other services of) the adversary.

- Flaws in implementations of A_3 and A_8: Briceno et al. [17] showed that the common implementations of both algorithms A_3 and A_8 by the keyed-hash algorithm, *COMP128*, has serious security flaws. Briceno et al. [17] launched a chosen-challenge attack, which works as follows: the SIM card was queried for specially chosen challenges (the 128-bit rs). The SIM applied the *COMP128* implementation of both A_3 and A_8 to compute the 32-bit SR and the 64-bit K_{sess}. The 128-bit Subscriber Authentication key K_i was determined by analyzing the 32-bit signed responses SRs. The attack requires querying the smartcard about 150,000 times. The chosen smart card issued 6.25 queries per second and the attack was successfully completed in 8 hours.

- Weakness in session key K_{sess}: There is a weakness in the session key K_{sess} due to a flaw in the implementation of the A_8 algorithm, *COMP128*. The keyed-hash algorithm *COMP128* always sets the least significant 10 bits of the 64-bit session key K_{sess} to 0. This reduces the strength of the key to 54 bits.

- Weakness in the subscriber identity confidentiality: One of the goals of the GSM authentication protocol is to avoid the MS identifying itself to the network by sending the IMSI in plaintext over the wireless radio link. Since the IMSI uniquely identifies the mobile subscriber, an adversary eavesdropping on the wireless radio link can learn that a particular subscriber is in the area, from the plaintext IMSI. To prevent this from happening, the network sends the TMSI to the MS after the authentication protocol has been successfully run for the first time. The network maintains the mapping between the TMSIs and the IMSIs in a database in the VLR. However, the IMSI has to be sent when the SIM card is used for the first time and when (if) there is a data loss at the VLR.

Asymmetric Approach: The asymmetric approach is based on asymmetric-key cryptography (also called public key cryptography), which uses different keys, viz., public key and private key for encryption and decryption, respectively. Aziz et al. [6] proposed an authentication protocol using the asymmetric approach, assuming the existence of a public-key infrastructure (PKI) in the GSM architecture. Both the NSS and the MS have public key certificates signed by a trusted CA. The authentication protocol by Aziz et al. [6] works as follows: let alg_list denote a list of flags representing various symmetric-key encryption algorithms chosen by the mobile station; sel_alg denote the flag representing the symmetric-key encryption algorithm selected by the NSS; x_{NSS} and x_{MS} denote the contributions of the NSS and the MS, respectively, toward the session key that will be generated at the end of a successful run of the protocol; and PK_A and PK_A^- denote the public key and private key of any arbitrary entity A.

$$MS \to NSS : Cert(MS), n_{MS}, alg\_list$$
$$NSS \to MS : Cert(NSS), \{x_{NSS}\}_{PK_{MS}}, sel\_alg,$$
$$\{h(\{x_{NSS}\}_{PK_{MSS}}, sel\_alg, n_{MS}, alg\_list)\}_{PK_{NSS}^-}$$
$$MS \to NSS : \{x_{MS}\}_{PK_{NSS}}, \{h(\{x_{MS}\}_{PK_{NSS}}, \{x_{NSS}\}_{PK_{MS}})\}_{PK_{MS}^-}$$

The mobile station MS sends its certificate $Cert(MS)$, a random challenge n_{MS} and alg_list to the NSS. The certificate $Cert(MS)$ binds the identity of MS with MS's public key. The NSS responds with its own certificate $Cert(NSS)$, its contribution toward the session key encrypted by the public key of the MS, $\{x_{NSS}\}_{PK_{MS}}$, and the preferred symmetric-key encryption algorithm sel_alg. The symmetric-key encryption algorithm will be used together with the session key to encrypt and decrypt messages by the NSS and the MS after a successful run of the authentication protocol. The NSS also computes a hash digest (using a cryptographic hash function h) of $\{x_{NSS}\}_{PK_{MSS}}$, sel_alg, n_{MS} and alg_list, signs the digest with its private-key PK_{NSS}^{-} and appends the signed digest to the message. Similarly, MS responds to NSS with its contribution component for the session key x_{MS} and a signed hash digest. Both the NSS and the MS compute the session-key $K_{sess} = x_{NSS} \oplus x_{MS}$.

Meadows [19] exposed a security weakness in the above protocol by Aziz et al. [6]. Inspired by the attack by Meadows [19], Boyd et al. [16] constructed yet another attack on the above protocol. Interested readers are referred to [16, 19] for details of these attacks.

Hybrid Approach: The hybrid approach uses a combination of both symmetric- and asymmetric-key cryptography for designing authentication protocols. Beller et al. [10, 11] proposed a series of authentication protocols suited for the GSM architecture using the hybrid (a combination of symmetric and asymmetric) approach. The protocols by Beller et al. were built using the Rabin cryptosystem [22], which is based on the difficulty of computing a square root modulo a composite integer. The Rabin cryptosystem was specifically chosen because the encryption algorithm in the Rabin cryptosystem is very efficient and hence suited for implementations in mobile stations. The protocol suite due to Beller et al. includes three variants: Basic MSR (modulo square root) protocol, Improved MSR (IMSR) protocol, and MSR+DH (Diffie-Hellman) protocol. In this section, we briefly discuss the Basic MSR protocol. Interested readers are referred to [10, 11, 16] for further reading. The Basic MSR protocol works as follows:

$$NSS \rightarrow MS : "NSS", PK_{NSS}$$
$$MS \rightarrow NSS : \{K_{sess}\}_{PK_{NSS}}$$
$$MS \rightarrow NSS : \{MS, Cert(MS)\}_{K_{sess}}$$

The NSS (via the Base Transceiver Station (BTS)) sends its identity "NSS" and its public-key PK_{NSS} to the mobile station MS. The MS uses PK_{NSS} to encrypt the session key K_{sess} and sends it to the NSS. Only the NSS can decrypt the session key using its private key. At the same time, the MS also sends its identity "MS" and the certificate $Cert(MS)$ (issued by a trusted CA) encrypted with the session key K_{sess} for authenticating itself to the NSS.

Remarks: In the Basic MSR protocol, the term "certificate" has been used in a different sense than the certificates used in the public-key infrastructure

(PKI). In the PKI, a certificate provides an assurance to users regarding the binding of a public key to the identity of the user who holds the corresponding private key. The certificate is made public in the PKI. In the Basic MSR protocol, the certificate constitutes the "bootstrapping material" that the trusted CA distributes to the NSS and the mobile stations. The NSS authenticates the MS by means of the certificate that it receives from the MS. It should be noted that in the Basic MSR protocol, the certificate is kept secret from other mobile stations because an adversary having access to the certificate can impersonate the mobile station. Furthermore, the Basic MSR protocol does not provide mutual authentication between the NSS and the MS. Only the NSS authenticates the MS. The other variants in the protocol suit are constructed by improving the Basic MSR protocol and have varying complexity and security features.

5.4   Wireless Networks: The IEEE 802.11 Standard

The IEEE 802.11 wireless networks can be classified into two models, viz., infrastructureless (or ad hoc model) and infrastructured model. In this section, we describe the IEEE 802.11 infrastructured model and discuss various authentication approaches in the 802.11 wireless networks. Authentication in ad hoc networks is discussed in Section 5.5.1. The 802.11 wireless local area network (WLAN) architecture consists of cells, called Basic Service Sets (BSSs) and each cell is controlled by a Base Station or Access Point (AP). We used the term "Base Station" to refer to the entity that the cell phone connects to in the GSM architecture. Hence, to avoid confusion, we will use the term "Access Point (AP)" for referring to WLANs. Each WLAN can be constructed using a single AP or multiple APs. A client is a wireless device that wants to join a certain network and sends a request to the AP asking entry to the network "guarded" by that AP. Before the client can access the network resources, it has to establish a relationship or association with the AP [3].

The process of association consists of the following stages: (1) unauthenticated and unassociated, (2) authenticated but unassociated and (3) authenticated and associated. All APs in the WLAN transmit periodic beacon management frames, which the clients listen to for identifying the APs within the wireless range. In the first stage, the client is not authenticated by, and not associated with the AP. After identifying an AP, the client engages in an authentication protocol with the AP. If the authentication protocol completes successfully, the client moves to the second stage, i.e., the client is now authenticated by, but not associated to, the AP. In the transition from state 2 to state 3, the client sends an association request frame to the AP and the AP responds with an association response frame to the client. This places the client in the third stage, i.e., authenticated and associated.

5.4.1 Design Goals for WLAN Authentication

In Section 5.2.2, we discussed the general design goals of wireless authentication protocols. Different forms of wireless networking environments pose different and unique security issues. This implies that the general design goals for authentication (described in Section 5.2.2) need to be fine-tuned before they can be applied to a particular scenario, like the WLAN. The goals for designing an authentication protocol suited for the IEEE 802.11 WLANs are as follows [7]:

1. Mutual Authentication: An authentication is defined to be unilateral if after the successful completion of a run of the protocol, one entity (the verifier) is assured of the identity of the second (claimant) involved in the protocol and that the second has actually participated. The authentication is mutual if both entities successfully carry out unilateral authentication with each other. A WLAN authentication protocol should provide mutual authentication between the Authentication Server (the role of the Authentication Server (AS) is described in Section 5.4.3.) and the client.

2. Resisting Dictionary Attacks: A dictionary attack takes place in the following scenario: Suppose an entity (claimant in an authentication protocol) is using a weak password that is chosen from a sample space small enough to be enumerated by an adversary. The password is then used to derive a secret key, $K = h(pwd)$, where h is a cryptographically strong hash function and pwd is the password. The secret key is used to encrypt a random challenge (R) sent by the verifier, to create the response $E_K(R)$. Let us assume the adversary eavesdrops on the channel and knows R and $E_K(R)$. He picks passwords from the small sample space, derives secret keys K', and encrypts the challenge to produce $E_{K'}(R)$. If $E_K(R) = E_{K'}(R)$, then the adversary has successfully guessed the password. If not, he picks another password and tries again. This kind of noninteractive (with the authentication server) attack is called a dictionary attack (more specifically, an offline dictionary attack). The WLAN authentication protocol should provide resistance to dictionary attacks.

3. Resisting Replay Attacks: The WLAN authentication protocol should use a time-variant parameter that serves to distinguish one protocol instance from another to thwart replay attacks. This time-variant parameter is usually generated by random numbers, sequence numbers, or timestamps. In this context, the term "random number" denotes a pseudorandom number that an adversary cannot predict. This is different from the phrase "choose a random number" used in the statistical sense in security protocols, which means "pick a number with uniform distribution on a given sample space" [20].

4. Known-key Security: Each run of the WLAN authentication protocol should result in a unique secret session key between the AP and the client. The compromise of the session key from one instance of the protocol should not compromise session keys of other instances.

5. Identity Privacy: A WLAN authentication protocol should provide a way to protect the client's identity from eavesdropping.

6. Proof of Security: Any relatively new WLAN authentication protocol is likely to contain flaws in the design of security attributes (as in the case of any security protocol). Before any efforts are made to standardize the protocol, a rigorous security analysis of the protocol needs to be done.

7. Fast Reconnect: When a client wants to join a WLAN, it has to authenticate itself to the Authentication Server (AS). After successful authentication between the client and the AS, the AP allows the client to join the WLAN. The clients in a WLAN have the freedom to move around from one AP to the other. When a client moves to a different AP that has not brokered the authentication process between the client and the AS, the client might lose network connection because the new AP may not be aware of the authentication between the client and the AS. The WLAN protocol should provide a lightweight version of the authentication protocol that the client can initiate for fast reconnect.

5.4.2 Authenticating 802.11 WLANs

The Wired Equivalent Privacy (WEP) protocol is used by the IEEE 802.11 standard to provide confidentiality for network traffic in a wireless local area network (WLAN). WEP aims to protect link-level data during wireless transmissions. In WEP, all clients of the wireless network share a secret key, called the WEP key, which is used by the client for authentication and for encryption/decryption of messages. This means the client uses the same key to do the following: (1) encrypt the challenge sent by the AS and create the response for authenticating itself to the AS and (2) encrypt messages. An obvious drawback of this protocol stems out of the fact that the WEP key is shared by the clients. The sharing of the WEP key becomes a bottleneck when the key needs to be updated because the updating requires changing the key that is stored in the clients' (wireless mobile devices). Several security flaws have been exposed in the WEP protocol [3, 15, 28] and new security solutions have been proposed by several researchers and vendors. To address the security weaknesses of WEP, the WiFi alliance (the international association of wireless device manufacturers) proposed a new standard for the WLAN security, called WiFi Protected Access (WPA). An IEEE 802.11i draft has also been finalized by IEEE and is known as Robust Security Network (RSN) or WPA2, which is designed to overcome the security flaws of WEP [7].

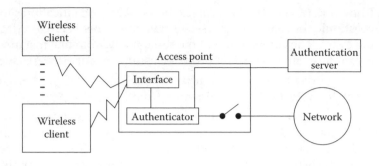

FIGURE 5.4
The WPA framework.

5.4.3   The WiFi Protected Access (WPA) Protocol

In the WPA protocol, the clients in the wireless network do not use the same symmetric key for both authentication and message encryption/decryption purposes. Instead, the clients use their own unique credentials (identities) for authentication and derive a cryptographically strong session key after the successful run of the authentication protocol. The session key is used to encrypt/decrypt subsequent messages. Moreover, every successful initiation of the authentication protocol generates a unique session key. This provides the known-key security. Also, the separation of the authentication protocol and the message protection makes the WPA scalable, allowing dynamic-key management.

The IEEE 802.1x standard offers an effective framework for authentication and message protection in wireless networks. The WPA protocol and the IEEE 802.11i adopted the framework of authentication provided by IEEE 802.1x. Figure 5.4 [7] shows the authentication framework provided by WPA and IEEE 802.11i. The framework is modeled with three entities, viz., the client, which is a (potentially mobile) wireless device, the access point (AP), and the authentication server (AS). The client connects to the AP for accessing the network. The AS connects to the client database and is responsible for authenticating the client. The AP acts as a broker between the AS and the client and controls a switch that connects the client to the network. The Extensible Authentication Protocol (EAP) is initiated by the three-entity model to facilitate authentication.

The EAP Framework: The EAP provides a framework of interactions between the entities for authentication in WLANs. The EAP can encapsulate other authentication protocols in its framework [9]. The EAP works as follows: In Step 1, the client C sends an *EAP-init* message to the AP, requesting access to a certain network. The AP acts as a guard to the network. In Step 2, the AP replies with an *EAP-request-identity* message to the client. In Step 3, the client responds with an *EAP-response-identity* message. In Step 4, the AP forwards the *EAP-response-identity* message to the AS. In Steps 5–10, the AS triggers the authentication protocol, which the EAP encapsulates and engages in a series of *request* and *response* messages with the client C. A successful run of the authentication protocol (which the AS starts with the client C) produces a cryptographically strong session key. The client C will use this session key for encryption and decryption of subsequent messages with the AP. The details of how the AS authenticates the client is protocol specific. The various approaches of authentication, viz., symmetric, asymmetric, or tunneled approaches, are discussed below. After the completion of the authentication protocol, the AS sends a *success* or *failure* message to the client via the AP.

(1) $C \rightarrow AP$: *EAP-init*
(2) $AP \rightarrow C$: *EAP-request-identity*
(3) $C \rightarrow AP$: *EAP-response-identity*
(4) $AP \rightarrow AS$: *EAP-response-identity*
(5) $AS \rightarrow AP$: *request*
(6) $AP \rightarrow AS$: *request*
(7) $C \rightarrow AP$: *response*
(8) $AP \rightarrow AS$: *response*
$$\vdots$$
(9) $AS \rightarrow AP$: *success/failure*
(10) $AP \rightarrow C$: *success/failure*

So far, the AP had been acting as a broker between the AS and the client by relaying messages between the AS and the client. Now, the AP is responsible for connecting the client to the network, depending on the *success/failure* message that the AS sends to the client via the AP. It is important to note that the above framework does not provide the actual authentication mechanisms. In the WPA protocol, the EAP type specifies the authentication protocol that needs to be initiated within the EAP framework. The Transport Layer Security (EAP-TLS) protocol [9] and the EAP Tunneled Transport Layer Security (EAP-TTLS) protocol [9] are examples of two authentication protocols that are currently supported by the EAP framework. The details of the various EAP types are beyond the scope of this chapter. However, we provide a brief discussion on various approaches that can be followed.

Symmetric Approach: The client and the AS authenticate each other by proving the knowledge of a shared symmetric key. Encryption and decryption of messages are carried out using symmetric keys, which makes the authentication protocol computationally efficient. However, the shared symmetric key (K) is derived from a password: $K = h(pwd)$, where h is a cryptographically strong hash function and pwd is a password. As the authentication protocols are carried out on a wireless medium, eavesdropping is common and offline dictionary attacks can be launched on most protocols based on this approach. Authentication protocols using a symmetric approach inside the EAP framework include the Lightweight Extensible Authentication Protocol (LEAP) [9], Kerberos and the EAP over Secure Remote Password (EAP-SRP) [7].

Asymmetric Approach: The client and the AS authenticate each other using the public-key cryptography. The realm of public-key cryptography demands the necessity of providing an assurance to users regarding the binding of a public key to the identity of the user who holds the corresponding private key. The deployment and management of the underlying infrastructure is quintessential to providing such an assurance. Certificates play the role of providing such authenticity in traditional public-key infrastructures. The advent of identity-based public-key cryptosystems has made it possible to do away with the burden of certificate management and yet provide implicit authentication to users [2]. Shamir [25] first proposed the concept of an identity-based cryptosystem where the public key of a user could be generated from his identity information. A trusted third party called the Key Generation Center (KGC) was required to derive the private keys corresponding to the public keys of users. The KGC would also publish the public global system parameters needed for encryption, decryption, and signature algorithms that users need to execute. The use of bilinear pairings on elliptic curves has served as an exciting breakthrough in building various kinds of cryptographic schemes [13]. The EAP over Transport Layer Security (EAP-TLS) protocol [9] uses certificate-based authentication. Lee et al. [7] used the identity-based cryptosystem to build an EAP authentication protocol.

Tunneled Approach: This approach consists of two layers of protocols, viz., the outer layer and the inner layer. The goal of the outer-layer protocol is to authenticate the AS to the wireless client. After every successful run of the outer-layer protocol, a unique session key is created. The session key is used to encrypt messages of the inner-layer protocol. The outer-layer protocol creates a secure channel to protect the inner-layer protocol; hence the name, tunneled approach. The inner-layer protocol is used to authenticate the wireless client to the AS. Protocols using the tunneled approach, such as Protected EAP (PEAP) [9] and EAP-Tunneled TLS (EAP-TTLS) [9], are still being revised as Internet drafts. Asokan et al. [5] recently showed that a man-in-the-middle attack can be launched on the protocols using the tunneled approach.

This concludes the discussion of authentication protocols in infrastructured wireless networks in the GSM architecture and the IEEE 802.11 architecture. In the following section, we describe various authentication techniques in wireless ad hoc networks.

5.5   Wireless Ad Hoc Networks

Wireless ad hoc networks are self-organized networks that can be rapidly deployed for a special purpose in situations where no infrastructure exists and in harsh conditions where it would be impractical or infeasible to deploy an infrastructure [1]. Due to the lack of an infrastructure, nodes forming the ad hoc network are expected to cooperate in forwarding and routing data packets on behalf of other nodes in the network that need to send packets to nodes out of their transmission range. Ad hoc networks are dynamic in nature: The nodes are mobile and can arbitrarily join or leave the existing ad hoc network at any time. The nodes in an ad hoc network are typically resource constrained and their properties can be summarized as follows [26]: (1) small CPU—the processor in the node has limited computing power and large computations are either extremely slow or not feasible; (2) small memory; (3) limited battery power—transmitting (followed by receiving) data consumes most of the energy in a wireless device. Due to limited battery power, the nodes conserve energy by turning off their receivers and "going to sleep"; and (4) limited bandwidth.

Under the above-mentioned constraints, developing network security protocols that offer authentication, confidentiality, integrity, nonrepudiation and access control is a challenging task. The following example provides an insight to the authentication problem arising in ad hoc networks but not in other infrastructured wireless networks, like the GSM system. Consider a resource-constrained mobile wireless node trying to authenticate itself to a powerful base station in the GSM system. Authentication protocols in the GSM architecture can be designed so that the mobile node performs the cheap operations of encryption and verification and avoids the expensive tasks of decryption and signature generation. The heavy tasks are performed by the base station [26]. Recent authentication protocols in the GSM architecture using low-exponent RSA have been developed in the above-mentioned way [29]. In the realm of ad hoc networks, the two nodes wanting to authenticate each other are both resource constrained, which means tasks can no longer be classified as cheap and expensive tasks and delegated accordingly. Yet another security challenge special to ad hoc networking, stems from the unavailability of a central system administrator who can perform authorization between nodes.

5.5.1 Authentication in Ad Hoc Networks

Wireless networks present certain security issues that were not present in traditional wired networks. The fact that wireless networks are prone to eavesdropping is a significant problem in designing authentication protocols in wireless networks (as we discussed in Section 5.2.2). Wireless ad hoc networks, with their unique properties such as dynamically changing topology, resource-constrained nodes and reliance on node collaboration in routing packets, present unique challenges in designing authentication protocols. The absence of a central system administrator or an online trusted third party that can perform authorization between nodes makes the problem of authentication even worse. Several protocols, models, and frameworks proposed for authentication protocols in ad hoc networks exist in the literature [1, 8, 26, 31, 32]. In this chapter, we briefly discuss three existing approaches, viz., symmetric, asymmetric and hybrid approaches to address the problem of authentication in ad hoc networks.

Symmetric Approach: The process of authentication is composed of the bootstrapping phase and the authentication phase (discussed in Section 5.2.1). The ad hoc nodes wanting to authenticate each other would need to demonstrate the knowledge of the "bootstrapping material" in the authentication phase. In the symmetric approach, the "bootstrapping material" consists of a symmetric (shared) key that is distributed to all the nodes in the ad hoc network that wish to engage in the authentication protocol. An authentic and confidential channel needs to be established to distribute the "bootstrapping material" to the nodes. Several models exist that describe ways to achieve such a channel [26, 31]. The model suggested by Stajano et al. [26] works as follows: The symmetric keys are distributed to nodes by physical contact. This method, though it sounds trivial, turns out to be quite an effective, cheap, and simple way to construct the secure channel. This model assumes that the use of public keys is not viable in the resource-constrained nodes and the public-key infrastructure needed to distribute the public keys is absent in the domain of ad hoc networks. However, the requirement of physical contact among all nodes in the ad hoc network can pose severe restrictions in some applications. After the bootstrapping phase is completed, the authentication phase can be carried out by the nodes using a challenge-response type protocol [20]. The idea behind a challenge-response type protocol can be summarized as follows: The claimant initiates the protocol by sending a message to the verifier containing the claimant's identity. The verifier sends a random number, called a challenge. The claimant uses the challenge and the shared symmetric to perform some computation and sends the result, called a response, to the verifier. The verifier uses the symmetric key to perform the same computation and verifies the response. Since the verifier chooses a different challenge for every run of the protocol (called a time-variant challenge [20]), an adversary cannot eavesdrop, record messages, and resend them at a later time (a replay attack) for impersonating the claimant.

Asymmetric Approach: Several models have been built using public-key cryptography for authentication in ad hoc networks. One of the early works using this approach can be attributed to Zhou et al. [32] who used threshold cryptography to build a distributed Certification Authority (CA) functionality in ad hoc networks. Establishing a key management service with a single CA creates a single point of vulnerability in ad hoc networks. This observation led to the idea of distributing the functionality of the CA to $t+1$ special ad hoc nodes, called server nodes. The authentication protocol by Zhou et al. [32] is bootstrapped by splitting the system private key into n shares and assigning one share to each server. The splitting of the system private key is done in such a way that any $t + 1$ shares can collectively generate the system private key. However, it is infeasible for t servers to collude and generate the correct system private key. This is accomplished by using Shamir's secret sharing scheme [24], the mathematical details of which are beyond the scope of this chapter. When a node A wants to get an authentic copy of node B's public key, it broadcasts a "query" request to at least $t+1$ neighboring nodes. Each server nodes signs B's public key with its share of the system private key. Another special node, called the "combiner" node, collects all the partial signatures and sends the full signature on B's public key to the node A. The node A verifies the signature on B's public key with the system public key and either accepts or rejects.

Remarks: The protocol by Zhou et al. [32] assumes the presence of special nodes in the ad hoc network in the form of "server" nodes and the "combiner" node. The server nodes need to store the public keys of all the nodes in the ad hoc network, which demands quite a large memory. Also, the "server" node has to process a large number of request messages from the nodes that want authentic copies of public keys. The "combiner" node needs to combine the partial signatures of the server nodes for every request for an authentic copy of a public key. Clearly, these special nodes cannot be resource constrained. The assumption does not quite comply with the generic design of an ad hoc network.

Hybrid Approach: Following the hybrid approach, Asokan et al. [4] proposed a password-authenticated key agreement protocol suitable to wireless ad hoc networks. In the hybrid approach, passwords constitute the "bootstrapping material," meaning a password is shared by two nodes wanting to authenticate each other by an authentic and confidential channel. In the authentication phase, the entities engage in the authentication protocol constructed using the password. It should be noted that using passwords in a simple challenge-response type protocol would lead to offline dictionary attacks [30]. Hence, protocols that are stronger than simple challenge-response protocols are needed that can use these cryptographically weak passwords to securely authenticate entities. The first attempt to protect a password protocol against offline dictionary attacks was made by Bellovin and Merritt [12] who developed a password-based encrypted key exchange (EKE) protocol using a combination of symmetric and asymmetric cryptography. The protocol by Asokan et al. [4] was based on the EKE protocol.

The EKE protocol works as follows: let nodes A and B participate in a particular run of the protocol. In Step 1, node A generates a public/private key pair (E_A, D_A) and also derives a secret key K_{pwd} from its password pwd. In Step 2, node A encrypts his public key E_A with K_{pwd} and sends it to node B. In Steps 3 and 4, node B decrypts the message and uses E_A together with K_{pwd} to encrypt a session key K_{AB} and sends it to node A. In Steps 5 and 6, node A uses this session key to encrypt a unique challenge C_A and sends the encrypted challenge to node B. In Step 7, node B decrypts the message to obtain the challenge and generates a unique challenge C_B. In Step 8, node B encrypts both C_A and C_B with the session key K_{AB} and sends it to node A. Node A decrypts this message to obtain C_A and C_B and compares the former with its own challenge. If they match, the correctness of node B's response is verified. In Step 9, node A encrypts node B's challenge C_B with the session key K_{AB} and sends it to node B.

$$(1) \quad A \; : \; (E_A, D_A)$$
$$(2) \quad A \rightarrow B : A, K_{pwd}(E_A)$$
$$(3) \quad B \; : \; E_A = K_{pwd}^{-1}(K_{pwd}(E_A)); \; K_{AB}$$
$$(4) \quad B \rightarrow A : K_{pwd}(E_A(K_{AB}))$$
$$(5) \quad A \; : \; K_{AB} = D_A(K_{pwd}^{-1}(K_{pwd}(E_A(K_{AB})))); \; C_A$$
$$(6) \quad A \rightarrow B : K_{AB}(C_A)$$
$$(7) \quad B \; : \; C_A = K_{AB}^{-1}(K_{AB}(C_A)); \; C_B$$
$$(8) \quad B \rightarrow A : K_{AB}(C_A, C_B)$$
$$(9) \quad A \rightarrow B : K_{AB}(C_B)$$

The EKE protocol results in a session key (stronger than the shared password), which the nodes can later use to encrypt sensitive data.

5.6   Summary

Authentication in wireless networks is a challenging task. In a wireless environment, the radio medium can be accessed by anyone who has the proper wireless equipment. In such an "open" environment, authentication of communicating wireless devices becomes crucial for the secure exchange of sensitive data. Different kinds of wireless networks pose unique security issues in the design of authentication protocols. In this chapter, we described three different categories of wireless networks, viz., the GSM system, IEEE 802.11 WLANs, and wireless ad hoc networks.

We categorized the design of authentication protocols in the above-mentioned wireless networks according to three different approaches, viz., symmetric, asymmetric, and hybrid approaches. In the symmetric approach, the authentication procedure is carried out using symmetric-key cryptography (also called secret-key cryptography), which relies on the same key for both encryption and decryption. The asymmetric approach is based on asymmetric-key cryptography (also called public-key cryptography), which uses different keys, viz., public key and private key for encryption and decryption, respectively. The hybrid approach uses a combination of both symmetric- and asymmetric-key cryptography for designing authentication protocols.

Different designs of authentication protocols deliver unique security guarantees and no single approach can provide a complete solution for authentication in a particular kind of wireless network. Any authentication protocol should undergo thorough cryptanalysis before implementation, to verify the security that the protocol guarantees. Wireless authentication protocols need more than heuristic arguments to provide guarantees of security and formal methods should be applied (wherever possible) to analyze and validate security issues for building "acceptable" authentication protocols.

Bibliography

1. Nidal Aboudagga, Mohamed Tamer Refaei, Mohamed Eltoweissy, Luiz A. DaSilva, and Jean-Jacques Quisquater. Authentication protocols for ad hoc networks: taxonomy and research issues. In *Q2SWinet '05: Proceedings of the 1st ACM international workshop on Quality of Service and Security in Wireless and Mobile Networks*, pages 96–104, New York, NY, 2005. ACM Press.

2. Sattam S. Al-Riyami. *Cryptographic Schemes Based on Elliptic Curve Pairings*, Ph.D. Thesis, Royal Holloway, University of London, 2005.

3. W.A. Arbaugh, N. Shankar, Y.C.J. Wan, and Kan Zhang. Your 80211 wireless network has no clothes. *Wireless Communications, IEEE*, 9:44– 51, 1992.

4. N. Asokan and Philip Ginzboorg. Key-agreement in ad-hoc networks. *Computer Communications*, 23(17):1627–1637, 2000.

5. N. Asokan, V. Niemi, and K. Nyberg. Man-in-the-middle in tunnelled authentication protocols, 2003. In *11th Security Protocols Workshop*, Cambridge, U.K., April 2003. Springer-Verlag.

6. Ashar Aziz and Whitfield Diffie. Privacy and authentication for wireless local area networks. *IEEE Personal Communications*, 1(1):25–31, 1993.

7. Kwang-Hyun Baek, Sean W. Smith, and David Kotz. A Survey of WPA and 802.11i RSN Authentication Protocols. Technical Report TR2004-524, Dartmouth College, Computer Science, Hanover, NH, November 2004.

8. D. Balfanz, D. Smetters, P. Stewart, and H. Wong. Talking to strangers: Authentication in adhoc wireless networks. February 2002. In *Symposium on Network and Distributed Systems Security (NDSS '02)*, San Diego.

9. Andrew Balinsk, Darrin Miller, Krishna Sankar, and Sri Sundaralingam. *Cisco Wireless LAN Security*. Cisco Press, 2004.

10. M. Beller and Y. Yacobi. Fully-fledged two-way public key authentication and key agreement for low cost terminals. *Electronic Letters*, 30:999–1001, 1993.

11. Michael J. Beller, Li-Fung Chang, and Yacov Yacobi. Privacy and authentication on a portable communications system. *IEEE Journal on Selected Areas in Communications*, 11(6):821–829, 1993.

12. Steven M. Bellovin and Michael Merritt. Encrypted key exchange: Password-based protocols secure against dictionary attacks. In *SP '92: Proceedings of the 1992 IEEE Symposium on Security and Privacy*, page 72, Washington, DC, 1992. IEEE Computer Society.

13. Ian F. Blake, G. Seroussi, and N. P. Smart. *Elliptic Curves in Cryptography*. Cambridge University Press, New York, July 1999.

14. Simon Blake-Wilson, Don Johnson, and Alfred Menezes. Key agreement protocols and their security analysis. In *Proceedings of the 6th IMA International Conference on Cryptography and Coding*, pages 30–45, London, U.K., 1997. Springer-Verlag.

15. Nikita Borisov, Ian Goldberg, and David Wagner. Intercepting mobile communications: The insecurity of 802.11. In *MobiCom '01: Proceedings of the 7th Annual*

International Conference on Mobile Computing and Networking, pages 180–189, New York, 2001. ACM Press.

16. Colin Boyd and Anish Mathuria. Key establishment protocols for secure mobile communications: A selective survey. In *ACISP '98: Proceedings of the Third Australasian Conference on Information Security and Privacy*, pages 344–355, London, U.K., 1998. Springer-Verlag.

17. Marc Briceno, Ian Goldberg, and David Wagner. An implementation of the gsm a3a8 algorithm. (specifically, comp128), April 1998. GSM Security White Paper, http://www.gsm-security.net.

18. Richard R. Joos and Anand R. Tripathi. Mutual authentication in wireless networks, 1997. Technical Report, Computer Science Department, University of Minnesota.

19. Catherine Meadows. Formal verification of cryptographic protocols: A survey. In *ASIACRYPT*, pages 135–150, 1994.

20. Alfred J. Menezes, Scott A. Vanstone, and Paul C. Van Oorschot. *Handbook of Applied Cryptography*. CRC Press, Inc., Boca Raton, FL, 1996.

21. Jeremy Quirke. Security in the gsm system, May 2004. GSM Security White Paper, http://www.gsm-security.net.

22. M. O. Rabin. Digitalized signature and public-key functions as intractable factorization. Technical Report MIT/LCS/TR-212, Laboratory for Computer Science, M.I.T. Cambridge, 1979.

23. John Scourias. *Overview of GSM: The Global System for Mobile Communications*. University of Waterloo, March 1996.

24. Adi Shamir. How to share a secret. *Commun. ACM*, 22(11):612–613, 1979.

25. Adi Shamir. Identity-based cryptosystems and signature schemes. In *Proceedings of CRYPTO 84 on Advances in cryptology*, pages 47–53, New York, 1985. Springer-Verlag.

26. Frank Stajano and Ross J. Anderson. The resurrecting duckling: Security issues for ad-hoc wireless networks. In *Proceedings of the 7th International Workshop on Security Protocols*, pages 172–194, London, U.K., 2000. Springer-Verlag.

27. Douglas R. Stinson. *Cryptography: Theory and Practice*. CRC Press, Boca Raton, FL, 1995.

28. Adam Stubblefield, John Ioannidis, and Aviel D. Rubin. Using the fluhrer, mantin, and shamir attack to break wep. In *Proceedings of the Network and Distributed System Security Symposium, NDSS 2002, San Diego*, 2002.

29. Shiuh-Jeng Wang. Anonymous wireless authentication on a portable cellular mobile system. *IEEE Trans. Comput.*, 53(10):1317–1329, 2004.

30. Thomas D. Wu. The secure remote password protocol. In *Proceedings of the Network and Distributed System Security Symposium, NDSS 1998, San Diego*, 1998.

31. Yang Xiao, Xuemin Shen, and Ding-Zhu Du (Eds.). *Wireless/Mobile Network Security (Network Theory and Applications S.)*. Springer-Verlag, New York, 2006.

32. Lidong Zhou and Zygmunt J. Haas. Securing ad hoc networks. *IEEE Network*, 13(6):24–30, 1999.

33. Sencun Zhu, Shouhuai Xu, Sanjeev Setia, and Sushil Jajodia. Lhap: A lightweight hop-by-hop authentication protocol for ad-hoc networks. In *ICDCSW '03: Proceedings of the 23rd International Conference on Distributed Computing Systems*, page 749, Washington, DC, 2003. IEEE Computer Society.

6

Intrusion Detection in Wireless Sensor Networks

Fereshteh Amini, Vojislav B. Mišić and Jelena Mišić

CONTENTS

Abstract   Security is rapidly replacing performance as the first and foremost concern in many networking scenarios. This includes wireless sensor networks, which are becoming increasingly popular for many environmental, logistics, engineering, health, and military applications. While security prevention is important, it cannot guarantee that attacks will not be launched and that, once launched, they will not be successful. Therefore, detection of malicious intrusions forms an important part of an integrated approach to security. In this chapter, we review the basic tenets of intrusion detection in wireless sensor networks. We present the main differences between wireless sensor networks and other similar networks such as ad hoc networks, and discuss the manner in which these differences limit and guide the analysis and development of viable and effective approaches to intrusion detection.

We also present a survey of current research in this area, and outline main challenges for future research.

6.1   Introduction

Wireless sensor networks (WSNs) consist of a large number of tiny sensor devices or nodes with sensing, computational, and communication capabilities. Sensor nodes monitor some physical phenomena in their environment, record the values of appropriate variables, and send them using wireless transmission toward one (or, in some cases, several) network sinks. Along the way, data may pass through a number of intermediate nodes where some filtering and aggregation may be performed. Network sinks act as gateways that collect the data, possibly aggregate it, and pass it on to the sensing applications that requested it. Sensor nodes are small and possess limited energy, memory, bandwidth, and processing power. They can be deployed in inhospitable places, with little or no human intervention thereafter. A sensor network is (or should be) able to operate autonomously, from the moment sensor nodes are deployed in the space of interest to the time when batteries are exhausted and sensor nodes stop working. This generic scenario may be applied in many situations, and it should come as no surprise that wireless sensor networks are becoming increasingly popular in many environmental, business, engineering, healthcare, military, surveillance, and other applications [2].

However, the intrinsic characteristics of WSNs make them vulnerable to attacks by malicious intruders. In military and surveillance applications, sensor networks can provide crucial data to their operators, and degrading their performance or even subverting them may offer generous benefits to an adversary. Therefore, security issues are of primary concern for the design and deployment of wireless sensor networks.

Typically, security implies intrusion prevention through physical protection of the system, advanced cryptographic techniques, and appropriate security policies. However, wireless sensor networks are often expected to operate unattended for prolonged periods of time. On account of that, the importance of security policies is much less pronounced than for systems that include a significant human component, such as information systems and online applications. For that same reason, and because the sensor network often operates in places where an adversary can easily access the actual devices, physical protection of sensor devices is often impossible. As a result, we have to rely on cryptographic techniques for attack prevention, and this is not enough. Or, in other words, despite our best efforts in devising secure protocols and communication techniques to protect against attacks, we cannot really expect that the network will be able to resist all possible attacks.

Consequently, we have to consider not just intrusion prevention techniques as the first line of defense, but also the techniques to detect ongoing attacks and techniques to eliminate or, at least, diminish the impact of such attacks. The former techniques, which are collectively known as *intrusion detection*, form the second line of defense and they are the focus of this chapter; the latter belong to a wider range of security response policies that are not discussed here, although occasional recommendations will be made in relationship with particular intrusion detection techniques.

The chapter is organized as follows. First, we review the main characteristics of wireless sensor networks in more detail, and outline the reasons for which intrusion detection techniques developed for other types of wireless networks are not readily applicable to wireless sensor networks. Then, we present the possible criteria for classification of intrusion detection techniques and discuss their advantages and shortcomings. A brief overview of some of the techniques proposed in the literature follows. Finally, we outline the challenges for research in this area and discuss some promising avenues for future work.

6.2 Why Wireless Sensor Networks Are Difficult to Protect

As mentioned above, wireless sensor nodes are typically small, battery-operated devices with three main subsystems:

- The sensing subsystem consists of one or more sensors or transducers that convert the monitored physical variable to an electrical, possibly digital, signal.
- The computational subsystem is a small microcontroller with integrated memory; it controls the operation of the other two subsystems.
- The communication or radio subsystem enables the node to communicate with other nodes in its vicinity through wireless transmissions.

Three main problems that make wireless sensor networks difficult to protect and secure against intrusions can be readily identified. The first problem is the very nature of the wireless communication medium, which makes wireless communication inherently insecure. Unlike wired networks, where a device has to be physically connected to the medium, the wireless medium is open and accessible to anyone. Moreover, the range in which the impact of an intruder can be felt primarily depends on the characteristics of the intruder's equipment; an intruder with a strong transmitter can easily produce interference from a distance that makes any physical response infeasible or, in some applications, plain impossible.

The second problem is the absence of any fixed infrastructure; in particular, there is no central or master controller to monitor the operation of the network and analyze the data to detect intrusions. While most such networks have a designated network sink, its role is typically restricted to data collection and query distribution, and does not include any form of actual control. As a result, any intrusion detection technique has to be implemented as a cooperative, distributed effort of many among the nodes in the sensor network, or even all of them together. An added difficulty stems from the unstable topology of the network, which may be due to battery exhaustion or (in some cases) node mobility.

Yet other wireless networks exist that have both of these problems: wireless ad hoc networks. In those networks, wireless communication medium is used, and they operate with little infrastructure or none at all. A number of intrusion prevention techniques have been proposed for such networks [10], and also a few techniques for intrusion detection [5, 17, 28, 29]. Such techniques are a combination of several approaches, including use of cooperating mobile agents [6, 15], possibly combined with the analysis of audit logs [13], a game-theoretic approach [1], and a number of others.

However, the main problem with wireless sensor networks lies elsewhere: in their limited computational and communication resources; namely, wireless sensor networks need to operate autonomously for prolonged periods of time, and they have to run on battery power. To cater to those goals, the energy consumption of sensor nodes has to be minimized; this necessitates both the power efficiency of the hardware (and its small size) and the efficiency of communications protocols and the software that implements those protocols. The processing subsystem is invariably implemented with a small microprocessor with limited resources, which runs at low clock speeds, and thus offers only modest computational and memory capabilities. The results are as follows.

- The processing power of such subsystems is generally insufficient to run a full-scale software agent dedicated to intrusion detection [29].
- Even if sufficient computational capability were available, the low data rate of typical communication channels—250kbps for IEEE 802.15.4 networks operating in the ISM band at 2.4 GHz, but only 20 or 40 kbps when operating in other bands [12]—simply does not suffice for the rather intense communication that those agents need.
- By the same token, any substantial computation is infeasible.
- Moreover, since memory capacity is of the order of hundreds or, at best, thousands of bytes, an audit log of realistic size cannot be maintained.
- Simple and efficient protocols mean that individual layers that are traditionally observed in wired networks (but also in other wireless networks) [24] must be integrated; after all, a wireless sensor network is a highly specialized network for a limited class of

applications, and such integration makes perfect sense in view of the inherent limitations of wireless sensor networks [2]. The important implication is that existing techniques that focus on one layer only—for example, routing [3, 18] or media access control (MAC) [26]—cannot readily be applied.

Further problems pertinent to wireless sensor network include the following:

- Sensor networks have a large number of nodes, which may exceed hundreds or even thousands [2]. Security architectures developed for small-scale ad hoc networks are infeasible for resource-limited large-scale sensor networks.
- Sensor networks exhibit comparatively stable communication patterns as opposed to ad hoc networks. In ad hoc networks, nodes are assumed to communicate among themselves and traffic patterns are reasonably random. On the contrary, in sensor networks most of the traffic is created as many-to-one nearly periodic transmission, as nodes have to report sensor readings to a central, more capable node.
- In ad hoc networks, communications are generally of the point-to-point, and often of multihop variety. There is no fixed source or destination of packets; instead, roles change over time. The only exception might be slightly increased traffic to and from nodes, which act as access points to the wired network. In sensor networks, data flow is directional and there is a single common destination for most, if not all, traffic flows.
- Sensor devices are physically vulnerable—they are susceptible to being damaged, captured and subverted (perhaps through reprogramming), or simply destroyed by the attacker.

The inescapable conclusion is that existing solutions for intrusion detection cannot be reused directly; instead, they have to be adapted to the characteristics of wireless sensor networks [6, 14, 22]. In particular, intrusion detection, like other security-related challenges, requires an integrated and comprehensive approach; if added as an afterthought, it cannot be as effective [21].

That makes it particularly hard to design an ideal security architecture for the whole layers. In practical applications, we should design our protocols in each layer with security in mind. Before security considerations, there exist several protocols in every layer. But when it comes to the security, we should incorporate the security method into already existing protocols or cooperate with them. The consequence is that the original architecture works inefficiently or otherwise should be redesigned.

6.3   Security Considerations

As is well known [4], main aspects of security include the following:

- Authentication is necessary to enable sensor nodes to detect maliciously injected or spoofed packets. It enables a node to verify the origin of a packet and ensure data integrity. Almost all applications require data authentication. In many applications, military as well as civilian ones, an adversary has clear incentives to join the network in order to inject false information such as fake data or routing information. Although authentication tries to prevent outsiders from injecting or spoofing packets, it does not solve the problem of compromised nodes. Since an attacker may have access to the secret keys of a compromised node, it can authenticate itself to the network. However, we may be able to use intrusion detection techniques to find the compromised nodes and revoke their cryptographic keys networkwide.

- Confidentiality or secrecy of data communications prevents unauthorized users from learning the contents of the messages. To that end, we can use standard encryption functions that might include secret keys shared among the communicating parties. (Note that the use of public-private key cryptography, while much more resilient to attacks, is out of the question on account of limited computational resources of sensor nodes.) However, encryption itself is not sufficient for protecting the privacy of data, as an eavesdropper can perform traffic analysis on the overheard cipher text, and this can release sensitive information about the data. In addition to encryption, privacy of sensed data also needs to be enforced through access control policies at the base station to prevent misuse of information.

- Availability requires that the sensor network is functional throughout its lifetime. Denial-of-Service (DoS) attacks result in a loss of availability [26]. In practice, loss of availability may have serious impacts. In a manufacturing monitoring application, loss of availability may cause failure to detect a potential accident and result in financial loss; in a battlefield surveillance application, loss of availability may open a back door for enemy invasion. Various attacks can compromise the availability of the sensor network. When considering availability in sensor networks, it is important to achieve graceful degradation in the presence of node compromise or benign node failures.

- Integrity of services is another security requirement. Above the networking layer, the sensor network usually implements several application-level services. Data aggregation is one of the most important sensor network services. In data aggregation, a sensor node collects readings from neighboring nodes, aggregates them, and sends them to the base station or another data processing node. The goal of secure data aggregation is to obtain a relatively accurate estimate of the real-world quantity being measured, and to be able to detect and reject a reported value that is significantly distorted by corrupted nodes.

6.4 Classifying the Intrusions

Intrusion attacks can be categorized according to different criteria.

6.4.1 Location of the Attacker with Respect to the Network

According to this criterion, attacks can be classified into insider and outsider attacks. In an outsider attack, the attack node is not an authorized participant of the sensor network. As the sensor network communicates over a wireless channel, a passive attacker can easily eavesdrop on the used frequency range to steal private or sensitive information. The adversary can also alter or spoof packets to attack the authenticity of communication or inject interfering wireless signals to jam the network. Another form of outsider attack is to disable sensor nodes. An attacker can inject useless packets to drain the receiver's battery, or he can capture and physically destroy nodes. A failed node is similar to a disabled node.

Unlike outsider attacks, insider attacks are performed by compromised nodes in the WSN. With node compromise, an adversary can perform an insider attack. In contrast to disabled node, compromised node generally seeks to disrupt or paralyze the network. A compromised node may be a subverted sensor node or a more powerful device, like laptop, with more computational power, memory, and powerful radio. It may be running some malicious code and seek to steal secrets from the sensor network or disrupt its normal functions. It may have a radio compatible with sensor nodes such that it can communicate with the sensor network.

6.4.2 Networking Layer in Which the Attack Takes Place

Attacks on wireless sensor networks can occur in different networking layers such as application, data link, network and physical layers, or in two or more of these layers simultaneously.

Attacks on the physical layer are, in fact, the easiest to launch. Since wireless sensor networks can be deployed in hostile environments or densely populated areas, physical access to individual nodes is possible. Even casual passers-by may be able to damage, destroy, or tamper with sensor devices.

Destruction of the node could cause gaps in sensor or communication coverage. Better-equipped attackers can interrogate a device's memory, stealing its data or cryptographic keys. The code can be replaced with a malicious program that is potentially undetectable to neighboring nodes. The capability profile of the subverted node becomes a fully authorized insider.

Attacks on the data link layer, including the media access control layer, are also comparatively simple. Many data link protocols in wireless sensor networks just consider the efficiency and fairness of utilizing the common channel. In these protocols, all the nodes in the network follow the same set of rules to access the media. For these reasons, many data link protocols are very vulnerable. Currently known attacks on the data link layer are mainly focused on the channel access. That is to say, the malicious node could randomly access the link and transmit or eavesdrop messages from the channel. More seriously, this node may inject and alter transmitted data. These attacks can be organized in three categories: collision attack, unfairness attack, and exhaustion attack.

- Collision Attack: Each node could inform its neighbors that he has some data to send or receive by exchanging RTS (Request To Send)/CTS (Clear To Send) control packets. Neighbor nodes could detect that the public channel is busy, and they would back off their sending even if they have some data packets to send. Using this mechanism, the collision only happens in the exchanging period of RTS and CTS packets, which means the data-packet-sending process is a noncollision process. In addition, each node will check whether the channel is busy or idle before sending RTS and CTS packets. That is why the probability of collision is very low. Under the condition when there is a packet transmitting on channel, adversaries can easily conduct attacks through sending out some packets to disrupt it (such as data packs, control packets sent by normal nodes).

- Unfairness Attack: For most RTS/CTS-based data link protocols, each node has the same priority to get the common channel. The rule is that the first tried node gets hold of the channel. Besides, all other nodes have to wait for a random length of time before trying to transmit packets. This rule could ensure that every node accesses common channels fairly. Adversaries could utilize these characteristics to attack the network. They send out packets just waiting for a very short time or without waiting. This causes the common channel used more by adversaries than by normal nodes.

- Exhaustion Attack: RTS/CTS-based data link protocols are sender invitation data link protocols. That is, when a sender sends out an RTS control packet to start a transmission, the receiver has to acknowledge the invitation with a CTS control packet if it is available. Since adversaries are also normal nodes, the receiver cannot exactly distinguish whether the RTS packet was sent by normal nodes or by adversaries. Under this condition, adversaries can attempt to retransmit RTS control packets to normal nodes repeatedly, forcing the receiver to acknowledge them incessantly. These kinds of abnormal retransmissions could result in the exhaustion of battery resources of receivers.

Attacks on Network Layer   It is not enough to secure our sensor networks by only using the data link layer security countermeasures. Those countermeasures can only protect against the outsider attacks. Some insider attacks which cannot be defended against in the link layer involve the routing protocols in the sensor networks [14]. These attacks can be categorized into the following kinds: selective forwarding, sinkhole attacks, wormhole attacks, Sybil attacks, and HELLO flood attacks:

- In sensor networks, each node can act as a router, that is to say, it could forward messages received. In **selective forwarding** attacks, once a middle node is captured by a malicious node, this node may refuse to forward certain messages and simply drop them. This behaves like a black hole. In practical applications, the malicious nodes use the attack to modify the packets. The neighboring nodes will conclude that the compromised node has failed and decide to seek another route skipping this node.
- In **sinkhole** attacks, the malicious node's goal is to lure all the traffic from a particular area to gain the entire message from the inspect area. The motivation of a sinkhole attack is that it makes selective forwarding trivial. By transmitting all traffic to the base station, the adversary can easily modify packets origination from any node in the area.
- In **wormhole** attacks, the powerful adversary is usually close to a base station. Remote powerful nodes are often colluded to establish an artificial link to transmit packets the remote nodes collected. Since these packets are originated at the base station, all the packets may be captured by the adversary. So the wormhole usually happens with the sinkhole. The sinkhole and wormhole attacks can be difficult to detect.
- In **Sybil** attack, the adversary presents multiple identities to other nodes in the network. So if other nodes are fooled, the data flow will be transmitted through the adversary and the control of substantial fractions of the network system will be at risk [7].
- In **HELLO** attacks, all nodes have to send HELLO packets to neighbor nodes before the network is established. A powerful adversary could use this characteristic to send HELLO packets to all nodes, thus destroying the network.

Some or all of these attacks can be combined to attack the current routing protocols; for example, TinyOS beaconing protocol is used to construct the topology through a broadcast message from the base station and the rebroad-cast message from the node that received the message. An adversary with the ability of powerful transmission may replace the base station. If authentication is introduced, another adversary that is situated near the base station can launch a wormhole and sinkhole attack. Also, the adversary can use a HELLO flood to make itself a parent of another node in the network.

Attacks on Application Layer The most common kind of application level attack is the Denial-of-Services (DoS) attack [16, 26]. A DoS attack is any event that diminishes or eliminates a network's capability to perform its expected functions. It is the general result of any action that prevents any part of a WSN from functioning correctly or in a timely manner. Hardware failures, software bugs, resource exhaustion, environmental conditions, or any complicated interaction between these factors can cause a DoS.

6.5   Intrusion Detection

As noted above, intrusion prevention techniques (which typically use encryption and authentication) are generally insufficient to ensure security, and must be complemented with intrusion detection [10]. However, close collaboration of those techniques would allow the latter to make use of the information provided by the former and vice versa, and thus improve the efficiency of both [11].

Detection technique An Intrusion Detection System (IDS) may be classified on the basis of its detection technique [4]. The main techniques include:

- A potential intrusion is reported by **Misuse or Signature-Based** detection if a sequence of events within a system matches a set of known security policy violations. In order to detect an intrusion by a Misuse model a knowledge of potential vulnerabilities of the system should be available. The intrusion detection system then applies this rule set to the sequences of data to determine a possible intrusion. This technique may exhibit low false positives, but does not perform well at detecting previously unknown attacks. Subhadrabandhu et al. [25] present a robust intrusion detection using misuse detection techniques. Anjum et al. [3] deal with the ability of various routing protocols to facilitate intrusion detection techniques when the attack signatures are completely known in the network.

- **Anomaly** detection uses a set of expected values to compare with a system's behavior. If the computed statistics do not match the expected values, an anomaly is reported. Anomaly-based detection defines a profile of normal behavior and classifies any deviation of that profile as an intrusion. The normal profile is updated as the system learns the subject's behavior. This technique may detect previously unknown attacks but may exhibit high false positives. Zhang et al. [28], presents an anomaly detection model. They use trace data that describes the normal updates of routing information since the main concern is that false routing will be used by other nodes. The generated trace data will then bear evidence of normality or anomaly. High false positive rates are reported based on their simulation results.

Anomaly detection may be used to detect attacks against a network daemon or a SetUID program by building a normal profile of the system calls made during program execution. If the process execution deviates significantly from the established profile, an intrusion is assumed. Okazaki et al. [19] have proposed a lightweight approach using profiles consisting of the type of system call and its frequency occurrence, in which speech recognition methods are used to calculate the optimal match between a normal profile and a sample profile.

- Compared to the Misuse modeling, **specification** modeling takes the opposite approach; it looks for specification of how a system or program executes and marks a sequence of instructions as a potential intrusion if it violates the specification. This technique may provide the capability to detect previously unknown attacks, while exhibiting a low false positive rate. For example, Snort [23] is an open source network intrusion prevention and detection system utilizing a rule-driven language, which combines the benefits of signature-based and anomaly-based detection methods.

Location of the Intrusion Detection System   A second distinction can be made in terms of the placement of the IDS. In this respect IDSs are usually divided into host-based and network-based systems and once again, both systems offer advantages and disadvantages:

- Host-based systems are present on each host that requires monitoring, and collect data concerning the operation of this host, usually log files, network traffic to and from the host, or information on processes running on the host. Host-based systems are able to determine if an attempted attack was indeed successful, and can detect local attacks, privilege escalation attacks and attacks which are encrypted. However, such systems can be difficult to deploy and manage, especially when the number of hosts needing protection is large. Furthermore, these systems are unable to detect attacks against multiple targets within the network.

- Network-based IDSs monitor the network traffic on the network containing the hosts to be protected, and are usually run on a separate machine termed a sensor. Network-based systems are able to monitor a large number of hosts with relatively little deployment costs, and are able to identify attacks to and from multiple hosts. However, they are unable to detect whether an attempted attack was indeed successful, and are unable to deal with local or encrypted attacks.

Hybrid systems, which incorporate host- and network-based elements, can offer the best protective capabilities, and systems to protect against attacks from multiple sources are also under development.

6.6   Approaches to Intrusion Detection

Although wireless sensor networks belong to the general family of wireless or Mobile Ad Hoc NETworks (MANETs), they have their own distinctive features. Many works [5, 11, 28] have investigated various aspects related to security and intrusion detection in MANETs but few in WSNs. The main differences between the MANETs and sensor networks from the security viewpoint can be summarized as follows:

- **Simpler device characteristics:** Sensor nodes are small and inexpensive devices with restricted transmit power (short range) and energy supplies. Due to low computation and communication capabilities authentication- and encryption-based security solutions are difficult to implement in a large-scale sensor network. Unlike typical mobile devices, sensor nodes spend a considerable amount of energy not only while sending and receiving data but also in the listening mode. Thus, sensor networks are more vulnerable to resource depletion attacks.
- **Lack of mobility:** In most applications, sensor nodes are stationary. They stay put wherever they are deployed. This decreases routing overhead. Most important, in sensor networks, route request broadcasts of reactive routing protocols and periodic updates of proactive routing protocols either do not occur or occur much less frequently.
- **Large network size:** Sensor networks consist of large numbers of nodes.

These differences make the IDS solutions proposed for MANETs unsuitable for WSNs. The challenges for an IDS in WSN are mainly due to the lack of resources. Besides, methods developed to be used in traditional networks cannot be applied directly to WSNs, since they demand resources not available in sensor networks.

WSNs are typically application oriented, which means they are designed to have very specific characteristics according to the target application. The intrusion detection assumes that the normal system behavior is different from the behavior of a system under attack. The several possible WSN configurations make the definition of the usual or expected system behavior difficult.

Since common nodes are designed to be cheap and small, they have limited hardware resources. Thus, the available memory may not be sufficient to create a detection log file. Moreover, a sensor node is designed to be disposed after being used by the application and it makes it difficult to recover a log file due to the possible dangerous environment in which the network was deployed. The software stored in the node must be designed to save as much energy as possible in order to extend the network lifetime.

Finally, another challenge to the design of an IDS is the frequent failures of sensor nodes when compared to processing entities found in wired networks. Given all these characteristics, it is important to detect the intrusions in real time. In this way, we could hold the intruder and minimize the possible damages.

6.6.1 Intrusion Detection Architecture for WSN

The optimal intrusion detection architecture for a WSN demands to be both distributed and at the same time, hierarchical for the special characteristics of this kind of network that we mentioned earlier. Distributed architecture allows detecting distributed attacks and provides scalability and robustness since it has different views of the network. Using this architecture, we can also distribute the process of detecting an intrusion over several nodes in the network. Because this architecture relies on cooperative work of nodes and is not centralized, it can be very fault tolerant, and that is to say if a node is removed from the network for any reason, the intrusion detection can still work properly.

Figure 6.1 shows a hierarchical architecture for intrusion detection in WSN. The intrusion detection architecture mimics the hardware architecture of WSN in which intrusion detection is done in three levels. The first level, which includes sensor nodes, is responsible for collecting application data, monitoring the behavior of neighboring nodes, and some responding to intrusions locally (e.g., by isolating relevant nodes). The second level is the coordinator's level and is responsible for aggregating the application data from the sensor nodes

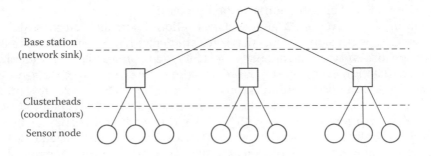

FIGURE 6.1
Hierarchical architecture for intrusion detection in WSN.

nearby and monitoring the behavior of the network or individual nodes. Identification of intrusions also will be done at this level by analyzing the aggregated data. Finally the intrusion reaction engine will react to the intrusions like done in the first level. Level three of the architecture detects intrusions by analyzing the application data from the coordinators. Similar to the access point level the base station level also has a monitoring and identification engine and will react to the intrusions that are detected at this point.

As described above, the functions of sensing, computation, and data delivery may be distributed across the tiers, with the lowest tier performing all sensing, the middle tier performing all computation, and the top tier performing all data delivery. Alternatively, each layer can perform a specialized role in computation.

6.7   Intrusion Detection Solutions for WSN

In this section we review some of the systems and algorithms that have been proposed for intrusion detection in WSNs. Some general approaches are presented as well as the algorithms that are based on the Markov model. Some other solutions utilize mobile agents in order to detect and respond to intrusions.

6.7.1   General Approaches

Silva et al. [6] propose a decentralized intrusion detection for WSNs. Function of the IDS component is loaded into some nodes called "monitor" nodes. The detection system is specification based, since the WSN may vary depending on the application goal. When deploying the sensor network, monitor nodes are distributed all over the network in a way that every node is covered by at least one monitor node. Their algorithm consists of three phases: in phase one, which is the data acquisition phase, messages are collected in a promiscuous mode and the important information is filtered before being stored for subsequent analysis. In the processing phase, the intrusion detection rules are applied to the stored data. Finally the last phase or intrusion detection phase will determine if an intrusion detection is raised.

Du et al. [9] propose a general Localization Anomaly Detection (LAD) scheme. They consider the fact that some anomalies happen in the process of location discovery (localization). For instance, deploying a Global Positioning System (GPS) in every sensor, in order to determine location of the sensors, is costly. A number of solutions consider deploying GPS to just a few numbers of nodes in the network. The remaining nodes will verify their location using the location of sensors with GPS. We can see that this approach may result in localization anomalies by adversaries. The proposed scheme (LAD) takes advantage of the deployment knowledge and the group membership of its neighbors, and uses such knowledge to find out whether the estimated location is consistent with its observations. If they are inconsistent, LAD will report an anomaly. They formulate the problem as an anomaly intrusion detection problem, and introduce a localization anomaly detection phase after the localization phase. In the localization phase, sensors derive their locations. Then in the detection phase, sensors verify whether the derived locations are correct. A failure of the verification indicates an anomaly.

Mittal and Vigne [18] describe a signature-based intrusion detection technique that is for detecting routing-based attacks. Detecting these kinds of attacks is difficult because malicious routing behavior can be identified only in specific network locations. They use the characteristics of the Routing Information Protocol (RIP), the network topology, and the positioning of the intrusion detection sensors to automatically determine both the signature configuration of the sensors and the messages that the sensors have to exchange to detect attacks against the routing infrastructure. The approach uses a set of sensors that analyzes routing traffic in different locations within a network. An algorithm to automatically generate both the detection signatures and the intersensor messages needed to verify the state of the routing infrastructure has been devised for the case of the RIP distance-vector routing protocol.

In another work ([20]), intrusion detection functions are distributed to all the nodes in the network. The authors introduce a novel anomaly-based intrusion detection method for wireless sensor networks suited to their simple and resource-limited nature. This detection-based security scheme, which is for large-scale sensor networks, exploits network stability in its neighborhood information. In many attacks against sensor networks, the first step for an attacker is to establish itself as a legitimate node within the network. If each node can build a simple statistical model of its neighbors' behavior, these statistics can later be used to detect changes in them. The authors have shown that, by looking at a relatively small number of received packet features, a node can effectively identify an intruder impersonating a legitimate neighbor.

6.7.2 Markov-Based Approaches

Doumit and Agrawal [8] propose an anomaly approach based on self-organized criticality (SOC), which is meant to link the multitude of complex phenomena observed in nature to simplistic physical laws and/or one underlying process. Hidden Markov models are used to detect data inconsistencies. This approach is developed based on the structure of naturally occurring events. With the acquired knowledge derived from the self-organized criticality aspect of the deployment region, a hidden Markov model is then applied. The proposed approach lets the sensor network adapt to the norm of the dynamics in its natural surrounding so that any unusual activities can be singled out. The work is focused on the fact that sensor nodes in WSNs are limited in resource and tries to minimize the resource consumption.

A new technique for handling security in WSNs is presented by Agah et al. [1]. They formulate the attack-defense problem by game theory and use the Markov Decision Process to predict the most vulnerable sensor nodes. Their approach formulates attack-defense problem as a two-player, nonzero-sum, noncooperative game between an attacker and a sensor network. In a noncooperative game unlike cooperative ones, no outside authority assures that players stick to the same predetermined rules, and binding agreements are not feasible. Each player (attacker and sensor network) tries to maximize its own payoff. The sensor network tries to defend the sensor nodes against intrusions. The algorithm is nonzero-sum in the sense that the increase in one player's payoff implies the decrease in the other player's payoff. The work shows that this game achieves Nash equilibrium, thus leading to a defense strategy from the network. Then, it uses the Markov Decision Process to predict the most vulnerable sensor node.

6.7.3 Mobile Agent Utilization

One solution to perform distributed intrusion detection is by using mobile agent technology [15]. Agents can be seen as guards that protect a network by moving from host to host and performing random sampling. Instead of monitoring each host at any time, agents only visit machines from time to time to conduct their examinations. When any anomaly is detected, a more comprehensive search is initiated. Although the idea of patrolling guards seems appealing at first, this approach has the disadvantage of leaving hosts vulnerable while no agents are present. On the other hand, random sampling definitely reduces the average computational load at each machine.

Kachirski and Guha [13] have proposed a distributed intrusion detection based on mobile agent technology. By efficiently merging audit data from multiple network sensors, their scheme analyzes the entire network for intrusions at multiple levels. There are three major agent categories: monitoring, decision-making, and action agents. Some are present on all mobile hosts, while others are distributed to only selected groups of nodes. The monitoring agents look for suspicious activities on the host node. If some anomalous activity is detected, the node is reported to the decision agent of the same cluster. The decision agent, based on these reports, will then decide whether the node has been compromised. When a certain level of threat is reached for the node in question, the decision agent dispatches a command that an action must be undertaken by local agents on the node.

Mobile agents introduce some advantages such as reducing network load, overcoming network latency, and scalability. On the other hand they may also result in some problems like securing the agent itself and a huge amount of code size.

6.8 Conclusion

Research in intrusion detection has been conducted for the past 20 years; however, its application to wireless sensor networks is fairly recent. We have argued that any secure network will have vulnerability that an advisory can exploit. This is specially true for WSN. Intrusion detection can complement intrusion prevention techniques to improve the network security. A number of research efforts concentrated on developing solutions for intrusion detection in WSNs in order to adapt (with special characteristics) these kinds of networks. Current solutions suggest distributed and cooperative intrusion detection to try to minimize false positives. Further research efforts are needed to explore new methods to detect attacks against WSNs.

Bibliography

1. A. Agah, S. K. Das, K. Basu, and M. Asadi. A non-cooperative game approach for intrusion detection in sensor networks. In *Third IEEE International Symposium on Network Computing and Applications*, pages 343–346, 2004.
2. I. Akyildiz, W. Su, Y. Sankarasubramaniam, and E. Cayirci. A survey on sensor networks. In *IEEE Communication Magazine 40 (8)*, 2002.
3. F. Anjum, D. Subhadrabandhu, and S. Sarkar. Signature-based intrusion detection for wireless ad-hoc networks: A comparative study of various routing protocols. In *Vehicular Technology Conference, Wireless Security Symposium, Orlando, FL*, 2003.
4. M. Bishop. *Computer Security: Art and Science*. Addison-Wesley, Pearson Education, Inc., Boston, 2004.
5. P. Brutch and C. Ko. Challenges in intrusion detection for wireless ad-hoc networks. In *SAINT: Symposium on Applications and the Internet*, pages 368–373, 2003.
6. A. P. R. da Silva, M. H. T. Martins, B. P. S. Rocha, A. A. F. Loureiro, L. B. Ruiz, and H. C. Wong. Decentralized intrusion detection in wireless sensor networks. In *Proceedings of the 1st ACM International Workshop on Quality of Service and Security in Wireless and Mobile Networks*, pages 16–23, 2005.
7. J. R. Douceur. The Sybil attack. In *International Workshop on Peer-to-Peer Systems*, pages 251–260, 2002.
8. S. S. Doumit and D. P. Agrawal. Self-organized criticality and stochastic learning-based intrusion detection system for wireless sensor networks. In *MILCOM: IEEE Military Communications Conference*, pages 609–614, 2003.
9. W. Du, L. Fang, and P. Ning. Lad: Localization anomaly detection for wireless sensor networks. In *IPDPS: 19th IEEE International Parallel and Distributed Processing Symposium*, 2005.
10. F. Hu and N. K. Sharma. Security considerations in ad hoc sensor networks. *Ad Hoc Networks*, 3(1):69–89, 2005.
11. Y. Huang and W. Lee. A cooperative intrusion detection system for ad hoc networks. In *SASN: Proceedings of the 1st ACM Workshop on Security of Ad Hoc and Sensor Networks*, pages 135–147, 2003.
12. IEEE. Standard for part 15.4: Wireless MAC and PHY specifications for low rate WPAN. IEEE Std 802.15.4, IEEE, New York, Oct. 2003.
13. O. Kachirski and R. K. Guha. Effective intrusion detection using multiple sensors in wireless ad hoc networks. In *Proceedings of the 36th Annual Hawaii International Conference on System Sciences*, pages 57–65, 2003.

14. C. Karlof and D. Wagner. Secure routing in wireless sensor networks: Attacks and countermeasures. In *First IEEE International Workshop on Sensor Network Protocols and Applications*, pages 113–127, May 2003.
15. C. Kruegel. Applying mobile agent technology to intrusion detection. In *Distributed Systems Group*, Technical University of Vienna, 2002.
16. J. Mirkovic, S. Dietrich, D. Dittrich, and P. Reiher. *Internet Denial of Service: Attack and Defense Mechanisms*. Prentice Hall, 2005.
17. A. Mishra, K. Nadkarni, and A. Patcha. Intrusion detection in wireless ad hoc networks. *IEEE Wireless Communications*, Vol. 11, No. 1, pp. 48–60, February 2004.
18. V. Mittal and G. Vigna. Sensor-based intrusion detection for intra-domain distance-vector routing. In *ACM Conference on Computer and Communications Security*, pages 127–137, 2002.
19. Y. Okazaki, I. Sato, and S. Goto. A new intrusion detection method based on process profiling. In *SAINT: Symposium on Applications and the Internet*, pages 82–91, 2002.
20. I. Onat and A. Miri. An intrusion detection system for wireless sensor networks. In *IEEE International Conference on Wireless and Mobile Computing, Networking and Communications*, pages 253–259, 2005.
21. A. Perrig, J. Stankovic, and D. Wagner. Security in wireless sensor networks. *Communications of the ACM*, 47(6):53–57, 2004.
22. R. Roman, J. Zhou, and J. Lopez. Applying intrusion detection systems to wireless sensor networks. In *Proc. CCNC 2006*, pp. 640–644.
23. http://www.snort.org, Date visited: April, 17, 2006.
24. W. Stallings. *Wireless Communications and Networks*. Prentice Hall, Upper Saddle River, NJ, 2002.
25. D. Subhadrabandhu, S. Sarkar, and F. Anjum. Rida: Robust intrusion detection in ad hoc networks. In *Proceedings of the 4th International IFIP-TC6 Networking Conference*, pages 1069–1082, 2005.
26. A. D. Wood and J. A. Stankovic. Denial of service in sensor networks. *IEEE Computer*, 35(10):54–62, 2002.
27. D. Xiao, C. Chen, and G. Chen. Intrusion detection-based security architecture for wireless sensor networks. *IEEE International Symposium on Communications and Information Technology*, 2:1412–1415, 2005.
28. Y. Zhang and W. Lee. Intrusion detection in wireless ad-hoc networks. In *MobiCom '00: Proceedings of the 6th Annual International Conference on Mobile Computing and Networking*, pages 275–283. ACM Press, 2000.
29. Y. Zhang, W. Lee, and Y.-A. Huang. Intrusion detection techniques for mobile wireless networks. *Wireless Networks*, Vol. 9, pp. 545–556, 2003.

7

False Data Detection and Secure Data Aggregation in Wireless Sensor Networks

Hasan Çam and Suat Ozdemir

CONTENTS

Abstract   In wireless sensor networks, sensor nodes are vulnerable to node compromise attacks that threaten their security and efficient utilization of resources. A compromised sensor node can inject false data during data forwarding and aggregation to forge the integrity of sensor data. It is highly desirable for sensor nodes to detect and drop false data as soon as possible in order to avoid depleting their limited resources such as battery power and bandwidth. In addition, the false data detection algorithms should be designed with data aggregation and confidentiality in mind. Data aggregation is used to reduce the redundancy in transmitted data and to improve the data accuracy. This chapter reviews the existing false data detection, data aggregation, secure data aggregation, and key establishment schemes for wireless sensor networks. It also addresses how false data detection can be integrated with data aggregation and confidentiality.

7.1   Introduction

Recent advances in low-power computing and communication technologies have given rise to the proliferation of wireless sensor networks having low-cost sensor nodes with limited processing capacity and battery power. Wireless sensor networks can be used in a wide range of applications such as environmental and patient monitoring, surveillance of critical areas and structures, and target tracking. For these applications, network security is usually an essential requirement and, therefore, the lack of proper security can curtail the widespread deployment of sensor networks. However, wireless sensor networks are prone to many types of security attacks, some of which do not even occur in traditional networks. For example, in node compromise attacks intruders gain the control of sensor nodes and threaten the security of the network by injecting false data, forging relayed data, or disturbing data transmission [1]. By injecting false data, compromised sensor nodes can distort data integrity, cause false alarms, and reduce the limited battery, computational and communication resources of sensor nodes.

Data aggregation is employed to eliminate data redundancy, reduce the amount of data transmitted to the base station, and/or improve data reliability. Data aggregation helps improve the utilization of resources such as bandwidth and battery power. Because approximately 70% of the total energy consumption in a wireless sensor network is due to communication, implementing data aggregation enhances the network's lifetime [2, 3]. Although any sensor node is usually capable of doing data aggregation, some sensor nodes are designated dynamically as data aggregators to aggregate data. Data aggregation is potentially vulnerable to security attacks because a compromised data aggregator can inject false data or forge the aggregated data. That is, since data are usually altered during data aggregation, it is difficult to determine whether the data are changed due to a proper data aggregation or a security attack of false data injection or data forging. This indicates that it is critical and challenging to provide secure data aggregation along with false data detection. In addition to secure data aggregation, data confidentiality is needed by many applications, including military and patient monitoring. But, data aggregation and confidentiality techniques unfortunately have conflicts in their implementation. Data confidentiality prefers data to be encrypted at the source node and decrypted at the destination. However, data aggregation techniques require any encrypted data to be decrypted at data aggregators, so that all the data arriving at a data aggregator can be aggregated based on the correlation between them.

To the best of our knowledge, none of the existing false data detection algorithms [4, 5, 6, 7] is performed along with data aggregation and confidentiality. Although these false data detection algorithms can be modified easily to support data confidentiality, it is a challenge for them to support data aggregation. For instance, the basic idea behind the false data detection scheme in [5] is to form pairs of sensor nodes, as shown in Figure 7.1, such that one

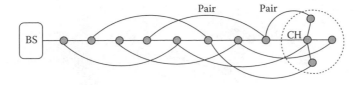

FIGURE 7.1
An example of false data detection scheme given in [5].

pairmate computes a message authentication code (MAC) of forwarded data and the other pairmate verifies later the data using the MAC. This scheme does not work if an intermediate node between the pairmates alters data in case of data aggregation. Thus, because data aggregation usually results in alterations in data, the false data detection scheme cannot be implemented when a data aggregator between two pairmates changes the data during aggregation.

In this chapter we review the existing schemes for false data detection, secure data aggregation, and key establishment. We also discuss how false data detection can be integrated with data aggregation and confidentiality in wireless sensor networks. The remainder of the chapter is organized as follows. Section 7.2 discusses the existing false data detection schemes. Sections 7.3 and 7.4 review the data aggregation and secure data aggregation protocols, respectively. Section 7.5 presents the existing key establishment schemes. The integration of false data detection with data aggregation and confidentiality is described in Section 7.6. The concluding remarks are made in Section 7.7.

7.2 False Data Detection

Consider the scenario presented in Figure 7.2 where compromised nodes inject false data about a fake border crossing in order to deplete the energy of sensor nodes and to mislead the border patrol. In general, it is not possible to prevent the injection of false data because sensor nodes are vulnerable to node compromise attacks. But, false data can be detected and dropped soon after its injection using message authentication codes (MACs) in data authentication schemes [4, 5, 6, 7]. For instance, the statistical en-route detection scheme [4], called SEF, enables relaying nodes and the base station to detect false data with a certain probability. To detect and filter out forged messages, SEF relies on the collective decisions of multiple sensor nodes as follows: (*i*) when an event occurs in an area of interest, the surrounding sensor nodes generate a legitimate report that carries multiple MACs, (*ii*) intermediate forwarding nodes detect incorrect MACs and filter out false reports with some probability, and (*iii*) the base station verifies the correctness of each MAC and eliminates

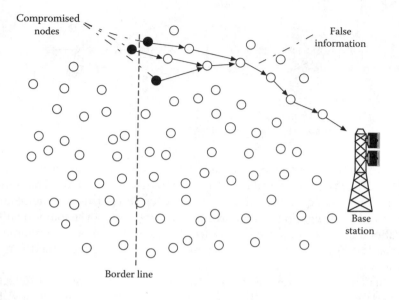

FIGURE 7.2
Compromised nodes inject false data about a fake border crossing to deplete the energy of sensor nodes and to mislead the border patrol.

the remaining false reports that cannot be detected by en-route filtering. In 10 hops, SEF is able to drop 80 to 90% of the false reports and the communication overhead is shown to be 14 bytes on average per report.

The interleaved hop-by-hop authentication scheme [5] guarantees that the base station detects any packet containing false data if at least $t + 1$ sensor nodes agree upon a report when there are at most t colluding compromised nodes. The value of the security threshold t is determined based on the network node density and the security requirements of the application under consideration. All those nodes that are involved in forwarding a report to the base station authenticate the report in an interleaved hop-by-hop fashion. The interleaved hop-by-hop scheme is implemented basically in five phases. In the first phase, each node establishes a one-hop pairwise key with each of its neighbors. In the second phase, each node discovers the ID numbers of its associated nodes to form pairs for data verification. In the third phase, when $t + 1$ sensor nodes in a cluster detect the same event of interest, each of them computes two MACs using the shared keys with the base station and its pairmate of forwarding node. When the clusterhead collects all these $2t + 2$ MACs along with the sensed data, it obtains one compressed MAC by XOR-ing those $t + 1$ MACs computed for the base station, so that $t + 2$ MACs are transmitted by the clusterhead. In the fourth phase, every forwarding node verifies the MAC computed by its pairmate, and either drops the data if the verification fails or removes the MAC and attaches a new MAC if the verification succeeds. In the fifth phase, the base station verifies the data based on the compressed MAC.

The Commutative Cipher-Based En-Route Filtering scheme (CCEF) [6] drops false data en route using the public key cryptography instead of sharing symmetric keys. In CCEF, the source node establishes a secret association with the base station on a per-session basis, while the intermediate forwarding nodes are equipped with a witness key. The base station first prepares two keys, namely, session and witness keys, for each source node per session. Then, the base station sends these keys to the respective source nodes. Although session keys are encrypted prior to their transmission to source nodes, the witness keys are transmitted in plaintext so that all intermediate nodes between the base station and source nodes can also use them later for data verification. Source and intermediate nodes employ a commutative cipher [8], which allows a forwarding intermediate node to use the witness key for data verification. If the verification fails, the data must be false and, therefore, it is dropped.

In the dynamic en-route filtering scheme [7], legitimate data are endorsed by multiple sensor nodes using their distinct authentication keys from one-way hash chains. Each clusterhead uses *hill climbing* approach for disseminating the authentication keys of sensing nodes along multiple paths toward the base station. The hill climbing approach guarantees that the nodes closer to a clusterhead store more authentication keys than those nodes farther from the clusterhead. This leads the number of authentication keys stored at each forwarding node to be balanced. False data are filtered in a probabilistic nature, similar to SEF [4]. If a forwarding node has the authentication key, it can validate the authenticity of the reports and drop the false reports. This scheme is shown to be more suitable for dynamic topology of sensor networks than SEF [4].

7.3 Data Aggregation

Data aggregation mitigates the data redundancy and the amount of transmitted data, leading to a smaller number of data transmissions. Data aggregation also combines several unreliable data measurements to produce more accurate data with meaningful and useful information for end-users. Data aggregation results in better bandwidth and battery utilization [1, 2], and helps improve the network's lifetime. Therefore, data aggregation is essential for improving the efficient utilization of limited resources in wireless sensor networks. A substantial number of data aggregation algorithms are presented in the literature [10, 11, 12, 13, 14]. Any sensor node that aggregates data is called a data aggregator, as shown in Figure 7.3.

In [10], a cluster-based protocol, called LEACH, is introduced to integrate data aggregation with the routing protocol, so that the amount of data that must be transmitted to the base station is reduced. In LEACH, each cluster head compresses data arriving from the sensor nodes of its cluster, and sends

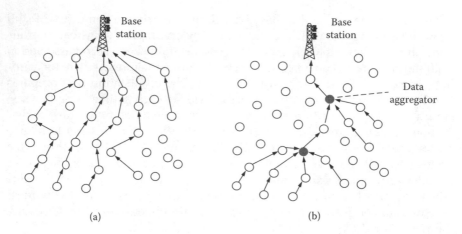

(a)   (b)

FIGURE 7.3
(a) Data are sent to base station without performing data aggregation at any sensor node. (b) Data are aggregated by *data aggregators*.

an aggregated packet to the base station. A localized coordination among sensor nodes is used for dynamic selection of clusterheads.

In PEGASIS (Power-Efficient Gathering in Sensor Information Systems) [11], sensor nodes form chains so that each node transmits to and receives from a nearby neighbor. Building a chain minimizes the total length that data travel. Sensed data move from node to node, get aggregated, and are eventually transmitted to the base station. Sensor nodes take turns transmitting to the base stations, thereby reducing the average energy spent by each node per round.

In [12], the tiny aggregation (TAG) service is presented to aggregate data. TAG provides a declarative interface for data collection and aggregation that is similar to functions in database query languages, such as max, min, count, sum, average, median, and histogram. TAG consists of two phases, namely, distribution and collection phases. In the distribution phase, aggregate queries are pushed down into the network. In the collection phase, the aggregated data are continually forwarded from children nodes to parent nodes toward the base station.

Directed diffusion [13] is introduced for the coordination of performing distributed sensing of an environmental phenomenon. Directed diffusion achieves energy savings by performing data aggregation and selecting empirically good paths. Attribute-value pairs are used to name task descriptions, and a sensing task is disseminated throughout the sensor network as an interest for the named data. This dissemination sets up gradients within the network designed to "draw" events. Specifically, a gradient is a direction state created in each node that receives an interest. The gradient direction is set toward the neighboring node from which the interest is received. As the sensed data are transmitted along multiple gradient paths, intermediate nodes on these paths aggregate the data.

In [14], a family of negotiation-based information dissemination protocols, called SPIN, is presented, where sensor nodes negotiate with each other before transmitting data. Negotiation helps to ensure that only useful information will be transferred. To negotiate successfully, however, nodes must be able to describe or name the data they observe. The descriptors used in these negotiations are referred to as *meta-data*. The negotiation process that precedes actual data transmission eliminates transmission of redundant data messages. The use of meta-data descriptors eliminates the possibility of overlap because it allows nodes to name the portion of the data that they are interested in obtaining.

7.4 Secure Data Aggregation

Security is a key requirement for many applications (e.g., surveillance, health-care, battlefield) in wireless sensor networks. In order to avoid security problems as much as possible, it is desirable to have end-to-end security by encrypting data at source nodes and decrypting them at the base station only. However, data aggregation requires data aggregators to examine the plain data coming from different neighboring sensor nodes and, therefore, any encrypted data should be decrypted at data aggregators. These two conflicting goals of secure communication and data aggregation necessitate data aggregation algorithms to be designed together with the secure communication algorithms. This has led many researchers [2, 15, 16, 17, 20, 21] to study secure data aggregation problems in wireless sensor networks.

In [15], a secure data aggregation (SDA) protocol is presented to detect node misbehaviors such as dropping or forging messages and transmitting false aggregate values. The main idea behind this protocol is the delayed aggregation based on the μTESLA protocol [3] that achieves asymmetry through delayed disclosure of symmetric keys as shown in Figure 7.4. In SDA, the nodes are organized into a tree-based hierarchy such that the internal nodes act as aggregators. A parent node is not able to immediately verify the authenticity

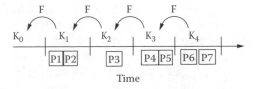

Time

FIGURE 7.4
The μTESLA one-way key chain where the sender generates the one-way key chain right to left by repeatedly applying the one-way function F. The sender associates each key of the one-way key chain with a time interval. Because time runs left to right, the sender uses the keys of the key chain in reverse order, and computes the MAC of the packets of a time interval with the key of that time interval [3].

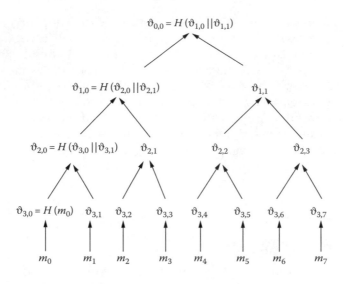

FIGURE 7.5
A sample Merkle hash tree. The ϑ at a parent node is obtained by hashing two ϑ's at its children nodes. If any of the values in the tree is changed, the ϑ value at the root changes as well.

of its children's data. SDA does not guarantee that nodes and aggregators provide correct values.

In [16], data aggregators aggregate information requested by a query. By constructing efficient random sampling mechanisms and interactive proofs, it is shown that a user can verify whether the aggregated data provided by an aggregator is a good approximation of the true value by implementing the aggregate-commit-prove technique in three phases. In the aggregate phase, the aggregator collects data from the sensors and computes the aggregation result according to a specific aggregate function. Each sensor node shares a key with the aggregator, and the data aggregator verifies the authenticity of collected data using the shared keys. In the commit phase, the aggregator commits to the collected data by constructing a Merkle hash tree [26] from sensor data; a sample Merkle tree is illustrated in Figure 7.5. This commitment ensures that data aggregator uses the data collected from the sensor nodes. The commitment is the root value of the tree. A hash function is used to ensure that the aggregator cannot change any input values after they are hashed. In the final prove phase, the aggregator is charged with proving the results to the base station. The aggregator first sends its aggregation result and the commitment to the base station, and then uses an interactive proof to prove the correctness of the results.

The energy-efficient secure data aggregation protocol (ESPDA) that we introduce in [17] uses small data representatives, called *data patterns*. After sensing an event, each sensor node first generates data patterns corresponding to the sensed data to represent its main characteristics, and then sends the patterns to the clusterhead in a cluster-based sensor network. After receiving

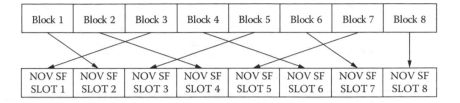

FIGURE 7.6
An example of mapping data blocks to time slots in the NOVSF block-hopping technique. Data blocks are transmitted in the order of block 3, block 1, block 5, block 2, block 7, block 4, block 6, and block 8.

data patterns from all active sensor nodes, a clusterhead determines a representative sensor node for each group of those sensor nodes that generate the same data patterns. Then, the clusterhead requests only the representative sensor nodes to send the actual sensed data. Hence, data redundancy is mitigated by examining the data patterns of sensor nodes. This also enables ESPDA to work in conjunction with the security protocol in the sense that sensor nodes can send the actual data in encrypted form, without any need for decryption in cluster heads because data aggregation is already implemented based on the data patterns. To strengthen data transmission security and to improve the spectral efficiency of wireless sensor networks, ESPDA employs the NOVSF Block-Hopping (NOVSF-BH) technique to interleave data blocks. An example of mapping data blocks to time slots in NOVSF codes [18] is illustrated in Figure 7.6.

In [19], we present a secure differential data aggregation (SDDA) protocol to support secure communication and to reduce the amount of redundant data transmitted from sensor nodes to clusterheads. In SDDA, every sensor node compares the raw sensed data with the reference data, and then only the differential data are transmitted, where the reference data are obtained by taking the average of previously transmitted data over a number of transmissions. For example, let 102°F denote the temperature measurement in a sensor node. If 100°F is considered as the reference temperature by the clusterhead, the sensor node sends only the difference (i.e., 2°F) of the current measurement from the reference value. The basic motivation behind this differential data aggregation is that significant changes in sensor measurements occur occasionally when an important event (e.g., a fire event for temperature sensors) happens in the environment. As shown in Figure 7.7, simulation results show that SDDA improves energy efficiency significantly by reducing the number of transmitted packets.

In Reference [20], sensor nodes use the cryptographic algorithms only when a cheating activity is detected. Topological constraints are introduced to build a secure aggregation tree (SAT) that facilitates the monitoring of data aggregators. In SAT, any child node is able to listen to the incoming data of its parent node. When the aggregated data of a data aggregator are questionable, a weighted voting scheme is employed to decide whether the data aggregator

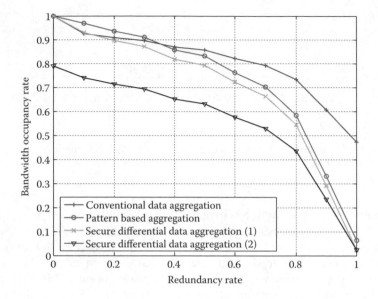

FIGURE 7.7
Comparison of bandwidth usage of SDDA with conventional and pattern-based data aggregation (ESPDA).

is properly behaving or is cheating. If the data aggregator is a misbehaving node, then SAT is rebuilt locally so that the misbehaving data aggregator is excluded from the aggregation tree.

In Reference [21], witness nodes are used for secure data aggregation. In addition to data aggregators, their witness nodes also perform data aggregation and compute the corresponding MACs. But, the witness nodes do not send their aggregated data to the base station. Instead, each witness node sends its MAC to its data aggregator for forwarding the MAC toward the base station. Those MACs that are computed by the witness nodes are used at the base station for verifying the correctness of the data aggregated by data aggregators. This enhances the assurance of data aggregation. In order to prove the validity of the aggregated data, every data aggregator provides proofs from several witnesses.

7.5   Key Establishment

Public key cryptography [34, 35] is computationally very expensive [1], and small-size sensor nodes usually do not have sufficient resources such as computational power, memory space to implement it. Therefore, symmetric key cryptography is preferred over public key cryptography in wireless sensor networks for data encryption and authentication. Symmetric key cryptography requires pairwise keys to be distributed or established between pairmates. This section describes some key distribution and establishment schemes proposed in the literature.

The simplest key distribution protocol is having a networkwide shared key. However, this solution is vulnerable to node compromise attacks because a single compromised node can reveal the networkwide key, which results in decryption of all encrypted data by intruders. Therefore, it is highly desirable to establish a distinct shared key between the members of every pair. To establish shared keys in the initial deployment of sensor nodes, a single networkwide master key may be used for a very short time, and is removed from the memory of sensor nodes after its usage is complete. But, in this case, the addition of new nodes to the network after initial deployment is very difficult. Another approach is to predistribute a unique symmetric key to each pair of nodes. Although this technique provides excellent security against node compromise attacks, it has the drawback of not being scalable because each sensor node has to store $n - 1$ keys to guarantee having a shared key between any two neighboring nodes in a network of n nodes.

To reduce the memory requirement of $n - 1$ keys in each node, several random key predistribution schemes are introduced [27, 28, 29]. The basic idea behind these schemes is to store some secret keys in sensor nodes to increase the probability that any two sensor nodes can establish a shared symmetric key. The random key predistribution scheme in [27] selects randomly a pool of keys from a key space to allow each sensor node to receive a random subset of keys from the key pool prior to network deployment. Any two nodes that are able to find a common key within their respective subsets can use the key as their shared secret to initiate the communication. This scheme is improved in [28, 29]. In [28], the estimated location information of sensor nodes is used to reduce memory space and computational overhead due to key distribution. The key distribution scheme in [29] is very similar to the one in [27], except that it requires any pair of sensor nodes to have q common keys within their key sets. Although random key predistribution schemes provide a balanced communication and memory overhead, they can be used in only those networks where the random graph model for connectivity holds. For example, if the node density of a network is nonuniform, these schemes could result in a disconnected network because some sensor node pairs may not be able to successfully perform key establishment. In addition, even if a small number of sensor nodes are compromised by a single intruder, then the amount of compromised keys could be significantly high, which reduces resilience against node compromise attacks.

In [30], we introduce a different type of key establishment protocol, called Event-Based Reconciliation (ERS). ERS takes advantage of overlapping sensing regions, distributed source coding, and secret key reconciliation. It enables sensor nodes with overlapping sensing regions to establish pairwise keys using distributed source coding and the sensed data of the events detected in overlapping sensing regions. The key establishment in ERS is achieved in three steps: (i) node authentication and seed agreement, (ii) key generation and transmission of the encoded/decoded key, and (iii) reconciliation of the

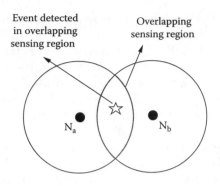

FIGURE 7.8
Sensor nodes N_a and N_b with an overlapping sensing region establish their secret key via ERS.

secret pairwise key. The first step of ERS is performed just after the initial network deployment. In this step, using a temporary master key, each sensor node authenticates its neighboring nodes and establishes a *seed* with each one of them. Neighboring nodes use those *seeds* to improve the confidentiality of the key establishment process. The second and third steps of ERS are implemented as follows: (i) a pair of neighboring sensor nodes with an overlapping sensing region possibly, as shown in Figure 7.8, agree to use the next event's sensed data as side information in the distributed source coding, (ii) one member of the pair generates an initial key, and then expands it to enhance confidentiality and better recovery of the key in the presence of bit errors, (iii) the expanded key is first encoded using the sensed data of the next event as side information, and then is sent to the other member of the pair, (iv) the other member of the pair decodes the expanded key using its own sensed data for the same event, and then tries to recover the initial key from the expanded key, and (v) because the initial key of the first member may not be recovered correctly by the second member of the pair, these pair-mates *reconcile* a common bit string by finding and correcting discrepancies between their keys. Finally, the common bit string that the pairmates agree upon is used as a shared key between them. These second and third steps are explained below in more detail using Figure 7.9.

Let N_a and N_b denote two neighboring sensor nodes. In the second step of ERS, N_a and N_b first agree on the event whose sensed data are used as side information to encode/decode the initial key. Once they detect the event, N_a first generates an initial n-bit key S, then expands S to $(m \times n)$-bit key S_a using the *seed* that it shares with N_b. As seen in Figure 7.9, to expand S, N_a maps and copies each bit of S to m locations in the expanded key S_a. The mapping locations are determined by a collusion-free one-way hash function that uses the *seed* as its key. Since the *seed* between N_a and N_b is only known to N_a and N_b, expanded S_a guarantees the confidentiality of S as long as N_a or N_b is not compromised. In addition to confidentiality, the expanded key is used to correct the bit errors at N_b. After expanding S, N_a encodes the expanded key S_a with respect to its sensed data D_a and sends the encoded data $H(S_a|D_a)$

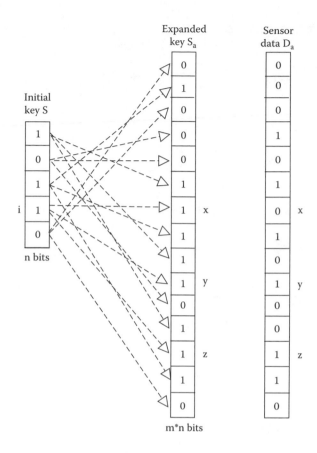

FIGURE 7.9
Expansion of the initial key S to S_a. The ith bit of S is copied to the xth, yth and zth bit positions of S_a. D_a is correlated to S_a.

to N_b. Upon receiving the data sent by N_a, N_b decodes $H(S_a|D_a)$ given D_b as side information and obtains S'_a. Then, N_b reverses the key expansion process and recovers its initial key S' from S'_a.

In the third step of ERS, N_a and N_b reconcile their pairwise shared key by finding the discrepancies between S and S'. N_b expands its initial key S' to S_b and sends S_b to N_a. Then, N_a obtains S' from S_b by reversing the key expansion process and compares S' with its initial key S to find out the common bits of these two keys. Once the common bits of S and S' are determined, to improve the secrecy of the final pairwise key, N_a randomly selects some of those common bits. After random selection of the bits, N_a sends indices of those bits to N_b, then N_a and N_b use the selected bits for their final pairwise key.

The security of ERS is evaluated with respect to its resilience against node compromise attacks. The simulations also aim to find an answer to the question: Given that a symmetric key K is used for communication between noncompromised nodes N_a and N_b, what is the probability that the key K

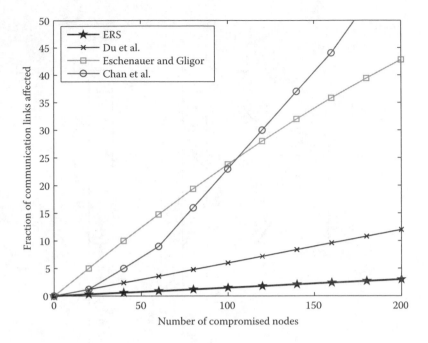

FIGURE 7.10
Comparison of ERS with some key establishment schemes in terms of resilience against node capture.

belongs to the set of those keys obtained by an intruder through compromising sensor nodes? Therefore, ERS is compared with the random key predistribution schemes proposed by Du et al. [28], Eschenauer and Gligor [27], and Chan et al. [29], as shown in Figure 7.10. As opposed to the random key predistribution schemes, each sensor node in ERS carries only the keys that it uses to communicate with its neighboring nodes. Hence, a compromised node affects its own communication links only, which significantly increases the network's resilience against node compromise attacks.

7.6   Integration of False Data Detection and Data Aggregation

When a data aggregator aggregates its incoming data, the data are usually altered. Therefore, false data detection and data aggregation protocols should be designed together to enable sensor nodes to determine whether the data are changed due to data aggregation or false data injection. This section presents briefly some operations of our data aggregation and authentication protocol (DAA) [36] that provides false data detection together with data aggregation and confidentiality. To the best of our knowledge, DAA is the first of its kind to allow data aggregation in false data detection. DAA can be used effectively

TABLE 7.1

Comparison of DAA with Existing *False Data Detection* and *Secure Data Aggregation* Schemes

Protocol	En-route False Data Detection	False Data Detection at Base Station	En-route Data Aggregation	Data Confidentiality
DAA	√	√	√	√
SEF [4]	√	√		
Interleaved hop-by-hop [5]	√	√		
CCEF [6]	√	√		
A dynamic en-route [7]	√	√		
SIA [16]		√		
Witness-based aggregation [21]		√		
Secure aggregation [15]		√	√	
ESPDA [17]		√		√

for various applications. As an example, consider the border surveillance sensor networks, where it is critical to distinguish a correct target report from a false report or a false alarm caused by various factors such as intruders, animals, faulty sensor nodes, or incorrect sensor readings. Because sensors have difficulty in differentiating between illegal activity and legitimate events, border patrol agents spend many hours investigating legitimate activities [38]. To reduce the false alarm rate in border surveillance sensor networks, the accuracy of data should be improved by aggregating the sensed data of multiple sensor nodes for the same event. Because DAA is able to aggregate the data and to detect false data injections, it can reduce false alarms, in addition to mitigating the amount of transmitted data.

In DAA, $2T + 1$ pairs of sensor nodes cooperate for data authentication against T compromised nodes, where the value of T depends on security requirements, node density, packet size, and tolerable overhead. Table 7.1 presents the comparison of DAA with existing false data injection and secure data aggregation schemes that are mentioned in this chapter. As seen from the table, only DAA offers false data detection, data confidentiality, and data aggregation during data forwarding. The terms that are used in DAA are defined next.

Definition 7.1

(*current data aggregator, backward data aggregator, forward data aggregator, forwarding node*). Let R={$A_0, A_1, \ldots, A_n$} represent the set of all data aggregators on a path P from a sensor node to base station BS. The aggregator that we currently consider is called the *current data aggregator*, denoted by A_u, for $1 \leq u \leq n - 1$. The previous and next data aggregators of A_u are referred

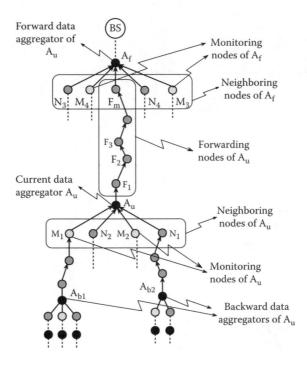

FIGURE 7.11
The system architecture of DAA.

to as its *backward data aggregator* A_b and *forward data aggregator* A_f, respectively, where $b = u - 1$ and $f = u + 1$. Sensor nodes that are located between A_u and A_f on the path P are called the *forwarding nodes* of A_u, as shown in Figure 7.11.

Definition 7.2
(*subMAC, FMAC*). For a given T and data D, let S denote the size of $MAC(D)$ in bits. A small-size MAC of D, called $subMAC(D)$, is a set of $S/(T + 1)$ bits that are randomly selected from $MAC(D)$. A full-size MAC of D, called $FMAC(D)$, is composed of $T + 1$ unique $subMAC(D)$s computed by $T + 1$ nodes.

The basic idea behind DAA can be explained using the system architecture presented in Figure 7.11. In order for DAA to aggregate data securely and detect false data injections, each data aggregator is monitored by some of its neighbors, called *monitoring nodes*. Monitoring nodes are selected randomly to perform data aggregation and compute subMACs for the aggregated data. To detect false data injections, A_u and its monitoring nodes form pairs with some other sensor nodes as follows. The monitoring nodes of a data aggregator A_u form T pairs with the forwarding nodes of A_u, called

MF-type pairs, and form another T pair with the neighboring nodes of the forward data aggregator A_f, called *MN-type* pairs. A_u and A_f form an *AA-type* pair. The incoming data of A_u are aggregated by A_u and its monitoring nodes. After aggregating the data, to provide data confidentiality, A_u encrypts the aggregated data and broadcasts it. Each monitoring node of A_u computes subMACs of the encrypted and plain aggregated data using the keys it shares with its pairmates so that the pairmate nodes can verify the authenticity of the aggregated data. A_u collects subMACs from its monitoring nodes and prepares two FMACs for the encrypted and plain data using those subMACs. These FMACs are forwarded along with the encrypted data. To detect false data injected during data forwarding, the encrypted data are verified by those forwarding nodes that are the pairmates of A_u's monitoring nodes. The plain data are verified by the neighboring nodes of the forward data aggregator that are the pairmates of A_u's monitoring nodes to detect the false data injected by A_u. If the verification of data fails at any node, data are dropped immediately. This results in better utilization of bandwidth and battery power.

DAA achieves integration of false data detection with secure data aggregation and confidentiality in two steps. In the first step, the monitoring nodes of data aggregators are selected, and then *MN-type, MF-type, and AA-type* pairs are formed. Also, to use an authentication key, each node pair that does not share a key establishes a symmetric key. In the second step of DAA, algorithm SDFC is executed for secure data aggregation, false data detection, and data confidentiality. In order to perform secure data aggregation, each data aggregator is monitored by its T neighboring nodes. Therefore, in the first step of DAA, T neighbors of a data aggregator A_u are selected as the *monitoring* nodes to perform data aggregation and to compute sub MACs of the aggregated data. Monitoring nodes are selected collaboratively by A_u and its neighboring nodes in such a way that a compromised node cannot affect the selected monitoring nodes. After the monitoring nodes are selected, the following $2T + 1$ pairs of sensor nodes are formed as shown in Figure 7.12.

- The current data aggregator A_u and A_u's forward data aggregator A_f form an *AA-type* pair.
- The monitoring and forwarding nodes of A_u form T *MF-type* pairs.
- The monitoring nodes of A_u and the neighboring nodes of A_f form T *MN-type* pairs.

To form the *MF-type* pairs, the neighboring and forwarding nodes of A_u send their ID numbers to each other. Each monitoring node selects a distinct forwarding node to form a pair. Similarly, to form the *MN-type* pairs, each monitoring node selects a distinct neighboring node of A_f to form a pair. If a monitoring node decides to form a pair with a node such that they happen to not have a shared key, then they establish a shared key. In what follows,

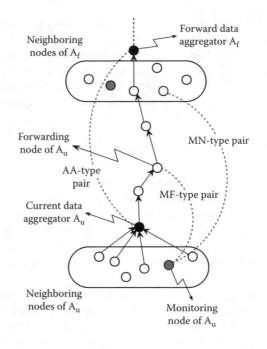

FIGURE 7.12
Sensor node pairs, for $T = 1$.

we present the system model for DAA, its algorithm SDFC, and simulation results.

7.6.1   System Model for DAA

We consider a large sensor network with densely deployed sensor nodes. Due to dense deployment, sensor nodes have overlapping sensing ranges, so that an event may be detected by multiple sensor nodes, thereby necessitating the aggregation of the correlated sensed data at neighboring sensor nodes. Some sensor nodes are dynamically designated as data aggregators to aggregate data from their neighboring sensor nodes, although every sensor node is assumed to be capable of doing data aggregation. To balance the energy consumption of sensor nodes, the role of data aggregator is rotated among sensor nodes based on their residual energy levels. We assume that data aggregators are chosen in such a way that (i) there are at least T nodes on the path between any two consecutive data aggregators, and (ii) each data aggregator has at least T neighboring nodes, so that they can form $T + 1$ pairs with the forwarding nodes on the path between two consecutive data aggregators. Network links are bidirectional. Transmission ranges of data aggregators can be adjusted depending on the number of their neighboring nodes. For instance, if a data aggregator needs more neighboring nodes, its transmission range may be increased. Sensor nodes have limited computation and communication capabilities. For example, the Mica2 motes [22] have a 4-Mhz 8-bit Atmel microprocessor, and are equipped with an instruction memory of 128 KB and a RAM of 4 KB.

Only data aggregators are allowed to encrypt and decrypt the aggregated data. The forwarding nodes first verify data integrity using MACs, and then relay the data if it is not false. The TinySec data packet structure [9] includes a 29-byte payload and a 4-byte MAC. Although a DAA packet contains the same size of payload (i.e., 29 bytes), it has two 4-byte MACs rather than one 4-byte MAC, leading to a 4-byte increase in the packet length. In DAA, each of these 4-byte MACs is called a full-size MAC, denoted as FMAC. An FMAC consists of $T + 1$ small-size MACs, called subMACs, such that one of them is computed by a data aggregator and the remaining T subMACs are computed by its T monitoring nodes. A subMAC is constructed by selecting some bits of a MAC. To form a subMAC by selecting some bits of a MAC, we assume that each sensor node has the same pseudo-random number generator (PRNG) [33] that generates random numbers ranging from 1 to 32. After the pairs are formed and their shared keys are established, the sensor nodes of each pair initiate their PRNGs using their shared key as the seed. In order for a neighboring sensor node N_i of the current data aggregator A_u to generate a $subMAC(D)$ for its pairmate F_j of data forwarding node, the sensor node N_i first computes the $MAC(D)$ of the data D using the key $K_{i,j}$ that it shares with F_j. Then, assuming that S denotes the size of $MAC(D)$ in bits, N_i pseudo-randomly selects $S/(T + 1)$ bits from $MAC(D)$ and forms the $subMAC(D)$. To select the bits from $MAC(D)$, N_i runs its PRNG $S/(T + 1)$ times, which results in $S/(T + 1)$ random numbers ranging from 1 to 32. Each random number indicates the index of a bit location in $MAC(D)$, and the bits of those selected locations constitute the $subMAC(D)$. To verify this subMAC computed by N_i, its pairmate similarly computes the $MAC(D)$ and runs its PRNG $S/(T + 1)$ times to generate its $subMAC(D)$. If the subMACs of a pair match, the data D are said to be verified by the pairmate of N_i.

A monitoring node establishes a pairwise shared key with its pairmate node using an existing pairwise key establishment scheme such as the one in [23, 24, 30]. Similarly, each data aggregator A_u and its neighboring nodes establish a group key, called K_{group}^u using an existing group key establishment scheme [32]. The group key is used for selecting the monitoring nodes of the data aggregator, and protecting data confidentiality while data are transmitted among the data aggregator and its neighboring nodes for data verification and aggregation. Intruders can compromise sensor nodes via physical capturing or through the radio communication channel. Once a sensor node is compromised, all information of the node becomes available to the intruder and the compromised node can inject false data into the network. Compromised nodes can collaborate for false data injection. Although compromised nodes can perform many types of attacks to degrade the network's security and performance, DAA considers only false data injection and eavesdropping attacks.

7.6.2 Algorithm SDFC

In DAA, algorithm SDFC provides false data detection, secure data aggregation, and data confidentiality. The basic idea behind SDFC is that monitoring nodes of data aggregators also perform data aggregation and compute subMACs of the encrypted and plain aggregated data for their pairmates. Then, to detect false data injections, each data aggregator forms two FMACs for the aggregated data: one FMAC for the encrypted data, and the other FMAC for the plain data. The FMACs of encrypted and plain data are forwarded along with the encrypted data. Pairmates of the monitoring nodes verify these FMACs to detect false data injections. Each FMAC consists of $T + 1$ subMACs computed by T monitoring nodes and the data aggregator. Assuming that the monitoring nodes of a current data aggregator are indexed in ascending order starting with 1, the current data aggregator forms each FMAC by concatenating subMACs of monitoring nodes in the order of their indices. Because the forwarding nodes of the current data aggregator and the neighboring nodes of the forward data aggregator are assumed to know the indices of their monitoring pairmates of the current data aggregator, they can easily locate the subMACs computed by their monitoring pairmates. When a pairmate node fails to verify its subMAC, the data are dropped immediately. Figure 7.13 presents the steps of algorithm SDFC.

As seen from Figure 7.13, algorithm SDFC's main steps are: (i) whenever some data are received by a data aggregator, the authenticity of data is verified by the data aggregator and its neighboring nodes, (ii) the data aggregator and its monitoring nodes aggregate the data independently of each other, (iii) each monitoring node computes one subMAC for the encrypted data and the other subMAC for the plain data, (iv) the data aggregator collects these subMACs from its monitoring nodes to form the FMACs of the encrypted and plain data, appends the FMACs to the encrypted data, and transmits them, (v) the forwarding nodes verify the data integrity of the encrypted data, and finally (vi) the neighboring nodes of the next aggregator verify the integrity of the plain data.

Now, we explain algorithm SDFC using the following example. Consider the sensor nodes illustrated in Figure 7.15 where A_u receives data D_1, D_2, and D_3 from N_1, N_2, and N_3, respectively. Note that D_1, D_2, and D_3 are indeed sent by A_{b1}, A_{b2}, and A_{b3}, respectively, and that N_1, N_2, and N_3 are their last forwarding nodes. In line 1 of SDFC, the neighboring node N_i of A_u sends D_i and its two FMACs to A_u. A_u first verifies D_i using the subMAC computed by A_{bi}, and decrypts D_i using the symmetric key that it shares with A_{bi}, for $1 \le i \le 3$. If the verification of D_1 is successful for $i = 1$, A_u encrypts D_1 using the group key K^u_{group}, and broadcasts the encrypted D_1 along with the FMAC of plain D_1 (line 2 of SDFC). If N_1 and N_3 are the *MN-pairmates* of monitoring nodes of A_{b1}, then the FMAC of plain D_1 that consists

Algorithm SDFC

Input: The current data aggregator A_u, its forward data aggregator A_f, k backward aggregators $\{A_{b1}, \cdots, A_{bk}\}$, n neighboring nodes $\{N_1, \cdots, N_n\}$ for $n \geq T$, T monitoring nodes $\{M_1, \cdots, M_T\}$, and z forwarding nodes $\{F_1, \cdots, F_z\}$ for $z \geq T$.

Output: Any false data that are injected during data aggregation or forwarding by up to T compromised nodes are detected and dropped by either A_u's data forwarding nodes or A_f's neighboring nodes. Data confidentiality is provided.

1: **for** (i=1 to k) **do**
2:     When the neighboring node N_i of A_u, which also serves as the last forwarding node of A_{bi}, receives the data $\{E_{K_{bi,u}}(D_i), FMAC(E_{K_{bi,u}}(D_i)), FMAC(D_i)\}$ from A_{bi}, the node N_i sends the data to A_u.
3:     **if** A_u successfully verifies $E_{K_{bi,u}}(D_i)$ **then**
4:         A_u decrypts $E_{K_{bi,u}}(D_i)$, obtains the plain data D_i, encrypts D_i using the group key K_{group}^u, and broadcasts $\{E_{K_{group}^u}(D_i), FMAC(D_i)\}$. Those neighboring nodes of A_u that are the *MN-pairmates* of A_{bi}'s monitoring nodes verify the integrity of D_i; if the verification fails, A_u discards D_i.
5:     **else**
6:         A_u discards D_i and informs A_{bi} about the unsuccessful verification of D_i.
7:     **end if**
8: **end for**
9: A_u and each monitoring node M_i aggregate all the verified data sent by the backward aggregators, for $1 \leq i \leq T$. Let D_{agg} denote the aggregated data.
10: A_u first encrypts D_{agg} using the symmetric key $K_{u,f}$ that it shares with A_f and then broadcasts the encrypted D_{agg} denoted by $E_{K_{u,f}}(D_{agg})$.
11: Each monitoring node M_i first computes two subMACs: 1) $subMAC(E_{K_{u,f}}(D_{agg}))$ using the key it shares with its *MF-pairmate* forwarding node, and 2) $subMAC(D_{agg})$ using the key it shares with its *MN-pairmate* that is a neighboring node of A_f. Then, each M_i sends its two subMACs to A_u.
12: A_u also computes its own two subMACs, namely, $subMAC(E_{K_{u,f}}(D_{agg}))$ and $subMAC(D_{agg})$, using the same key $K_{u,f}$ that it shares with A_f.
13: A_u forms two full-size MACs, namely, $FMAC(D_{agg})$ and $FMAC(E_{K_{u,f}}(D_{agg}))$. $FMAC(D_{agg})$ is formed by concatenating the A_u's $subMAC(D_{agg})$ with the $subMAC(D_{agg})$s of the T monitoring nodes. Similarly, $FMAC(E_{K_{u,f}}(D_{agg}))$ is formed by concatenating the A_u's $subMAC(E_{K_{u,f}}(D_{agg}))$ with the $subMAC(E_{K_{u,f}}(D_{agg}))$s of the T monitoring nodes.
14: A_u first forms a packet containing $E_{K_{u,f}}(D_{agg})$, $FMAC(D_{agg})$, and $FMAC(E_{K_{u,f}}(D_{agg}))$ and then sends it to the first forwarding node.
15: Call VERIFY(z)
16: Relabel A_f and A_u as A_u and A_{bz}, respectively, so that the old A_u becomes the *z*th backward aggregator of the new A_u. Go to Line 1, where the *z*th iteration of the "for" loop (in lines 1 to 8) determines whether D_{agg} contains any false data injected by the old A_u during data aggregation.

FIGURE 7.13
Algorithm SDFC.

Procedure VERIFY(z)

Input: The forward data aggregator A_f and the z forwarding nodes $\{F_1, \cdots, F_z\}$ of A_u.

Output: Data forwarding nodes verify the data integrity by computing and comparing their subMACs.

1: **for** (j=1 to z) **do**
2:      **if** F_j is not the *MF-pairmate* of a monitoring node of A_u **then**
3:          F_j just forwards the incoming packet to the next forwarding node or A_f.
4:      **else**
5:          F_j verifies the $subMAC(E_{K_{u,f}}(D_{agg}))$ computed by its pairmate of monitoring node.
6:          **if** the verification is successful **then**
7:              F_j forwards the packet to the next forwarding node if $j < z$. If $j = z$, the forwarding node F_j, which is also a neighboring node of A_f, sends the entire packet to A_f.
8:          **else**
9:              F_j drops $E_{K_{u,f}}(D_{agg})$ and informs A_u about it.
10:          **end if**
11:      **end if**
12: **end for**

FIGURE 7.14
Procedure VERIFY.

of three subMACs need to be verified by A_u, N_1, and N_3. Therefore, N_1 and N_3 decrypt D_1 and verify it using their associated subMACs. Similarly, D_2, and D_3 are also verified by A_u and its neighboring nodes. Once D_1, D_2 and D_3 are verified, each of A_u and its monitoring nodes N_1 and N_2 aggregate them to obtain the aggregated data D_{agg} (line 9 of SDFC). In line 10 of SDFC, A_u encrypts D_{agg} using the key that it shares with A_f, and broadcasts the encrypted D_{agg}. In line 11 of SDFC monitoring nodes compute subMACs for the encrypted and plain aggregated data. Monitoring node N_1 computes the subMAC for the encrypted D_{agg} using the key that it shares with its *MF-pairmate* F_1. N_1 also computes the subMAC for the plain D_{agg} using the key that it shares with its *MN-pairmate* N_7. Similarly, N_2 computes the subMAC for the encrypted D_{agg} using the key that it shares with its *MF-pairmate* F_3, and computes the subMAC for the plain D_{agg} using the key that it shares with its *MN-pairmate* N_4. In line 12 of SDFC, A_u computes two subMACs for the encrypted and plain D_{agg} using the key that it shares with A_f. A_u collects subMACs from N_1 and N_2, forms two FMACs for the encrypted and plain D_{agg}, and finally sends the encrypted D_{agg} along with two FMACs (line 14 of SDFC). Line 15 of algorithm SDFC calls Procedure VERIFY (Figure 7.14) for verification of encrypted data. In Procedure VERIFY, the FMAC of the encrypted D_{agg} is verified by F_1, F_3, and A_f. Once D_{agg} is verified by A_f, A_f is relabeled as A_u and the FMAC of plain D_{agg} are verified by the new A_u's

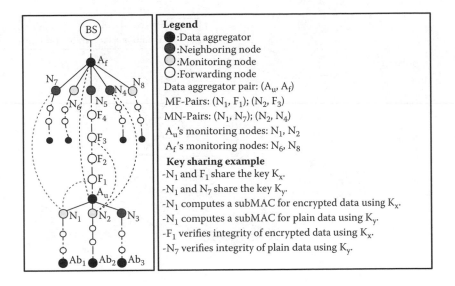

FIGURE 7.15
An example for algorithm SDFC, for $T = 2$. For the sake of simplicity, some pair relations are not illustrated.

neighboring nodes N_4 and N_7 that are the pairmates of old A_u's monitoring nodes.

7.6.3 Simulation Results

DAA is simulated using QualNet [25] network simulator for an area of $100\,m \times 100\,m$ and 100 sensor nodes with a transmission range of $15\,m$. Sensor nodes are distributed uniformly and the base station is located at one corner of the network. TDMA is used as the media access control scheme. Some nodes are designated as data aggregators. Data are assumed to be generated mainly by the nodes located at the edges of the network, although any node is allowed to sense events and generate data. The performance of DAA is compared with the *traditional data authentication* scheme where a source node computes a MAC of its data and sends the data and its MAC to a destination node that is usually a base station.

Computational Overhead The computational overhead of DAA is evaluated in terms of the number of MAC computations required for false data detection, secure data aggregation, and data confidentiality. Figure 7.16 compares the number of MAC computations in a network using DAA and traditional data authentication where a source node computes the MAC of the data and its destination node verifies this MAC [9]. The number of MAC computations in the network is shown as a function of security parameter T and percentage of data redundancy. Percentage of data redundancy is defined as

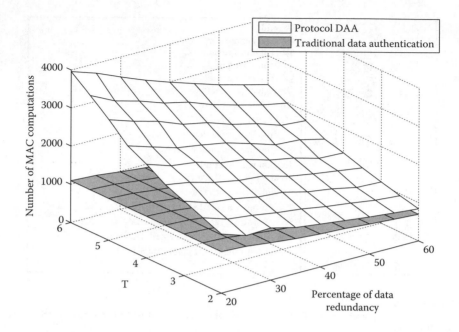

FIGURE 7.16
The overhead of MAC computations versus T for data with variable redundancy.

being the ratio of redundant data to the total generated data by sensor nodes. The data redundancy is included in the simulations to show the benefit of data aggregation in a network of densely deployed sensor nodes.

Figure 7.16 illustrates that as T increases, the number of MAC computations in DAA increases as well. Consequently, the network becomes more secure against false data injections because sensor network's ability to detect and eliminate false data increases. Hence, the value of T trades off between security and computation overhead of the network. Figure 7.16 also shows that as the percentage of data redundancy increases, the number of MAC computations decreases because data aggregation reduces significantly the amount of data to be transmitted by eliminating redundant data. As seen from Figure 7.16, DAA has more computational overhead than the traditional data authentication scheme. However, DAA can still result in energy savings because (i) the data aggregation in DAA significantly reduces the data transmission in the network, and (ii) the transmission of a bit can consume as much energy as the execution of 900 instructions [1].

Communication Overhead   The communication overhead of DAA occurs as follows: (i) two FMACs are transmitted during data forwarding, and (ii) those data are transmitted from data aggregators to their neighboring nodes for aggregation and subMAC computation. Because DAA detects and

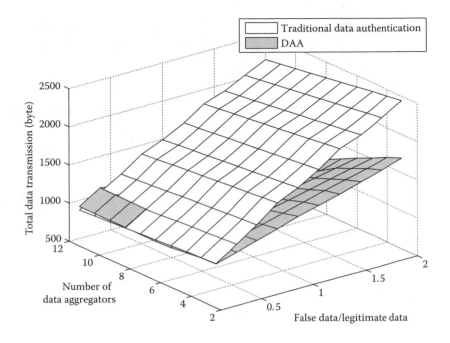

FIGURE 7.17
The total transmitted data in DAA is compared with that of the traditional data authentication scheme, as the number of data aggregators and the ratio of false data to legitimate data β/α vary. Because DAA reduces the amount of overall data using data aggregation and false data detection, the amount of data transmitted in DAA is up to 60% less than the traditional data authentication scheme.

eliminates false data between two consecutive data aggregators, simulations are performed for various number of data aggregators in the network. The percentage of data redundancy in the network is assumed to be 30% on average. Because most of the energy in sensor networks is consumed due to data transmission, it is critical to mitigate data redundancy and to detect false data as early as possible.

The total data transmission of the network with DAA and with traditional data authentication are shown in Figure 7.17. When $\beta/\alpha = 2$ and the network has 12 data aggregators, DAA results in 60% less data transmission as compared with the traditional data authentication. This data reduction of up to 60% occurs due to two reasons: (i) the 30% data redundancy is reduced significantly by data aggregation, and (ii) those false data that could be twice as much as the legitimate data (i.e., β/α could be equal to 2) are detected and dropped as early as possible. Hence, even though there exists communication overhead, implementing data aggregation and false data detection in DAA still reduces the amount of overall data transmission in the network. As the number of data aggregators increases, the number of hops between data aggregators decreases and DAA detects false data earlier, thereby leading to less

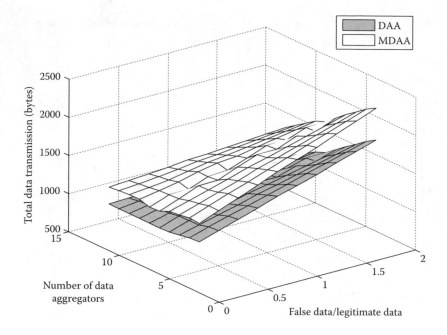

FIGURE 7.18
The comparison of total transmitted data for DAA and MDAA, as the number of data aggregators and the ratio of false data to legitimate data β/α vary.

data transmission over the network. However, in case of the network with traditional data authentication, the amount of transmitted data is not affected by the number of data aggregators simply because false data are detected only at the base station.

Impact of Data Aggregation   To show the importance of data aggregation in a densely deployed sensor network, we assume that MDAA is a modified version of DAA such that it is the same as DAA, except that MDAA does not perform any data aggregation at all. That is, DAA mitigates the redundant data at data aggregators by implementing data aggregation, whereas MDAA transmits all of the redundant data to the base station. But, both of them drop false data as soon as it is detected. We compare the performance of DAA with MDAA with respect to the total amount of data transmitted over the network, where β/α ranges from 0.2 to 2, and the number of data aggregators ranges from 2 to 12. The data aggregators are assumed to be distributed uniformly over the network. Simulations also assume that 10 compromised sensor nodes are spread over the network to inject false data, and that the percentage of data redundancy is 30% on average. Figure 7.18 shows that DAA results in up to 25% less data transmission than MDAA. It is worth mentioning that the impact of data aggregation in DAA grows as the percentage of data redundancy increases.

7.7 Conclusion

This chapter has reviewed the basic concepts and the existing algorithms for false data detection, data aggregation, secure data aggregation, and key establishment in wireless sensor networks. We have also addressed the importance of integrating false data detection with data aggregation. In this regard, we have briefly described the protocol DAA.

Bibliography

1. I.F. Akyildiz, W. Su, Y. Sankarasubramaniam, and E. Cayirci, "A survey on sensor networks," *IEEE Communications Magazine*, 40(8), pp. 102–114, Aug. 2002.
2. C. Intanagonwiwat, D. Estrin, R. Govindan, and J. Heidemann, "Impact of network density on data aggregation in wireless sensor networks," *Proc. of the 22nd International Conference on Distributed Computing Systems*, pp. 575–578, July 2002.
3. Adrian Perrig, Robert Szewczyk, Doug Tygar, Victor Wen, and David Culler "SPINS: Security protocols for sensor networks," *Wireless Networks Journal (WINE)*, September 2002.
4. F. Ye, H. Luo, S. Lu, and L. Zhang, "Statistical en-route detection and filtering of injected false data in sensor networks," *Proc. of the IEEE INFOCOM 2004*, Hong Kong, China, March 7–11, 2004.
5. S. Zhu, S. Setia, S. Jajodia, and P. Ning, "An interleaved hop-by-hop authentication scheme for filtering false data in sensor networks," *Proc. of IEEE Symposium on Security and Privacy*, pp. 259–271, Oakland, CA, May 2004.
6. H. Yang and S. Lu, "Commutative cipher based en-route filtering in wireless sensor networks," *Proc. of IEEE VTC Fall 2004*, 2004.
7. Z. Yu and Y. Guan, "A dynamic en-route scheme for filtering false data in wireless sensor networks," to appear in *Proceedings of IEEE INFOCOM 2006*, Barcelona, Spain, April 23–27, 2006.
8. W. Diffie and M. Hellman, "New directions in cryptography," *IEEE Trans. on Information Theory*, 22(6):644–654, November 1976.
9. C. Karlof, N. Sastry, and D. Wagner, "TinySec: A link layer security architecture for wireless sensor networks," *Proc. of the Second ACM Conference on Embedded Networked Sensor Systems*, November 3–5, Baltimore, MD, 2004.
10. W. Heinzelman, A. Chandrakasan, and H. Balakrishnan, "Energy-efficient protocol for wireless micro sensor networks," *Proc. of 33rd Hawaii International Conference on System Sciences*, 2000.
11. S. Lindsey and C. S. Raghavendra, "PEGASIS: Power efficient gathering in sensor information systems," *Proc. of IEEE Aerospace Conference*, 2002.
12. S. R. Madden, M. J. Franklin, J. M. Hellerstein, and W. Hong, "TAG: a tiny aggregation service for ad-hoc sensor networks," *Proc. of the 5th Symposium on Operating Systems Design and Implementation*, pp. 131–146, 2002.
13. C. Intanagonwiwat, R. Govindan, D. Estrin, J. Heidemann, F. Silva, "Directed diffusion for wireless sensor networking," *IEEE/ACM Transactions on Networking*, vol. 11, no. 1, February 2003.

14. W. Heinzelman, J. Kulik, and H. Balakrishnan, "Adaptive protocols for information dissemination in wireless sensor networks," *Proc. of 5th ACM/IEEE Mobicom Conference*, 1999.

15. L. Hu and D. Evans, "Secure aggregation for wireless networks," *Proc. of Workshop on Security and Assurance in Ad hoc Networks*, Jan. 28, Orlando, FL, 2003.

16. B. Przydatek, D. Song, and A. Perrig, "SIA: Secure information aggregation in sensor networks," *Proc. of SenSys'03*, Nov. 5–7, Los Angeles, 2003.

17. H. Çam, S. Ozdemir, P. Nair, D. Muthuavinashiappan, and H.O. Sanli, "Energy-efficient and secure pattern-based data aggregation for wireless sensor networks," *Special Issue of Computer Communications on Sensor Networks*, pp. 446–455, Feb. 2006.

18. H. Çam, "Nonblocking OVSF codes and enhancing network capacity for 3G wireless and beyond systems," *Computer Communications*, vol. 26, no. 17, pp. 1907–1917, 1 Nov. 2003.

19. H. Çam, S. Ozdemir, H.O. Sanli, and P. Nair, "Secure differential data aggregation for wireless sensor networks," *in Sensor Network Operations*, S. Phoha, T.F. La Porta, and C. Griffin (eds.), Wiley-IEEE Press, April 2006.

20. K. Wu, D. Dreef, B. Sun, and Y. Xiao, "Secure data aggregation without persistent cryptographic operations in wireless sensor networks," *Proc. of 25th IEEE International Performance, Computing, and Communications Conference, (IPCCC) 2006*, pp. 635–640, 2006.

21. W. Du, J. Deng, Y. S. Han, and P. K. Varshney, "A witness-based approach for data fusion assurance in wireless sensor networks," in *Proc. of IEEE Global Telecommunications Conference (GLOBECOM '03)*, pp. 1435–1439, 2003.

22. Crossbow Technologies Inc., http://www.xbow.com.

23. D. Liu and P. Ning, "Establishing pairwise keys in distributed sensor networks," *Proc. of the 10th ACM Conference on Computer and Communications Security (CCS)*, October 27–31, pp. 52–61, Washington, DC, 2003.

24. W. Du, J. Deng, Y. S. Han, and P. K. Varshney, "A pairwise key pre-distribution scheme for wireless sensor networks," *Proc. of the 10th ACM Conference on Computer and Communications Security (CCS)*, October 27–31, pp. 42–51, Washington, DC, 2003.

25. QualNet Network Simulator by Scalable Network Technologies, www.scalable-networks.com/

26. R. C. Merkle, "Protocols for public key cryptosystems," *Proc. of the IEEE Symposium on Research in Security and Privacy*, April 1980, pp. 122–134.

27. L. Eschenauer and V. D. Gligor, "A key-management scheme for distributed sensor networks," *Proc. of the 9th ACM Conference on Computer and Communications Security*, November 18–22, pp. 41–47, Washington, DC, 2002.

28. W. Du, J. Deng, Y. S. Han, S. Chen, and P. K. Varshney, "A key management scheme for wireless sensor networks using deployment knowledge," *Proc. of IEEE INFOCOM'04*, March 7–11, 2004, Hong Kong.

29. H. Chan, A. Perrig, and D. Song, "Random key predistribution schemes for sensor networks," *IEEE Symposium on Security and Privacy*, May 11–14, pp. 197–213, Berkeley, CA, 2003.

30. S. Ozdemir and H. Çam, "Key establishment with source coding and reconciliation for wireless sensor networks," *Proc. of IEEE IPCCC 2006*, pp. 407–414, April 2006.

31. K.-C. Nguyen, G. Van Assche, and N. J. Cerf, "Side-information coding with turbo codes and its application to quantum key distribution," *Proc. International Symposium on Information Theory and Its Applications*, 2004.

32. C. Blundo, A. Santis, A. Herzberg, S. Kutten, U. Vaccaro, and M. Yung, "Perfectly-secure key distribution for dynamic conferences," *Proc. of Crypto 92*, 1992.

33. D. Seetharam, S. Rhee, "An efficient pseudo random number generator for low-power sensor networks," in *Proc. of 29th Annual IEEE International Conference on Local Computer Networks*, pp. 560–562, 2004.

34. W. Diffie and M. E. Hellman, "New directions in cryptography," *IEEE Transactions on Information Theory*, vol. 22, pp. 644–654, November 1976.

35. R. L. Rivest, A. Shamir, and L. M. Adleman, "A method for obtaining digital signatures and public-key cryptosystems," *Communications of the ACM*, vol. 21, no. 2, pp. 120–126, 1978.

36. S. Ozdemir and H. Çam, "Integration of false data detection with data aggregation and confidentiality in wireless sensor networks," *submitted to IEEE Transactions on Mobile Computing*.

37. T. He, S. Krishnamurthy, J.A. Stankovic, T. Abdelzaher, L. Luo, R. Stoleru, T. Yan, and L. Gu, "Energy-efficient surveillance system using wireless sensor networks," *Proc. of the 2nd International Conference on Mobile Systems, Applications, and Services*, pp. 270–283, 2004.

38. M. Madden, "Sensors along border wasting agents' time," *http://www.azcentral.com/specials/special03/articles/0121border-sensors.html*.

8

Privacy and Anonymity in Mobile Ad Hoc Networks

Xiaoyan Hong and Jiejun Kong

CONTENTS

Abstract   Privacy in mobile ad hoc networks has new semantics in addition to the conventional notions for infrastructure networks. Mobility enabled by wireless communication has significantly changed privacy issues and anonymity research in many ways. In particular, mobility requires ad hoc routing schemes to transmit control packets frequently in an open wireless medium. The routing traffic facilitates adversaries in conducting various attacks threatening the network security and privacy. In this chapter we introduce new privacy demands associated with ad hoc networks and new privacy threats under passive routing attacks. We then investigate new routing design principles for defending against the new threats. The chapter also demonstrates through examples on how the effectiveness of the attacks can be quantified and how the attacks can visualize critical information about a network. Finally, a countermeasure, namely, an on-demand, anonymous, and untraceable routing protocol for a mobile ad hoc network is introduced.

8.1   Introduction

Mobile ad hoc networks (MANETs) are capable of establishing an instant communication infrastructure for many time-critical and mission-critical applications. Most routing protocols in MANETs fall into one of the two categories: proactive routing and reactive routing (also known as on-demand routing). In proactive ad hoc routing protocols like OLSR [7], TBRPF [29], and DSDV [31], mobile nodes constantly exchange routing messages, which typically include node identities and their connections to other nodes (Link State or Distance Vector), so that every node maintains sufficient and fresh network topological (or routes) information to allow them to find any intended destinations at any time. On the other hand, reactive routing has become a major trend in MANETs [5]. AODV [32] and DSR [19] are dominant examples. Unlike their proactive counterparts, reactive routing operation is triggered by the communication demand at sources. Typically, a reactive routing protocol has two components: *route discovery* and *route maintenance*. In the route discovery phase, the source seeks to establish a route toward the destination before sending the first data packet. The source floods a route request (RREQ) message, and the destination will respond with a route reply (RREP) message upon receiving an RREQ. The RREP traces backward along the path that the RREQ takes to the source, which pinpoints the on-demand route. In the route maintenance phase, nodes en route monitor the status of the forwarding path, and report to the source about link breakages. Optimizations could lead to local repairs of broken links.

Nevertheless, the innate characteristics of mobile wireless networks, such as node mobility and wireless transmissions, make MANETs very vulnerable to security threats. Among all forms of threats, we focus on passive routing attacks that threaten the privacy of mobile wireless networks. The open-air wireless communication can be explored by curious or malicious individuals or teams to gather information on mobile nodes and to further prepare counterattacks. The needed eavesdropping devices, such as sensors and portable computing devices, are all available off the shelf, on line, or from local electronic stores. Providing supports of identity anonymity, location privacy, and motion pattern privacy thus is critical in many MANET applications. This poses challenging constraints on secure routing and data forwarding.

In this chapter, we demonstrate that MANET routing protocols become a critical factor in network privacy, and more specifically, in anonymous communication. We identify new privacy requirements for mobile ad hoc networks by showcasing a set of passive routing attacks and defense strategies against these new threats. More specifically, we demonstrate that mobility enabled by wireless communication has changed privacy and anonymity issues in many ways compared to legacy privacy issues discussed in infrastructure network research (e.g., message privacy on the Internet, transaction anonymity in distributed banking systems). We define "mobile anonymity," the new privacy and anonymity aspects for mobile ad hoc networks. In addition to the conventional identity anonymity, the mobile anonymity has to address *venue anonymity, privacy of network topology,* and *privacy of motion pattern.* These new privacy and anonymity aspects have little significance in fixed infrastructures, but become critical issues in mobile networks. We then identify design principles for new countermeasures. Our study suggests that a hybrid approach of *identity-free routing* and *on-demand routing* assisted with *neighborhood traffic mixing* provides better mobile anonymity support than other approaches. We demonstrate through examples on how to quantify the effectiveness of the mobile anonymity attacks and how to visualize motion patterns. Finally, we introduce an on-demand anonymous and untraceable routing protocol as a countermeasure in mobile ad hoc networks.

8.2 Passive Attacks

Wireless communications can be protected by strong cryptographic methods at application (end-to-end) or MAC layer (hop-by-hop). However, these protections are not sufficient for privacy purposes. For example, MAC addresses are not encrypted by the standard MAC security protections. In addition, an eavesdropper assisted with radio detection devices can always detect a radio wireless transmission near its own location. With the help of localization algorithms [27] and GPS information, the eavesdropper can use its own coordinates and naming system to name all identified network members without knowing their real identities. Moreover, the reoccurrences of some payload patterns provide plenty of opportunities for analysis on the traffic contents and time instances. In a nutshell, besides denial-of-service threats, propagation of routing messages is challenged by traffic analysis as well.

Independent of whether and how the wireless transmissions are protected, traffic analysis leads to a passive type of attacks against the ad hoc routing schemes. The goal of such attacks is very different from other related routing security problems such as route disruption and "denial-of-service" attacks. In fact, the passive enemy will avoid such aggressive schemes, in the attempt to be as "invisible" as possible, until it traces, locates, and then physically destroys the assets. The attackers try to be *protocol compliant*, so they are harder to be detected before potential devastating physical attacks are launched.

We further characterize the passive adversary in terms of an escalating capability hierarchy.

- *Mobile eavesdropper and traffic analyst*: Such an adversary can at least perform eavesdropping and collect as much information as possible from intercepted traffic. It is mobile and equipped with GPS to know its exact location. The minimum traffic it can intercept is the routing traffic from the legitimate side. An eavesdropper with enough resources is capable of analyzing intercepted traffic on the scene. This ability gives the traffic analyst quick turnaround action time about the event it detects, and imposes serious physical threats to mobile nodes.

- *Mobile node intruder*: If adequate physical protection cannot be guaranteed for every mobile node, node compromise is inevitable within a long time window. A successful passive node intruder is protocol-compliant, thus hard to detect. It participates in collaborative network operations (e.g., ad hoc routing) to boost its attack strength against mobile anonymity; thus it threatens the entire network including all other uncompromised nodes. This implies that a countermeasure must not be vulnerable to a single point of compromise.

- *Mobile colluding attackers*: Adversaries having different levels of attacking ability can collaborate through separate channels to combine their knowledge and to coordinate their attacking activities. This realizes the strongest power at the adversary side.

Clearly, transmitted routing messages and cached routing tables, if revealed to the aforementioned adversary, will leak large amounts of private information about the network. When this happens, proactive protocols and on-demand protocols show different levels of damages by design. With the proactive routing, a compromised node has fresh topological knowledge about other mobile nodes during the entire network lifetime. The adversary can also translate the topological map to a physical map with the help from localization algorithms [27] and GPS. Thus, a single point of compromise allows the adversary to trace the entire network. On the other hand, with the on-demand routing, an adversary has reduced chances of breaking mobile anonymity in the sense that only active routing entries are in cache and in transmission, and the traffic pattern is probabilistic (with respect to communication needs) and expires after a while.

Secure ad hoc routing protocols, such as SEAD [14], Ariadne [15], and ARAN [37], focus on authentication rather than anonymity. Simple encryption of routing information [2] can stop less sophisticated eavesdroppers, but not traffic analysts. Using pairwise keys between neighbors in encryption can alleviate the damages, but cannot fully thwart intruders and traffic analysts; for example, a DSR route is traceable by a single intruder en route, while an AODV route is traceable by collaborative intruders.

8.3 Mobility Changes Anonymity

In this section, we describe various new anonymity threats and vulnerabilities in MANET routing protocols. On one hand, the locations and motion patterns of mobile nodes, standing venues, and even the varying network topology, become new interests of the adversaries. This brings in new privacy challenges in addition to conventional identity anonymity and message privacy. On the other hand, new vulnerabilities exist in current MANET routing protocols. Mobility requires an ad hoc routing protocol to transmit messages frequently in an open wireless medium. The routing traffic, if not protected from anonymity attacks, facilitates adversaries in conducting various attacks threatening the network security and privacy. We present extensive examples to illustrate the feasibility and effectiveness of these new privacy threats, and to present the new anonymity aspects for mobile wireless networks, namely, "mobile anonymity." The mobile anonymity includes *venue anonymity*, *privacy of network topology*, and *privacy of motion pattern*.

8.3.1 Conventional Concept of Anonymity

The concept of *anonymity* is defined as the state of being not identifiable within a set of subjects, namely, the *anonymity set* [33]. In conventional anonymity research, the anonymity set is the set of the *identities* of possible senders/recipients. Further, anonymity is defined in terms of *unlinkability*. *Unlinkability* describes the property that a sender/receiver not to be identified from the anonymity set, and the relationship of the sender and the receiver not to be identified. In this chapter, the notion of identity refers to a mobile node's routing and forwarding ID, such as an IP address or a MAC address, since our focus is on routing and data forwarding. Another aspect of anonymity is the *unobservability*, a property that states that transmissions are physically indiscernible from random noises. Discussions on the *unobservability* problem are not the intention of this chapter.

8.3.2 Venue Anonymity

Figure 8.1 illustrates an adversary's network, which is comprised of a number of eavesdropping cells. The dense grid of eavesdroppers presents a strong form of adversary that collaboratively gathers global knowledge of traffic. The figure helps to illustrate several possible attacks described in this section. For example, it characterizes the capability of a collection of colluding traffic analysts from multiple cells. And it also characterizes the capability of a mobile traffic analyst who can travel along the grid structure to launch anonymity attacks anywhere and anytime.

For a mobile node, we define its **venue** in terms of the *adversary*'s capability in positioning a wireless transmission, i.e., a **venue** is the smallest area

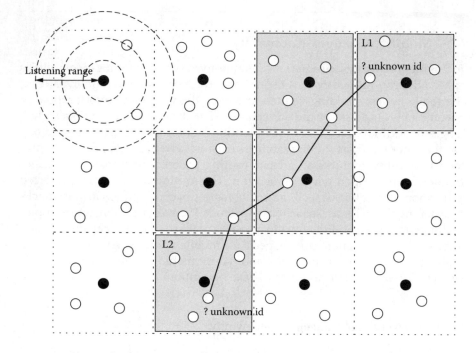

FIGURE 8.1
A network with a number of eavesdropping cells. Traffic analysts are depicted as solid black nodes. A sender in cell *L*1 is communicating with a recipient in cell *L*2. Active routing cells are depicted in shade.

to which the adversary can "pinpoint" the mobile node via wireless eavesdropping. In other words, a venue is a location perceived by an adversary. Therefore, a venue is at most as large as the radio receiving range of the eavesdropper (Figure 8.1). With a better positioning technique support, the adversary can improve the precision to a smaller area. The network is then comprised of many venues given all intercepted wireless transmissions. Several mobile nodes could be associated with the same venue. An undirected graph $G = \langle V, E \rangle$ can describe the adversarial network. For example, in Figure 8.1, a cell is a venue, it then becomes a vertex in graph G. Adversarial eavesdropping nodes form the vertex set V. The topological links among the vertexes, indicating the communication capability among the adversaries, form the edge set E. The **venue anonymity set** is the set of all the venues. The **venue anonymity** is defined as the state of the sender/recipient's venue being not identifiable within the venue anonymity set. The relationship between the sender's and recipient's venues should not be linked given the venue anonymity set. Clearly, the venue anonymity concept is defined in parallel to the identity anonymity. However, to ensure venue anonymity, the identities of the transmitting nodes must not be revealed.

Venue anonymity captures the subtle differences from the term *location privacy*, where the relationships between nodes' identities and their locations are concerned [3, 8, 10, 11, 18]. In the context of venue, the adversary's knowledge about legitimate wireless nodes is reasoned. This also sets aside the geographical positions used in georouting [20, 25] in the sense that locations are not used by any legitimate nodes in routing. Similar to location privacy, where the association between the venue and the node's identity is concerned, we can define the concept of *venue privacy*, which maps to location privacy directly. On the other hand, the venue anonymity can be compromised by the adversary through various routing attacks regarding the legitimate nodes. In a nutshell, the venue anonymity presents new semantics in describing the location-related privacy issue of mobile networks where routing presents major vulnerability.

Mobility differentiates venue anonymity from identity anonymity. In static networks (e.g., the public Internet), a sender (or recipient)'s identity and its venue are synonyms due to the rich semantics carried in the identity (e.g., an IP address, or a domain name). Thus, identifying a sender's (or recipient's) venue implies the compromise of the sender's (or recipient's) anonymity. But in mobile networks, a legitimate node is not locked in a vertex of the underlying graph. Thus a node's identity is dissociated from a specific venue. However, at each traffic analyst's vertex, the adversarial analyst can correlate a mobile node with its own exact location. Examples 8.1, 8.2, and 8.3 show that *identity anonymity* and *venue anonymity* are different concepts in mobile networks. While identity anonymity is still an issue, venue becomes a new anonymity problem that needs to be addressed separately.

Example 8.1
(Sender or recipient identity anonymity attack in on-demand route request flooding) In common on-demand ad hoc routing schemes like DSR and AODV, identities of the source/sender and the destination/recipient are explicitly embedded in route request (RREQ) packets. Any eavesdropper who has intercepted such a flooded packet can uniquely identify the sender's and the recipient's identities. However, he may not know the venue/vertex of the sender or the recipient. This example also verifies that neither sender nor recipient identity anonymity is protected in DSR and AODV.

Example 8.2
(Sender or recipient identity anonymity attack in on-demand route request flooding with per-hop encryption) A seemingly ideal cryptographic protection is to apply *per-hop* encryption using pairwise key agreement, i.e., a transmission is protected by an ideal point-to-point secure pipe between the two neighbors of a forwarding hop. The secure channel protects every packet including the packet header. This solution prevents eavesdroppers from understanding routing messages. But it does not prevent passive node intruders from identifying the sender's and the recipient's identities upon receiving an RREQ packet. Again, the intruder may not know the venue/vertex of the sender or the recipient.

Example 8.3

(Packet flow tracing attack)   Similar to anonymity attacks revealing the relationship between senders and recipients, the packet flow tracing attack can reveal the relationship between a sender's venue and its recipient's venue. Even protected by ideal encryption along a multihop forwarding path, timing correlation and content correlation analysis can be used to trace a packet flow. For example, by collusion or mobility, mobile traffic analysts can trace an ongoing packet flow to the sender's venue $L1$ and the recipient's venue $L2$ (Figure 8.1), thus breaking the sender's (or recipient's) venue anonymity. But they may not be able to see the identities.

8.3.3   Privacy of Ad Hoc Network Topology

Internet topology are mostly stable and can be viewed through various public tools. Routing protocols in Internet (e.g., BGP [36], OSPF [28], and RIP [12]) make no attempts to protect the privacy for network topology. However, in mobile networks, network topology constantly changes due to mobility. Once information about the network topology (or routes as partial topology information) is revealed, the adversary can launch further security breaches or locate positions of a few nodes given other out-of-band information like geographic positions and physical boundaries of the underlying mobile network. If the targeted ad hoc network has localization and positioning support, the topology privacy problem is aggravated when the localization results (locations) are revealed. Therefore, the privacy of network topology becomes a new anonymity requirement in mobile networks. Example 8.3 has shown a packet flow tracing attack to compromise relationship anonymity between a sender's venue and its recipient's venue. It is also an example of partial compromise of topology anonymity (the path connecting the sender and the receiver).

Example 8.4

(A mobile node intruder tries to locate where a specific node is)   In proactive ad hoc routing protocols, mobile nodes constantly exchange routing messages to ensure that each sender knows enough network topological information for any intended recipient at any moment. Such design indeed establishes a lot of single points of compromise in the network, i.e., a single-node intruder can break anonymity protection by seeing the topological map. This example shows that precomputing routing schemes, in particular proactive routing schemes, directly conflict with anonymity protection requirements in mobile networks. With on-demand routing, a node intruder can simply function as a source/sender to establish a route toward the victim, then position and move toward the next hop close to the victim. By continuously probing and moving, the attacker can shorten the route and finally reach the victim. If more attackers collude, locating a victim is easier. Thus, an anonymous routing protocol should prevent a sender from knowing a forwarding path toward any mobile node.

In infrastructure networks, a node's topological location and related physical location are determined *a priori*. Therefore, anonymity solutions proposed for infrastructure networks use neighborhood information for transmission. For example, a Chaumian MIX [6] knows its immediate upstream and downstream MIXes, a jondo in Crowds [35] knows its next jondo or the destination recipient. If directly ported to mobile networks, these schemes are vulnerable to attacks described in Example 8.5.

Example 8.5
(Neighborhood location privacy attack) Given any cell *L* depicted in Figure 8.1, a mobile traffic analyst or an intruder may gather and quantify (approximate) information about active mobile nodes within the transmission range. For example, it can: (a) enumerate active nodes in *L*; (b) get related quantities such as the size of the set; (c) perform traffic analysis against *L*, e.g., how many and what kind of connections in and out of the cell. Currently common ad hoc routing protocols [19, 32, 39] do not address this attack.

8.3.4 Privacy of Motion Pattern

Besides venues, the change of venues, or the nodes' motion patterns are very important information. For example, a network mission may require a set of legitimate nodes to move toward the same direction or a specific spot. Any inference of the motion pattern will effectively visualize the outline of the mission and may finally lead to the failure of the mission. Ensuring the privacy for mobile nodes' motion patterns is a new expression. If the network fails to ensure topological venue privacy, a mobile node's motion pattern can be inferred by a dense grid of traffic analysts, or even by a sparse set of node intruders under certain conditions [13], e.g., capable of knowing neighbors' relative positions (clockwise or counterclockwise), and capable of overhearing or receiving route replies (RREPs) of on-demand routing.

Example 8.6
(Motion pattern inference attack: dense mode) The goal of this passive attack is to infer (possibly imprecise) motion patterns of mobile nodes. In Figure 8.1, the omnipresent colluding intruders can monitor wireless transmissions in and out a specific mobile node, they can combine the intercepted data and trace the motion pattern of the node at the granularity of cell.

Example 8.7
(Motion pattern inference attack: sparse mode) When node intruders are sparse in the network, they may still be able to infer motion patterns from ongoing routing events, though the information gathered could be imprecise. Here we describe a probabilistic *H(op)-clique* attack. Figure 8.2 depicts the situation when a node intruder *X* finds from the routing packets that its next hop toward the node *Y* switches from node *V*1 to *V*2 (both are *X*'s neighbors). With high probability, this routing event indicates that either the target node *Y*

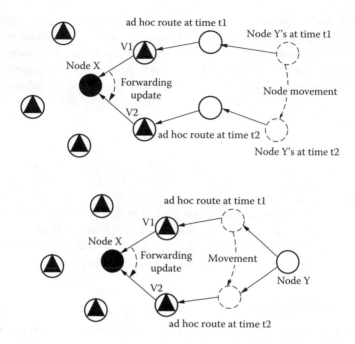

FIGURE 8.2
Sparse-mode motion pattern inference (H-clique attack). The solid black node is a protocol-compliant node intruder. The neighbors (denoted by circled triangles) are legitimate network members, but cannot detect a protocol-compliant node intruder. Left: target movement; right: forwarding node movement.

(left figure) or some intermediate forwarding nodes (right figure) have moved along the direction $V_1 \rightarrow V_2$ (clockwise). We assume that a node intruder can be furnished with basic ad hoc localization techniques (e.g., using Angle-of-Arrival, Receiver-Signal-Strength-Index, etc.). The H-clique is comprised of a single-node intruder and its gullible neighbors. Through colluding, multiple H-cliques can combine their knowledge to obtain more precise information on motion pattern. Figure 8.3 shows that a mobile node cutting through two H-cliques is detectable by the adversary. Figure 8.4 shows the case of three H-cliques. Therefore, a few node intruders can effectively launch motion pattern inference attacks against the entire network. Both proactive routing schemes and on-demand schemes are vulnerable to such passive attacks.

As a summary of this section, we point out that without security protections, all the listed privacy goals are violated by easy eavesdropping and traffic analysis. Further, while encryption and pseudonyms can be used for the mobile anonymity as a first defense as they have been widely used in Internet practice, problems such as the venue anonymity still exist. If coordinations among the attackers are possible, the motion pattern privacy and topology privacy are in great danger. With intrusions, the listed privacy goals are also mostly compromised. More design issues have to be addressed to

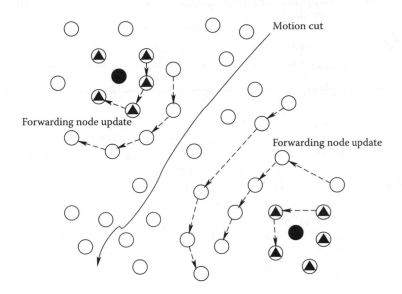

FIGURE 8.3
H-clique attack: A motion cutting through two H-cliques is detectable from forwarding node updates.

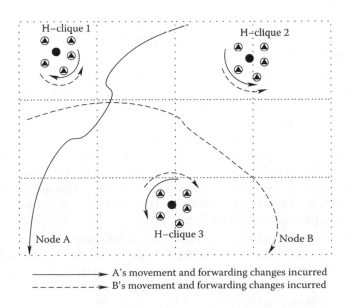

FIGURE 8.4
Composite H-clique attack: More H-cliques can obtain more precise motion patterns.

ensure the mobile anonymity for routing in a mobile wireless network facing various passive adversaries.

8.4   Designing Countermeasures

Being a member of MANET, a node must rely on at least one of its neighbors to forward its packets. When all the vulnerabilities exist, a node is facing a dilemma. On one hand, a node must forward packets to one of its neighbors, so that the neighbor can further forward the packets toward the destination. On the other hand, this node does not trust any of his neighbors. Given that the node has no way of knowing which neighboring node is adversarial (a passive attacker), the node must not reveal its identity and other identifiable information in its transmission. This dilemma becomes the major challenge in designing an anonymous routing protocol to meet all the network privacy goals for MANET. Given the limited dimensions in routing protocol design, the following directives are useful to serve the cause.

- *On-demand routing approach as a baseline to ensure privacy of network topology*: In on-demand routing, fresh network topology knowledge is gathered only when needed. Compared to proactive routing and any other proactive features (e.g., constant neighborhood beaconing), purely on-demand routing schemes reduce the node intruders' chances in knowing fresh network topology. In addition, on-demand routing generates less routing overhead and is more energy efficient. These features are highly desirable for many MANET applications.

- *Identity-free routing for strong identity protection*: The idea of identity-free routing is to hide a node's identity from its neighboring nodes in exchanging routing messages. This also implies *identity-free forwarding* for packets. In the design, usage of any identity/pseudonym of any node is not allowed in routing. Thus when the worst case presents, the adversary only knows the presence of neighboring nodes (by wireless transmissions) but not their identities (or any replacement pseudonyms) nor the associated relationship among identities. This design ensures perfect identity anonymity against strong passive node intruders.

- *Wireless neighborhood traffic mixing*: Without using identities directly in any routing message, traffic should be further mixed within a neighborhood where multiple nodes move in and out of the venue. Any counting or statistically meaningful analysis is difficult to obtain over a certain period of time. Thus the traffic from or to a venue is protected against strong passive traffic analysts. The venue anonymity is partly ensured.

Collectively, the directives illustrate critical design principles for building anonymous routing protocols for mobile ad hoc networks. It is possible to apply these principles to design various anonymous routing protocols that achieve different levels of protocol efficiency and anonymity protection. In Section 8.6 we introduce technical details on how the principles are used in routing protocols, e.g., how to establish a route, through ANODR routing protocol [22]. More work can be found in [4, 38, 39, 40]. The design choices, however, also depend on the cost paid by both the legitimate and the adversarial sides. While anonymity protection comes with cryptographic and routing overhead, we note that the passive adversary (e.g., Figure 8.1) also pays nontrivial deployment and communication cost for strong privacy attacks. When a costly attack is unlikely to occur, a balance between the performance and the degree of protection could be justified further. A choice on a protocol design is performance driven. Studies on the performance issue can be found in [23, 24, 26].

8.5 Threat Evaluation

This section aims at illustrating various issues discussed above through simulation. We present two sets of simulation study on the mobile anonymity attacks. First, we show how to quantify the effectiveness of mobile anonymity attacks; we use the packet flow tracing attack as the example. We then show a visual illustration of the mobile anonymity attacks; we use the sparse mode motion inference attack (SMIA) as the example.

8.5.1 Route Traceable Ratio

In order to realize *identity-free routing*, we have to employ a very different approach from common on-demand routing protocols [19, 30, 32]. Figure 8.5 depicts a typical active route established by different on-demand routing protocols. In Figure 8.5, common on-demand routing protocols use a node's identity to furnish packet forwarding, while an identity-free routing must use a *random* pseudonym shared between neighboring forwarders. This design bears a resemblance to virtual circuits used in Internet QoS [1]. We use a new metric called *traceable ratio* to quantify the degree of exposure of path segments. Such exposure leads to the violation of the motion pattern privacy and the topology privacy (a route contributing partially to topology knowledge).

In an identity-free routing, when node X is compromised, the adversaries can link two random pseudonyms together for each route passing node X. Thus, for each route, if F forwarding nodes are compromised and they are consecutive en route, then a route segment of $F + 1$ hops are linked together. If the compromised nodes are not consecutive en route, then the adversary

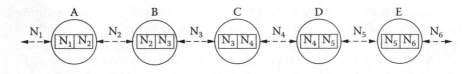

FIGURE 8.5
Identity-free routing (using random pseudonyms N_1, N_2, ...) vs. common routing (using node identity pseudonyms A, B, ...).

is able to construct multiple route segments, but not to link the multiple compromised segments together. For example, if A is the source and E is the destination in Figure 8.5, and A, B, D, E are intruded, then the adversaries can form traceable segments $\overline{ABC}$ and $\overline{CDE}$, but they have to intrude C to discover that $\overline{ABC}$ and $\overline{CDE}$ belong to the same route. For the same example, if an ordinary on-demand routing is used, comprising A, B, D, E leads to revealing the entire path $\overline{ABCDE}$.

Let us quantify the damage caused by node intrusion. Suppose a route has L hops in total, where K route segments are compromised. And suppose the hop count of the ith compromised segment is F_i,$1 \le i \le K$, we define the *traceable ratio R* of the route as

$$R = \frac{\sum_{i=1}^{K}(F_i \cdot W_i)}{L} = \frac{\sum_{i=1}^{K}(F_i \cdot \frac{F_i}{L})}{L}$$

where W_i is a weight factor. The weight W_i can be of form $(\frac{F_i}{L})^r$ where $r \ge 0$, so that the traceable ratio of a route is 100% when all forwarding nodes en route are intruded, or 0 when no forwarding node en route is intruded. Without loss of generality, we select $W_i = \frac{F_i}{L}$. In addition, the longer a compromised segment is, the larger the traceable ratio R is. This means that the victim being traced is in greater danger if the mobile intruders can get as far as possible to approach the victim. Using the same example in Figure 8.5, we have $L = 4$, the traceable ratio $R = \frac{2 \cdot \frac{2}{4} + 2 \cdot \frac{2}{4}}{4} = \frac{1}{2}$ when A, B, D, E are intruded, or $R = \frac{3 \cdot \frac{3}{4} + 1 \cdot \frac{1}{4}}{4} = \frac{5}{8}$ when A, B, C, E are intruded.

In our simulation, we compare the traceable ratio between DSR and the identity-free routing for identical scenarios. Figure 8.6 shows the traceable ratios over different path lengths. Longer paths are more likely to include intruded forwarding nodes. The figure shows that identity-free routing is not sensitive to the path length because the knowledge exposed to intruders is localized only in the intruded node. The traceable ratio of the identity-free routing remains at the percentage of the intruded nodes. In contrast, the traceable ratios of DSR increase quickly (note that DSR does not scale to long hops; thus data collected for the path length as long as 7 or more are not sufficient for statistically meaningful display).

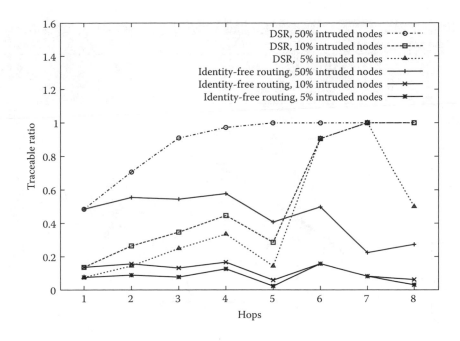

FIGURE 8.6
Traceable ratio evaluation.

8.5.2 Illustration: Sparse Mode Motion Inference Attack

We simulate a scenario where a target node moves straight across a network from the left side to the right. Figure 8.7 is a snapshot of the simulation. While moving, the target node periodically communicates with other nodes (two in the figure) using the routes established by the AODV routing protocol. In the figure, a routing path between the target and the destination is depicted by the linked solid lines. When the target moves, different paths are taken and the figure shows that the intermediate forwarding nodes have changed several times due to the target mobility. In the meantime, node intruders (two in the figure; shown also are their radio ranges) are presented in the network. They use the aforementioned radio techniques to obtain the relative positions of their neighbors. In addition, a node intruder is capable of launching worm-hole attacks [16] and rushing attacks [17] to place itself on the ad hoc routes with high probability. By analyzing the intercepted RREQs and the corresponding RREP packets, e.g., taking the source, destination and broadcast-id tuple from the RREQs and matching them with the later received/intercepted RREP, the attackers can detect that the next hop has switched from one neighbor to another for this target node. When encryption is not implemented to protect the routing messages, this H-clique attack is easier in the sense that no intrusion is needed.

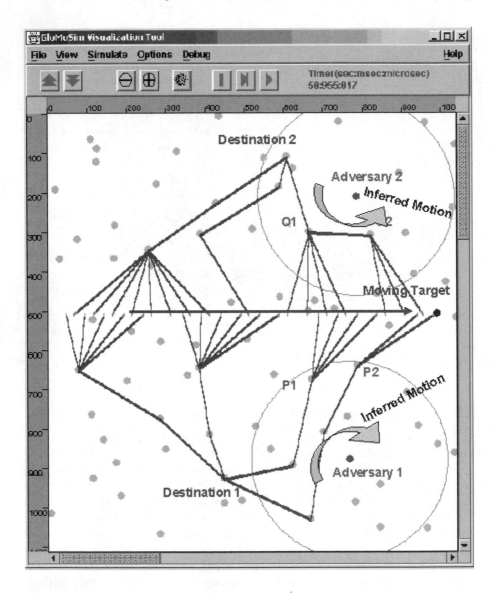

FIGURE 8.7
Illustration through simulation: 2 H-clique attacks (depicted nodes and ad hoc routes are from GloMoSim animation).

In Figure 8.7, *adversary 1* suggests a clockwise motion to its northwest, *adversary 2*, hearing the path migration from node Q1 to node Q2, figures out that the target is moving counterclockwise to its southwest. Combining these two pieces of information, the adversaries successfully discover that there is a motion cutting through between them. Through the case, we demonstrated that with a certain number of adversaries (which are capable of communicating with each other), in a bounded time, the motion pattern inference is possible.

8.6 A Countermeasure: The Anonymous On-Demand Routing Protocol

Many anonymous routing schemes have been proposed for MANET recently. Most of them adopt the on-demand routing approach [4, 21, 22, 38, 39, 40]. These protocols differ in the usage of cryptographic operations to anonymize both the transmission events and stored data, and in the mechanisms of establishing identity-free routes through route discovery.

The Anonymous On-Demand Routing protocol (ANODR) [21, 22] uses *anonymous virtual circuit* in routing and data forwarding to realize the principle of identity-free routing. In its design, no node identity is ever used in route discovery and data delivery. In addition, each ANODR node does not know its immediate upstream and downstream nodes. Instead, the node only knows the physical presence of neighboring ad hoc nodes. This is achieved by a special anonymous signaling procedure. The protocol is described below.

8.6.1 Route Discovery

The anonymous signaling procedure is implemented with the route discovery. The source creates an anonymous *global trapdoor* and the inner core of an *onion* [6, 34] in the route request (RREQ) packet. It then initiates the search for the destination by flooding the packet.

1. *Anonymous Global Trapdoor*: The global trapdoor is a (semantically secure [9]) encryption of a well-known tag message that can only be decrypted by the destination. Once the destination receives the flooded RREQ packet, it decrypts the global trapdoor and sees the well-known tag. But all other nodes see random bits after decryption. The design of the global trapdoor requires anonymous end-to-end key agreement between the source and the destination.

2. *Trapdoored Boomerang Onion (TBO)*: When the RREQ packet is flooded from the source to the destination, each RREQ forwarding node adds a self-aware layer to the onion (creating a trapdoor). Eventually the destination receives the RREQ with the multilayer onion. The destination broadcasts a route reply (RREP) packet with the onion. Only the right upstream node that produced the outermost layer of the onion is able to decrypt it (opening a trapdoor) and marks itself en route. This node strips off a layer of the onion and broadcasts the RREP with the updated onion. Eventually the RREP traces the onion layers and is forwarded back to the source and pings down the path. This signaling procedure resembles a boomerang bouncing back by the destination. Figure 8.8 depicts the creation and the use of the Trapdoored Boomerang Onion (TBO) in the signaling procedure, where $f_{K_A}(M)$ denotes encryption/decryption of

$$TBO_A = f_{K_A}(core)$$

$$TBO_B = f_{K_B}(f_{K_A}(core))$$

$$TBO_C = f_{K_C}(f_{K_B}(f_{K_A}(core)))$$

$$TBO_D = f_{K_D}(f_{K_C}(f_{K_B}(f_{K_A}(core))))$$

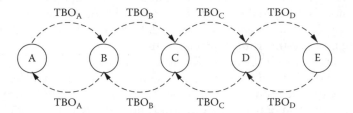

FIGURE 8.8
Trapdoored boomerang onion (TBO) between source A and destination E.

message M with symmetric key K_A using a symmetric encryption function f and "core" is a random nonce.

The actual ANODR route discovery design adds symmetric key agreement between two consecutive en route nodes together with TBO to produce one-time packet content in order to allow traffic mixing among the neighborhood. The packet formats are:

$$\langle RREQ, seq\#, global\_trap, onion, pk\_1time\rangle;$$

$$\langle RREP, \{K_{seed}\}_{pk\_1time}, f_{K_{seed}}(proof_{dest}, onion)\rangle.$$

Where, pk_1time is a one-time temporary public key, K_{seed} is a secret key shared between two consecutive RREP forwarders; hence two consecutive nodes en route, and $proof_{dest}$ is the cryptographic structure that shows the destination successfully opened the global trapdoor.

At RREQ phase, a RREQ upstream node (which is later the RREP downstream) puts its one-time public key pk_1time in the RREQ packet. The RREQ downstream node records this one-time public key for the source/destination session and overrides the field with its own one-time public key. Note that a node will reasonably generate many one-time public/private key pairs (pk_1time_X, sk_1time_X) at its idle time for different RREQ flooding.

At the RREP phase, the RREP upstream node (the RREQ downstream earlier) uses the stored one-time public key pk_1time to encrypt a pairwise per-hop session key K_{seed}, which is used to further encrypt some fields of the RREP packet. The next RREP receiver will be able to decrypt the encrypted contents using the recorded sk_1time_X and identify a unique route discovery session and get the per-hop session key K_{seed}. The session key K_{seed} also serves as the route pseudonym N, the identifier of the anonymous virtual circuit (Anonymous Circuit Identifier (ACI)) for this link. The need for the secret hop key between two neighboring RREP nodes is further justified in subsection 8.6.3. Each node records the incoming route pseudonym together with the outgoing route pseudonym and inserts the pseudonym pair to the route table (ACI table). The anonymous virtual circuit is established when the source receives the RREP with route discovery session information confirmed.

8.6.2 Anonymous Route Maintenance

Following the soft-state design, the routing table entries are recycled upon timeout T_{win} similar to the same parameter used in DSR and AODV. Moreover, when one link is broken due to mobility or node failures, the upstream node cannot forward the packet via the broken link. The upstream node can detect such anomalies when the retransmission count exceeds a predefined threshold. Upon anomaly detection, the node looks up the corresponding entry in its ACI table, finds the other ACI N' that is associated with the ACI N of the broken hop, and assembles an anonymous route error report packet of the format $\langle RERR, N' \rangle$. The node then recycles the table entry and locally broadcasts the RERR packet. A receiving node of the RERR packet looks up N' in its ACI table. If the lookup returns a result, the node concludes that it is on the broken route and should follow the same procedure to notify its neighbors. The RERR thus will eventually reach the source. The source will start a new route discovery.

8.6.3 Anonymous Data Forwarding

After the route discovery, an anonymous virtual circuit is established between the source and the destination. Intuitively, the route pseudonym N shared by the two ends of a link used as the Anonymous Circuit Identifier (ACI) in data packets:

$$\langle DATA, route\_pseudonym, payload \rangle.$$

Nodes hearing the packet must look up the route pseudonym in their ACI tables. A node discards the packet if the route pseudonym in the packet does not match any incoming ACI in its table. Otherwise, it changes the packet's route pseudonym field to the matched outgoing ACI, then acts as the current forwarder and locally broadcasts the modified packet. The procedure is then repeated until the data packet arrives at the destination.

A more sophisticated design is to use K_{seed} as the secret seed to generate cryptographically strong pseudorandom sequences and use the ith in the sequence as the route pseudonym ACI for the ith data packet. The ACI table updates itself for each sequence item. Such design ensures strong venue anonymity.

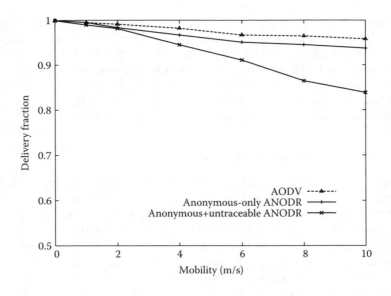

FIGURE 8.9
ANODR delivery fraction.

8.6.4 Analysis

The ANODR protocol has an impact on the performance of data delivery. AN-ODR uses public-key cryptographic operation for opening the global trapdoor and for processing RREP packets. It also introduces more bits in the control packets. In addition, ANODR cannot implement route optimization techniques specified in AODV and DSR standards, e.g., gratuitous route reply, etc. These facts will decrease the fraction of data packets that can be delivered to the destination successfully when a network is highly mobile. Figure 8.9 shows such a trend for mobile devices with relatively low CPU power. The comparison is made to the original AODV routing protocol. Two variants of ANODR are presented. Variant "Anonymous+untraceable ANODR" is the one described here, which achieves both ID anonymity and route untraceability at the cost of using public-key processing during RREP forwarding. Variant "Anonymous-only ANODR" is a simplified version that does not use a one-time public key and per-hop session key. It surely achieves ID anonymity. The figure shows the trade off between the performance and the degree of protection. Nevertheless, the design principles that ANODR follows allow it to achieve protection for the mobile anonymity. It hides the nodes' identities and also hides the relationship among nodes from each other, yet is able to successfully build up a route and forward packets. The anonymity protection it achieves is illustrated in Figure 8.6 through traceable ratio. With ANODR, the intruders cannot obtain more active route information other than themselves.

8.7 Summary

This chapter has presented an extensive study on privacy threats against mobile ad hoc networks. In addition to privacy required in an infrastructure network, a mobile ad hoc network should prevent its mobile members from being traced by a passive adversary. The network needs new mobile anonymity protections like (1) venue anonymity in addition to conventional identity anonymity, (2) privacy of ad hoc network topology, and (3) privacy of a node's motion pattern. Practical examples are given to illustrate the feasibility and effectiveness of many attacks that threaten the new privacy requirements. As a defense against the new anonymity threats in mobile ad hoc networks, on-demand and identity-free routing with neighborhood traffic mixing are presented as practical design principles. The Anonymous On-Demand Routing protocol is then described as an example of the countermeasures. The protocol protects not only nodes' identities but also the up- and downstream relationships. With ANODR, the intruders cannot obtain more active route information other than about themselves.

Bibliography

1. ATM Forum. Asynchronous Transfer Mode. http://www.atmforum.org, 2002.
2. S. Basagni, K. Herrin, E. Rosti, and D. Bruschi. Secure Pebblenets. In *MobiHoc*, pages 156–163, 2001.
3. A. R. Beresford and F. Stajano. Location Privacy in Pervasive Computing. *IEEE Pervasive Computing*, 2(1):46–55, 2003.
4. A. Boukerche, K. El-Khatib, L. Xu, and L. Korba. SDAR: A Secure Distributed Anonymous Routing Protocol for Wireless and Mobile Ad Hoc Networks. In *The 29th IEEE International Conference on Local Computer Networks (LCN04)*, 2004.
5. J. Broch, D. A. Maltz, D. B. Johnson, Y.-C. Hu, and J. Jetcheva. A Performance Comparison of Multi-Hop Wireless Ad Hoc Network Routing Protocols. In *ACM MOBICOM*, pages 85–97, 1998.
6. D. L. Chaum. Untraceable electronic mail, return addresses, and digital pseudonyms. *Communications of the ACM*, 24(2):84–88, 1981.
7. T. Clausen and P. Jacquet. Optimized Link State Routing Protocol(OLSR). Internet RFC 3626, http://www.ietf.org/rfc/rfc3626.txt, March 2005.
8. J. Deng, R. Han, and S. Mishra. Intrusion Tolerance and Anti-Traffic Analysis Strategies for Wireless Sensor Networks. In *IEEE International Conference on Dependable Systems and Networks (DSN)*, 2004.
9. S. Goldwasser and S. Micali. Probabilistic Encryption. *Journal of Computer and System Sciences*, 28(2):270–299, 1984.
10. M. Gruteser and D. Grunwald. Anonymous Usage of Location-Based Services Through Spatial and Temporal Cloaking. In *MobiSys03*, 2003.

11. Q. He, D. Wu, and P. Khosla. Quest for Personal Control over Mobile Location Privacy. *IEEE Communications Magazine*, 42(5):130–136, 2004.

12. C. Hedrick. Routing Information Protocol. http://www.ietf.org/rfc/rfc1058.txt, 1988.

13. X. Hong, J. Kong, and M. Gerla. A New Set of Passive Routing Attacks in Mobile Ad Hoc Networks. In *IEEE MILCOM*, 2003.

14. Y.-C. Hu, D. B. Johnson, and A. Perrig. SEAD: Secure Efficient Distance Vector Routing in Mobile Wireless Ad Hoc Networks. In *Fourth IEEE Workshop on Mobile Computing Systems and Applications (WMCSA'02)*, 2002.

15. Y.-C. Hu, A. Perrig, and D. B. Johnson. Ariadne: A Secure On-Demand Routing Protocol for Ad Hoc Networks. In *ACM MOBICOM*, pages 12–23, 2002.

16. Y.-C. Hu, A. Perrig, and D. B. Johnson. Packet Leashes: A Defense against Wormhole Attacks in Wireless Networks. In *IEEE INFOCOM*, 2003.

17. Y.-C. Hu, A. Perrig, and D. B. Johnson. Rushing Attacks and Defense in Wireless Ad Hoc Network Routing Protocols. In *ACM WiSe'03 in Conjunction with MOBICOM'03*, pages 30–40, 2003.

18. Y.-C. Hu and H. J. Wang. A Framework for Location Privacy in Wireless Networks. In *ACM SIGCOMM Asia Workshop*, 2005.

19. D. B. Johnson and D. A. Maltz. Dynamic Source Routing in Ad Hoc Wireless Networks. In T. Imielinski and H. Korth, editors, *Mobile Computing*, volume 353, pages 153–181. Kluwer Academic Publishers, 1996.

20. Y.-B. Ko and N. Vaidya. Location-Aided Routing (LAR) in Mobile Ad Hoc Networks. In *ACM MOBICOM*, pages 66–75, 1998.

21. J. Kong. *Anonymous and Untraceable Communications in Mobile Wireless Networks*. Ph.D. thesis, University of California, Los Angeles, June 2004.

22. J. Kong and X. Hong. ANODR: Anonymous On Demand Routing with Untraceable Routes for Mobile Ad-Hoc Networks. In *ACM MOBIHOC'03*, pages 291–302, 2003.

23. J. Kong, X. Hong, M. Sanadidi, and M. Gerla. Mobility Changes Anonymity: Mobile Ad Hoc Networks Need Efficient Anonymous Routing. In *The Tenth IEEE Symposium on Computers and Communications (ISCC)*, 2005.

24. J. Kong, J. Liu, X. Hong, and M. Gerla. Toward Efficient Solutions to Resist Mobile Traffic Sensors: How Much Performance Cost is Paid by On-Demand Anonymous Routing Protocols. In *International Workshop on Research Challenges in Security and Privacy for Mobile and Wireless Networks (WSPWN 06)*, pages 61–70, March, 2006.

25. J. Li, J. Jannotti, D. De Couto, D. Karger, and R. Morris. A Scalable Location Service for Geographic Ad Hoc Routing. In *ACM MOBICOM*, pages 120–130, 2000.

26. J. Liu, J. Kong, X. Hong, and M. Gerla. Performance Evaluation of Anonymous Routing Protocols in MANETs. In *IEEE Wireless Communications and Networking Conference 2006, Las Vegas*, NV, April, 2006.

27. S. Meguerdichian, F. Koushanfar, G. Qu, and M. Potkonjak. Exposure in Wireless Ad Hoc Sensor Networks. In *ACM Proc. of 7th Annual International Conference on Mobile Computing and Networking (MobiCom '01)*, 2001.

28. J. Moy. OSPF Version 2. http://www.ietf.org/rfc/rfc1131.txt, 2005.

29. R. Ogier, F. Templin, and M. Lewis. Topology Dissemination Based on Reverse-Path Forwarding (TBRPF). Internet RFC 3684, http://www.ietf.org/rfc/rfc3684.txt, March 2005.

30. V. D. Park and M. S. Corson. A Highly Adaptive Distributed Routing Algorithm for Mobile Wireless Networks. In *IEEE INFOCOM*, pages 1405–1413, 1997.
31. C. E. Perkins and P. Bhagwat. Highly Dynamic Destination-Sequenced Distance-Vector Routing (DSDV) for Mobile Computers. In *ACM SIGCOMM*, pages 234–244, 1994.
32. C. E. Perkins and E. M. Royer. Ad-Hoc On-Demand Distance Vector Routing. In *IEEE WMCSA'99*, pages 90–100, 1999.
33. A. Pfitzmann and M. Köhntopp. Anonymity, Unobservability, and Pseudonymity—A Proposal for Terminology. In H. Federrath, editor, *Designing Privacy Enhancing Technologies; Workshop on Design Issues in Anonymity and Unobservability (DIAU'00)*, June 2000.
34. M. G. Reed, P. F. Syverson, and D. M. Goldschlag. Anonymous Connections and Onion Routing. *IEEE Journal on Selected Areas in Communications*, 16(4), 1998.
35. M. K. Reiter and A. D. Rubin. Crowds: Anonymity for Web Transactions. *ACM Transactions on Information and System Security*, 1(1):66–92, 1998.
36. Y. Rekhter and T. Li. A Border Gateway Protocol 4 (BGP-4). http://www.ietf.org/rfc/rfc1771.txt, 2005.
37. K. Sanzgiri, B. Dahill, B. N. Levine, C. Shields, and E. Royer. A Secure Routing Protocol for Ad Hoc Networks. In *10th International Conference on Network Protocols (IEEE ICNP'02)*, 2002.
38. R. Song, L. Korba, and G. Yee. AnonDSR: Efficient Anonymous Dynamic Source Routing for Mobile Ad-Hoc Networks. In *ACM Workshop on Security of Ad Hoc and Sensor Networks (SASN)*, 2005.
39. Y. Zhang, W. Liu, and W. Lou. Anonymous communications in mobile ad hoc networks. In *IEEE INFOCOM'05*, 2005.
40. B. Zhu, Z. Wan, M. S. Kankanhalli, F. Bao, and R. H. Deng. Anonymous Secure Routing in Mobile Ad-Hoc Networks. In *29th IEEE International Conference on Local Computer Networks (LCN'04)*, pages 102–108, 2004.

9

Security Issues in the IEEE 802.15.1 Bluetooth Wireless Personal Area Networks

Yang Xiao, Daniel Kay, Yan Zhang, Tianji Li and Ji Jun

CONTENTS

Abstract   In this chapter, we provide a survey of security issues in the IEEE 802.15.1 Bluetooth wireless personal area network. Security aspects and security flaws are identified, and security enhancements are presented.

9.1   Introduction

In May 1998, the Bluetooth Special Interest Group (SIG) was formed with founding members such as Ericsson, Nokia, Intel, IBM, and Toshiba. Bluetooth SIG has continued to grow so that today, the SIG has over 4,000 members in the telecommunications, computing, automotive, music, apparel, industrial automation, and network industries [2]. Bluetooth's name was inspired by the Danish King Harald Bluetooth, known for unifying Denmark and Norway in the 10th century. Bluetooth was further standardized in the IEEE 802 working group to become the IEEE 802.15.1 standard [1], which defines the Physical (PHY) layer and Medium Access Control (MAC) layer, including the lower transport layers of Logical Link Control and Adaptation Protocol (L2CAP), Link Manager Protocol (LMP), baseband, etc.

Bluetooth wireless technology uses a radio link that is optimized for power-conscious, battery-operated, small size, and lightweight personal devices. Bluetooth operates at an unlicensed, 2.4-GHz industrial, scientific and medical (ISM) band, and adopts a fast frequency-hopping in order to minimize interference, handling voice and data communications between Bluetooth devices, with very limited range, usually less than 10 meters. A Bluetooth Wireless Personal Area Network (WPAN) supports a synchronous communication channel for voice-type communication, and an asynchronous channel for data communication. A Bluetooth device may be configured to handle both types of communication channels over the same time interval [1]. Data traffic is transmitted unidirectionally and is limited to 723.2 kb/s, and voice traffic is bidirectional and is limited to 64 kb/s [1]. Bluetooth packets are smaller than those higher-layer counterparts so that higher-layer packets must be broken into smaller packets before they can be transmitted [1]. A packet consists of three main parts: access code, header, and payload.

This chapter is a survey of security issues in the IEEE 802.15.1 standard (version 1.1), which is also referred to as Bluetooth standard. Before going into details of Bluetooth, we first introduce some concepts as follows [1]. The Bluetooth baseband is the layer that determines the MAC and PHY layers' procedures used to support the exchange of data between Bluetooth devices. Link establishment is a procedure for creating a link on the LMP. Piconet is a group of Bluetooth devices that share a common channel. A piconet includes one master and multiple slaves. When a master and a single slave share a point-to-point link in a piconet, it is called a Synchronous Connection-Oriented (SCO) link. Authentication is a process of verifying a device using a known procedure. Authorization is to grant access to a specific service to a Bluetooth device by a user or by a user-defined rule. Creation of a secure connection is a procedure for creating a connection that includes authentication and encryption. A link key is a 128-bit random number that is used in the authentication process and as a parameter when deriving the encryption key. LMP authentication is a link management-level procedure verifying and identifying a remote device via a challenge-response mechanism using a combination of a random number, a secret key, and the Bluetooth device address. LMP pairing is a procedure to authenticate two devices using a personal identification number (PIN), creating a common link key used as the basis for a secure connection. A paired device is a Bluetooth device with a link key that has been exchanged with another device.

The rest of the chapter is organized as follows. Section 3 introduces link layer security. Section 4 presents LMP. We provide control interface and a generic access profile in Section 5 and Section 6, respectively. Security flaws and enhancements are presented in Section 7 and Section 8, respectively. Finally, Section 9 outlines our conclusions.

9.2 Link Layer Security

9.2.1 Overview

To provide secure peer-to-peer communications, a Bluetooth system must provide security functionality at the application layer and the link layer. In order to provide usage protection and information confidentiality, the system uses four entities for maintaining security at the link layer, listed in Table 9.1 [1]. These entities are as follows: a public address that is unique for each user, two secret keys (private user keys), and a random number (RAND). The random number is different for each new transaction.

The Bluetooth device address (BD_ADDR) is unique for each Bluetooth device. It is a 48-bit IEEE address that is publicly known, and can be received using any of several methods [1]. One of the secret keys is for the authentication algorithm and is randomly generated with 128 bits. For the encryption algorithm, another of the secret keys is used and its key size may vary between 8 bits and the full 128 bits. There are two reasons to vary the encryption key size. First, there are many different restrictions imposed on cryptographic algorithms in various countries. Second, it allows strengthening security without the need for expensive replacement of encryption algorithms and encryption hardware. The lifetime of the encryption key is not necessarily the same as the lifetime of the authentication key. A new encryption key is generated every time encryption is needed [1].

The RAND is a number generated from a pseudorandom process. All Bluetooth devices have a random number generator, which is used for many security-related functions: challenge-response schemes, generating authentication keys, and encryption keys, etc. For Bluetooth, the requirements for a number to be classified as a random number are that it be nonrepeating and randomly generated. This means that the number should be highly unlikely to be generated more than once in the lifetime of the authentication key

TABLE 9.1

Link Layer Security Entities

Link Layer Security Entity	Size
BD_ADDR	48 bits
Private user key	128 bits
Private user key, encryption configurable length	8-128 bits
RAND	128 bits

and that it is not probable to predict its value with a probability significantly greater than 0.

The key size is a factory-set value. In order to prevent the changing of the permitted key size by a user, the Bluetooth baseband processing does not accept an encryption key from a higher software layer. The changing of a link key should only be done through a defined procedure. The procedure to change a key depends on the type of key [1].

9.2.2 Link Keys

A link key is a 128-bit random number that can be either semipermanent or temporary. It is shared among two or more parties. The link key is used in the authentication routine and is the basis of all secure transactions. A semipermanent link key can be used in the authentication of many connections between Bluetooth devices that share it. It is stored in a nonvolatile memory. A temporary link key has only the lifetime of the current session.

The five types of link keys are defined as:

- Combination key K_{AB}
- Unit key K_A
- Temporary key K_{master}
- Initialization key K_{init}
- Encryption key K_c

The encryption key is generated when the LM command is activated. The encryption key is derived from the current link key. Because there are restrictions on the strength of encryption algorithms but not on the strength of the authentication algorithms, the authentication and encryption keys are different. This allows the encryption key to be shorter without compromising the security of the authentication algorithm. The only difference between the combination key and the unit key, with regard to Bluetooth, is how they are generated. The unit key is generated when the Bluetooth device is installed. The combination key is generated when two Bluetooth devices are connected. The combination key is shared between each pair of Bluetooth devices. The key size of a combination key depends on the Bluetooth device [1].

The master key (i.e., temporary key), which is a link key, is only used during the current session and only used during initialization. The main purpose of this key is for when a master wants to use the same encryption key when transmitting to two or more slaves. The master key is generated by using a random number, an L-octet PIN code, and a BD_ADDR. The PIN can be either a fixed number provided with the Bluetooth device or set by the user. A PIN can be from 1 to 16 octets in length. The larger the PIN is, the more secure it is. The sharing of large PINs is usually handled by software at the application layer [1].

9.2.3 Security Procedures

In order for the Bluetooth device to use link keys during the authentication procedure, the link keys have to be generated and shared between Bluetooth devices. Link keys must be kept secret, so the only way to exchange link keys is through the initialization process. The initialization process consists of five parts:

- Generation of the initialization key
- Generation of the link key
- Link key exchange
- Authentication
- Generating of an encryption key in each unit (optional)

After a link key is shared between two Bluetooth devices, it does not have to be regenerated for the next connection between the two devices. A new encryption key will be generated upon the next connection; hence an encryption key is not reused between the two devices for a new connection. The LM command will automatically start the initialization procedure if no link key is available [1].

The initialization key is the link key used temporarily during the initialization process. The initialization key is generated by the E_{22} algorithm, shown in Figure 9.1, using a BD_ADDR, a PIN code, the length of the PIN code (in octets), and a random number. If one of the Bluetooth devices has a fixed PIN, the nonfixed PIN is used. If both devices have fixed PINs, they cannot

E_{21}: $\{0, 1\}^{128} \times \{0, 1\}^{48} \rightarrow \{0, 1\}^{128}$
   (RAND, address)$| \rightarrow A\hat{l}_r(X, Y)$

For mode 1:
$X = RAND[0...14] \cup (RAND[15] \oplus 6)$
$Y = \overset{15}{\underset{i=0}{!}} address[i(\bmod 6)]$
Let L equal the number of octets in the user PIN,
$PIN' = \begin{cases} PIN[0...L-1] \cup BD\_ADDR[0...\min\{5, 15-L\}, L < 16 \\ PIN[0...L-1, L = 16 \end{cases}$

For mode 2:
E_{22}:$\{0, 1\}^{8L'} \times \{0, 1\}^{128} \times \{1, 2, ..., 16\} \rightarrow \{0, 1\}^{128}$
$(PIN', RAND, L')| \rightarrow A'_r(X, Y)$
$\begin{cases} X = \overset{15}{\underset{i-0}{!}} PIN'[i(\bmod L')] \\ Y = RAND[0...14] \cup (RAND[15] \oplus L') \end{cases}$
$L' = \min\{16, L + 6\}$

FIGURE 9.1
Calculation of E_{21} and E_{22}.

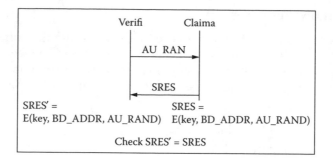

FIGURE 9.2
Challenge-response for symmetric key systems.

be paired. The output of E_{22} (Figure 9.1) is a 128-bit number. After the link keys have been exchanged, the initialization key is discarded. The application has the responsibility to take countermeasures against a Bluetooth device that tries to use a large number of PINs each time by using a different BD_ADDR [1].

Authentication in a Bluetooth network is handled by a challenge-response scheme, shown in Figure 9.2. In the challenge-response scheme, the knowledge of the secret key is verified through a two-step protocol using a symmetric secret key. In this scheme, the verifier challenges the device that wants to be authenticated, to authenticate a random input (AU_RAND). This device generates an authentication code (SRES) and sends it to the verifier. In order to generate the SRES, three pieces of information are needed: the AU_RAND, the Bluetooth device address (BD_ADDR), and the shared secret key. The shared secret key is the current link key. The BD_ADDR is used to protect against a simple reflection attack [1].

In Bluetooth authentication, the application determines who has to authenticate whom. The application may require only the claimant to authenticate, or it may require mutual authentication. If an authentication attempt fails, a given wait interval must pass before the verifier will start a new authentication attempt with the same claimant. The verifier will also adhere to this same wait period before it will respond to the authentication attempt by a device claiming the same BD_ADDR as the one that failed the authentication attempt.

Bluetooth uses a computationally secure authentication function. The encryption function used by Bluetooth is called SAFER+, which is an enhanced version of a 64-bit block cipher called SAFER-SK128. It consists of a set of eight layers, or rounds, and separate mechanisms for generating round keys. The output of this function is a 128-bit result, which is produced from a 128-bit pseudorandom string and a 128-bit key. The result consists of an encryption with a round key, substitution, encryption with the next round key, and a Pseudo-Hadamard Transform (PHT) [1].

The key used for authentication is derived using the function E_2. This function has two modes of operation. The first mode (E_{21}) produces a 128-bit link key when the input is a 128-bit random number and a 48-bit address. This mode is used when creating unit and combination keys. In the second mode (E_{22}), the input has to be a 128-bit random number, an L octet user PIN, and a 128-bit link key. The mode is used to create the initialization and master keys, when a master key is generated. The function of Ar is identical to SAFER+.

The combination key, if desired, is first generated during the initialization process. The combination key is the result of combining two numbers generated in the two devices that are involved in the initialization process. Each device produces a number that is used to make the combination key. The number produced by a device is generated using the E21 function using the random number and the device's BD_ADDR. The random numbers generated by the two devices are securely exchanged by XORing with the current link key. Then, each device computes the other's contribution to become the combination key. The devices combine the two numbers together to produce a 128-bit link key. The combining of the two parts is done with an XOR operation. The old link key is discarded, and the combination key becomes the new link key.

There are a few ways to use encryption keys in a point-to-multipoint configuration. In the first method, it is possible for the master to use a separate encryption key for each slave, but the master would have to encrypt, using an individual encryption key, and send a message to each slave device. That would be very inefficient and could allow an intruder access to a ciphertext/plaintext pair, which could be used for cryptanalysis. In the second way, the master and slave devices can share a common encryption key. However, since it is not possible for Bluetooth slave devices to switch encryption keys in real time, the master device can send a message to each slave device and tell it to use the common link key and therefore use a common encryption key [1].

In order to generate a master key, K_{master}, which will replace the link key during an initial session shown in Figure 9.3, two random numbers are generated and are passed as arguments to the E22 function. The output of this function is a 128-bit random number. The use of the function output and not a straight random number help to ensure against a poorly implemented random number generator.

$$K_{Master} = E_{22}(RAND1, RAND2, 16) \tag{9.1}$$

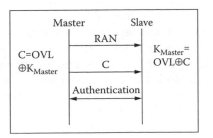

FIGURE 9.3
Master link key distribution.

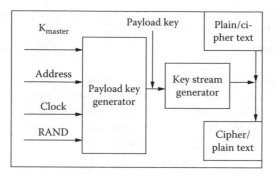

FIGURE 9.4
Stream ciphering for Bluetooth with E_0.

The slave device is sent a third random number that it uses in the E_{22} function with the current link key to calculate a 128-bit overlay. The output of the function, also computed by the master device, is XORed with the new master key. This output is then sent to the slave device, which does a bitwise XOR on it and now has the new master key. After a mutual authentication procedure using the new link key, the authenticated ciphering offset (ACO) values are not discarded, so that the master device may return to the previous link key. The new encryption key, KC, is then computed by each slave device. The value of the ciphering offset number (COF) is derived from the BD_ADDR of the master. This encryption key cannot be used for broadcasts until the master has ensured that the slave device needs it [1].

$$K_C = E_3(K_{Master}, EN\_RAND, COF) \qquad (9.2)$$

The stream cipher system E_0, shown in Figure 9.4, consists in the initialization of the payload key, the generation of the key stream bites, and the final encryption/decryption. The payload key generator combines and places the input bits in order and shifts them into the four linear feedback shift registers, LFSR, used in the key generation. The key stream bits are generated using the summation stream cipher by Massey and Rueppel [3].

Each Bluetooth device has a parameter, which defines the maximum key length, from 1 to 16 octets. For an application that uses the Bluetooth device, there is an acceptable minimum key length. A negotiation between two Bluetooth devices must occur in order for the key length to be decided on. The first step in this negotiation is that the master sends a suggested key length to the slave, which is always the largest key that the master can handle. If the slave can handle the key length, the slave would respond with the key length. If not, it should respond with the largest key length that it can handle. This key size will be used unless the application denies the key length because it is too small. An intruder could try to force the use of a weak protection by claiming it can only use a small maximum key length [1].

A stream cipher algorithm is used for the encryption routine. In the algorithm, the ciphering bits are bitwise modulo-2 and then added to the data stream. Each payload packet is encrypted separately using the master BD_ADDR, 26 bits of the master real-time clock and the encryption key K_C. As stated earlier, the encryption key is derived from the current link key, COF, and a random number EN_RAND. The random number EN_RAND is publicly known because it is broadcast as the plain text. The encryption key is modified to be the appropriate size. In order to ensure that at least one bit changes between transmissions, the real-time clock is incremented for each slot. The E_0 algorithm is reinitialized with each new packet for both the master and slave device. Since the cipher is symmetric, decryption is performed in exactly the same way [1].

9.3 Link Manager Protocol

The LMP is used for link setup and control. The LMP interprets and filters the signals received that are not propagated to higher layers. LMP message may be used for three things: link setup, security, and control. They are transmitted in the payload and have a higher priority than user data. The messages in LMP, since the link controller (LC) provides a reliable link, do not have to be acknowledged. The LC does not guarantee the time taken to deliver a message or the delay between the delivery of the message to the remote device and the reception of the corresponding ACK by the sender. It only guarantees that it will attempt to communicate once per poll interval slot [1].

The LM protocol data units (PDUs) are always single-slot packets so the payload header is 1 byte. The two least significant bits determine the logical channel. The channel determines whether the packet is the start of a L2CAP message, a continuation of an L2CAP message, or an LMP message. By examining the active member address (AM_ADDR) in the packet header, the source/destination of the PDUs is determined [1].

For authentication, shown in Figure 9.5, a challenge-response scheme is used. The verifier sends an LMP_au_rand PDU, which contains the challenge,

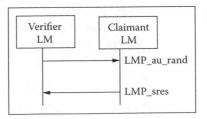

FIGURE 9.5
Authentication: Claimant has link key.

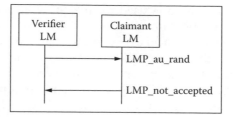

FIGURE 9.6
Authentication fails: Claimant has no link key.

a random number, to the claimant. The random number is input into a function, along with the claimant's BD_ADDR and a secret key, and the output is the response to the challenge. The response is sent back for verification. A successful calculation of the authentication response requires the prior knowledge of a shared secret key. A verifier can be either a master or a slave device. The claimant can challenge the verifier but only after a successful response to the verifier. The claimant can only calculate the correct challenge response, LMP_sres, if it has the current link key. If the claimant does not have a link key associated with the verifier, the claimant responds with an LMP_not_accepted with the *key missing* reason code that means that link key is missing. If the authentication fails, the scheme mentioned earlier is used to prevent an intruder from using a large number of keys in a relatively short period of time. This scheme will help prevent a denial-of-service attack [1].

When two devices do not have a common link key, another method of authentication is attempted. The initialization key (K_{init}) is created using a PIN, a random number, and a BD address. Once the initialization key is calculated, the link key is created. Only then can mutual authentication be made. This link key will be used for all future authentications until it is changed [2]. If the authentication fails after creation of a link key because of a wrong authentication response, the earlier-mentioned scheme is used to prevent an intruder from using a large number of PINs in a short period of time [3].

If the current link is a temporary link key and the link key is derived from combination keys, the link key may be changed. If the link key is a unit key, the unit goes through a pairing procedure, previously stated, in order to change the link key. If there is a successful change in the link key, the old link key is discarded and the new link key is used until it is changed. The link key can be changed permanently or it can be changed for the current session only.

After the authentication, the master and slave must decide whether to use encryption and if that encryption is for point-to-point packets or if it is for both point-to-point and broadcast packets. Once master and slave agree on the encryption mode, the master gives more details about the encryption. The next step is to determine encryption key size [1].

There are three steps to the start of encryption [1]:

- Master is set up to transmit unencrypted packets, but to receive encrypted packets.
- Slave is set up to transmit and receive encrypted packets.
- Master is set up to send and receive encrypted packets.

There are three steps in order to stop encryption [1]:

- Master is set up to transmit encrypted packets, but to receive unencrypted packets.
- Slave is set up to transmit and receive unencrypted packets.
- Master is set up to transmit and receive unencrypted packets.

In order to change the encryption mode, the encryption key, or random number, encryption must be first stopped and then restarted with new parameters as stated earlier. The change in encryption mode occurs as a device goes from point-to-point transmitting to point-to-broadcast transmitting [1].

9.4   Control Interface

The control interface for IEEE 802.15.1-2002 is based on the host controller interface (HCI) section of the Bluetooth specification. The HCI provides the command interface to the baseband controller and link management. It also provides the access to hardware status and control registers. The HCI section defines a basis for a physical interface for a Bluetooth external module. In addition, the control functions are necessary for all Bluetooth implementations. The IEEE 802 standards describe protocols and not implementation. The HCI provides a uniform command method for accessing the Bluetooth hardware capabilities. It provides the host with the ability to control the link layer connection to other Bluetooth devices. HCI Policy commands control the behavior of the local and remote LM. The HCI Command packet sends commands from the host to the host controller. The HCI Event packet is used to notify the host when an event occurs. It is sent by the host controller [1].

The Link Control commands allow the host controller to control connections to other Bluetooth devices. The Link Control commands allow the LM to create and modify the link layer connected with Bluetooth devices that are part of the Bluetooth piconets and scatternets [1]. A Bluetooth piconet is a network of Bluetooth devices in which one of the Bluetooth devices acts as the master and the remainder of the devices are slave devices. The frequency-hopping channel of a piconet is determined by the master of the piconet. A Bluetooth scatternet is a network of overlapping piconets. In a scatternet, a Bluetooth device in a scatternet can belong to one or more of the piconets and it can be the master in only one of the scatternets and a slave in another scatternet [1].

9.5   Generic Access Profile (GAP)

The generic procedures related to the discovery of Bluetooth devices are defined by the Generic Access Profile, GAP. GAP also defines the procedures to the use of different security levels. A user of a Bluetooth device should be able to connect to any other Bluetooth device, even if the two devices do not share a common application [1].

The GAP provides the following fundamental functions [1]:

- States the requirements on names, values, and coding schemes.
- Defines modes of operation that are not service or profile specific.
- Defines the general procedures that can be used for discovering identities, names, and basic capabilities of other Bluetooth devices. The other Bluetooth devices have to be in a mode such that they can be discovered.
- Explains the basic procedure for how to create a dedicated exchange of link keys between Bluetooth devices.
- Explains the basic procedure that can be used for establishing connections to other Bluetooth devices that are in a mode that allows them to accept connections and service requests.

The BD_ADDR is a 48-bit unique address of a Bluetooth device. It is received by a remote Bluetooth device during the discovery procedure. The Bluetooth PIN is used to authenticate two Bluetooth devices, if they do not already share a link key, to each other. The PIN can be entered at the user interface level, or it may be stored in the device. If the PIN can be entered at the user interface level, then an intruder could create an application that can change the PIN of a Bluetooth device in order to try to gain unauthorized access to a Bluetooth device or network [1].

9.6   Security Flaws

The popularity of Bluetooth devices makes the security flaws in Bluetooth more important. There are several flaws that have been detected in Bluetooth security using a schema called VERDICT [4]. These flaws can be broken down into four categories:

- Improper Validation
- Improper Exposure
- Improper Randomness
- Improper Deallocation

Improper validation can be broken down into five subcategories. These subcategories are:

- Device Address Validation
- Invalid State (Link Control)
- Invalid State (Encryption Modes)
- Encryption Keys
- Link Keys

The address for a Bluetooth device is similar in format to that of an 802.3 address. A Bluetooth address has to be unique for it to be secure, but if a user is allowed to change the address in hardware, then an individual address can be spoofed similar to the way IP addresses can be spoofed. Address spoofing can occur since there is no address validation. There is no checking of the address of a Bluetooth device. Using address spoofing, a spoofed device was able to create to a piconet with the authentic device and the master-slave switch was also made between the two devices with the same address [4].

In the Bluetooth controller, there are two major states, standby and connection. There are also seven substates: page, page scan, inquiry, inquiry scan, master response, slave response, and inquiry response. These seven substates are stored in three bits. Since three bits are capable of storing eight values, the Bluetooth device must ensure that the eighth state, which would be an invalid state, is never entered. However, if this invalid state is entered, the Bluetooth device's state machine must have a way to enter the correct state.

Once a slave device receives a master key, a Bluetooth device can be in one of three encryption states. These states are listed in Table 9.2.

A Bluetooth device's design could use two bits to hold the mode of encryption for the device. If a Bluetooth device entered state four, then broadcast transmissions would be encrypted and point-to-point traffic would not be encrypted. This state would allow any Bluetooth device in the receiving area to capture all data traffic and allow an intruder unrestricted access to all information transmitted [4].

According to the Bluetooth standard, the master cannot use different encryption keys for broadcast and individual transmissions. The Bluetooth device, which is the master, can request all slave devices to use the same link key. This ability could allow an intruder to decipher and use only one link key and hence one encryption key, which could be used to capture transmitted

TABLE 9.2

Possible Encryption Modes for a Slave with a Master Key

State Number	Broadcast Traffic	Unicast Traffic
1	No encryption	No encryption
2	No encryption	Encryption (master key)
3	Encryption (master key)	Encryption (master key)
4 (invalid state)	Encryption (master key)	No encryption

data for all devices in the piconet. Another weakness in encryption keys is noticeable when a Bluetooth device needs to know what encryption key size can be used by another device. An intruder could configure his device to claim it was only capable of using the minimum key size, so forcing other devices in the network to use the small key size [4].

Authentication and encryption operate on the assumption that the link key is a shared secret. However, if a Bluetooth device uses its unit key as a link key with more than one other Bluetooth device, one of these devices can compute the encryption key, shared between two other devices, by using one of those device's addresses [4]. Another vulnerability of link keys is the assumption that the link key is shared with only one other Bluetooth device. If an intruder's device has previously interacted with a device, the intruder's device can listen to the traffic between other Bluetooth devices in the range. For most devices, an intruder's device would have to be within 10 meters but there have been some devices built recently that are capable of eavesdropping on Bluetooth devices up to 1.1 miles away [5].

Another vulnerability of Bluetooth involves the switching of a master and slave device. When the master-slave switch starts, encryption, if used, on the old piconet is disabled. If an intruder poses as the new master, sends a signal saying that it wants to start and manage its own piconet, the intruder can eavesdrop on the piconet until the old master finishes transferring timing information [4].

A Bluetooth device uses a PIN in generating security keys. The PIN can be a fixed string of numbers set by the manufacturer or can be a user/application set value. The device could also not have a PIN; in this case, the device would use a default value of zero in generating the keys. If a PIN is small or zero, a brute force search for the initialization keys can be done [4].

Improper deallocation may occur if the Host Controller does not issue a disconnect command after an authentication failure occurs. An intruder would have an easier time gaining access because the intruder would not need to authenticate after spoofing an encryption link [4].

9.7   Security Recommendation

In Reference [6], two general recommendations are made: use combination keys instead of unit keys, and pair Bluetooth devices in an area as secure as possible and use long random passkeys. A unit that uses its unit key is only able to use one key for all its "secure" connections. The Bluetooth device has to share its unit key with other "trusted" devices, so any of these "trusted" devices would be able to eavesdrop on any communications. The calculating of Bluetooth keys is not computationally complex. If an intruder is able to capture all transmissions between two devices while these devices are pairing, the intruder would be able to compute all possible keys in order to compute the link key. With a longer key, the calculation of these keys will take longer. Since Bluetooth are vulnerable while pairing, it is recommended to only pair Bluetooth devices in a secure area [6].

9.7.1 Service Discovery Application Profile

The Service Discovery Application Profile describes the features and procedures used to determine what services other Bluetooth devices are offering while using the Bluetooth Service Discovery Profile (SDP) [6]. SDP is insecure because it does not require authentication or encryption for its transactions. The main problem with the way SDP works is that if a device offers the service of connecting to a LAN, the SDP would inform any user of this fact. This information could be used by an intruder to penetrate the Bluetooth device's security to gain access to the LAN. The service record only provides information about the services offered, not how to access these services [6].

9.7.2 Bluetooth Headset Profile

Bluetooth-enabled headsets, cell phones, and PDAs have been made possible because of the security options offered by the Bluetooth Baseband specifications. The use of the Bluetooth passkey is an important part of Bluetooth security [6]. The passkeys are used as part of all authentication and encryption. The proper use of passkeys can prevent the illegal use of a lost or stolen headset. To use a Bluetooth headset, an audio gateway is needed. An audio-gateway is a device that is capable of sending audio data to the headset for playback. It is recommended that both the headset and the audio gateway both store the necessary passkeys and link keys. These two items would allow for a secure channel for the audio data. The Bluetooth SIG recommends that security mode three is used for the headset. Security mode three requires authentication each time the headset has a connection to set up [6].

Since most headsets do not have a user interface that allows the changing of the headset's passkey, the headset passkey can be changed using a device with an adequate user interface (i.e., PDA, laptop, cell phone, etc.). The passkey should be changed only after making a secure connection using the manufacturer's randomly generated passkey that is stored in nonvolatile memory. This passkey must be highly secured, to prevent tampering. The original passkey would always remain stored in the headset, and would allow users to reset the passkey if they have lost or forgotten the passkey, but only if they have the original paperwork. The pairing of a headset with an audio gateway should only be allowed when the headset is in a pairing mode and when the user is in a secure location. A passkey length of 128 bits is recommended for maximum security. The Bluetooth SIG also recommends that the headset use combination keys for its connections. They also recommend that the audio gateway be capable of storing link keys in tamper-resistant, nonvolatile memory [6], shown in Figure 9.7.

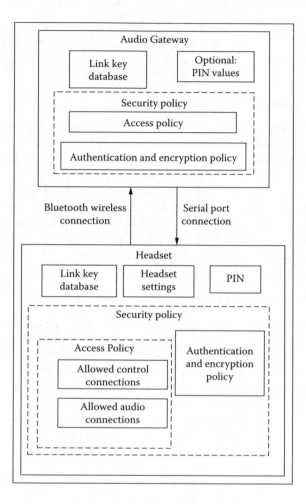

FIGURE 9.7
Headset security architecture [6].

9.7.3   Dial-Up Network Profile

The Bluetooth wireless technology allows two or more units to connect via authentication and encryption using mechanisms in the Bluetooth Baseband Specification. The link keys, shared between the devices, determine the security level of the connection. The passkey used in the connection must be entered by the user or be available to the devices by some other means. There are two types of devices defined in the dial-up networking profile: gateway and data terminals. A gateway device is a device that allows access to a public network. The data terminal is the device that uses the dial-up service of a gateway. The Dial-Up Network profile allows for only one connection between a gateway and a data terminal. The typical DNP is given in Figure 9.8 [6].

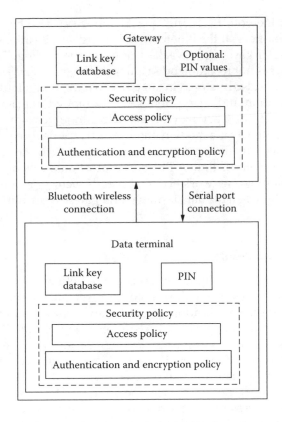

FIGURE 9.8
Dial-up networking security architecture[6].

The security configuration depends on the device being connected. A modem, which is not usually capable of advance security features, does not have a user interface, so its security settings are determined by using a data terminal. On the other hand, a cellular phone would have more robust security options and have its own user interface for setting security options. The Bluetooth Baseband authentication and encryption mechanisms should be used to secure the connection. To help ensure maximum security, both the gateway and the data terminal should store the link keys. Data terminals usually do not have fixed Bluetooth passkeys. Gateways can have fixed or nonfixed passkeys. It is recommended that the pairing of a data terminal and a gateway should only be possible while both devices are in a pairing mode. This pairing mode helps limit the time where an intruder could pair with one of the devices when it is not desired. The entering of passkeys in a data terminal and gateways can be handled by a few methods. If the gateway used a fixed passkey, the user would have to enter the passkey from the gateway into the data terminal. If the gateway is nonfixed, the user will have to generate a new passkey and enter it into the two devices. The passkey can be generated by an application on the data terminal by the user and then by the user into the gateway. The Bluetooth Baseband specifications can also be used to generate and store passkeys. If the Bluetooth Baseband specifications are used, the initial exchange of keys is the weakest point of the pairing. The use of unit keys as link keys is not recommended, but combination keys should be used. These keys should also be stored in tamper-resistant, nonvolatile memory [6].

For authentication and encryption, security mode three is recommended for the data terminal and gateway, but security mode two may be used for the gateway. If the gateway is a modem, then security mode three should be used for maximum security. If security mode two is used, the connection is only secure over the Bluetooth link. A PPP authentication mechanism can be used for authentication by a network access server. If security mode three is used, all connections between the data terminal and gateway are authenticated and encrypted [6].

9.7.4   LAN Access Profile

The Bluetooth wireless technology can provide both authentication and encryption for LAN access using the Bluetooth Baseband specifications [6]. The mechanisms that the specifications provide can be used to connect both personal and nonpersonal devices. A LAN access device, called a LAN access point (LAP), is the perfect example of a device that may be connected to many different types of devices. For a LAN access device, many different layers of security mechanisms will be used, most of them should be used to strengthen each other, not to replace one another. The Serial Access Profile and the General Access Profile are both used in the LAN Access Profile. The Bluetooth Baseband authentication and encryption are adequate to secure the link between the LAP and the data terminal (DT). Both the LAP and the DT must store the link keys. These link keys should be combination keys, as unit keys do not provide adequate security. If pairing of two devices occurs in an insecure location, which is normal for where a LAP would be located, then a long randomly generated passkey should be used. If that is not possible, then a key exchange should be done and applied at a higher layer using a more robust encryption mechanism, such as AES, Triple-DES, or Diffie-Hellman. These procedures would help protect the passkey from being intercepted by an intruder. If access to the LAP is restricted or managed, then a fixed passkey can be used. The passkey would have to be entered into the DT. If anyone is permitted access to the LAP, then the Bluetooth Baseband specification should be used. The LAP in this case should be physically inaccessible by everyone except the administrator.

For authentication and encryption, security mode three is recommended for the data terminal and LAP, but security mode two may be used for the LAP. If security mode two is used, the connection is only secure over the Bluetooth link. If security mode three is used, all connections between the data terminal and gateway are authenticated and encrypted. A PPP authentication mechanism can be used by a network access server, but is not required for a single DT and LAP [6].

9.7.5   Synchronization Profile

The Synchronization Profile defines requirements and mechanisms for synchronizing applications such as phone book, calendar, messages, and notes between Bluetooth devices [6]. The synchronization profile does not recommend any important security precautions. The authors in [6] suggest that the information used during synchronization is highly confidential and must be protected from misuse. The profile specification includes authentication and encryption from the Bluetooth Baseband. It is recommended that combination keys, instead of unit keys, are used for link keys. It is also recommended that the passkey be a long randomly generated passkey. If possible, the pairing for synchronization should be done in a "private area," in order to prevent eavesdroppers [6].

9.7.6   Summary

The information provided in [6] covers recommendations for the secure use of Bluetooth devices. The authors strongly recommend the use of combination keys, instead of unit keys. It is also recommended that the pairing of Bluetooth devices should be done in a "private area," this would be anywhere that the pair could be overheard by a casual bystander. The use of long randomly generated passkeys is recommended. The passkeys should be of the longest length that the devices can handle [6].

9.8   Conclusions

The main subject covered in this chapter is Bluetooth security. With any technology, security is only as good as how it is implemented and used. The most serious vulnerabilities occur during the pairing of Bluetooth devices. This happens when an intruder, in the area that the pairing occurs, could eavesdrop on the pairing sequence when link keys are exchanged over an unencrypted channel. The use of a unit key as a link key is another major vulnerability. If a Bluetooth device uses its unit key as the link key, then any device that pairs with this device could have access to all communications from and to the device. This happens because the link key is used as the basis of all authentication and encryption. The use of a combination key, a link key generated by both devices, is what is recommended.

The length of the encryption keys is vulnerable. The fact that a device can claim that it can only use the smallest key size usable by the application, can and will be abused. The use of a small encryption key allows for the easier cryptoanalysis of communications and could make communications at the application layer vulnerable. An intruder with the authentication and encryption keys does not just want the keys, but wants access to the communications of the application layer that use the Bluetooth device.

It appears to us that the standard in [1] assumed that Bluetooth security would be adequate for personal use. If users wanted to use a Bluetooth device in a more secure way, these users would use an application, at the application layer, which would encrypt all communication through the Bluetooth device.

With the popularity of Bluetooth devices such as headset, PDAs, and cell phone, the need for security has never been more important. The users of these devices are usually unaware of these security vulnerabilities and are open to various attacks. These users could have personal and/or business information revealed. The security of distance is no longer available because people have recently built a device that can "sniff out" Bluetooth devices up to 1.1 miles away. The users attacked by these intruders will probably not even know that they are being attacked because they would not see the intruder.

If a BD_ADDR can be associated with a particular user, that user's activities can be logged and their privacy could be compromised. There should be a two-way challenge-response authentication, instead of the one-way; this would prevent man-in-the-middle attacks. There is also no user authentication, only device authentication; this would allow an intruder to use an authenticated device to gain access to information he should not have [7].

Bluetooth technology is still in its infancy. As with any new technology, there will be problems. The Bluetooth devices that people would use in their homes are most likely as secure as they need to be. The use of Bluetooth devices in the public area is not as secure as it should be. A user of Bluetooth devices should secure all sensitive data with a password and if possible an encryption application.

Bibliography

1. Part 15.1: Wireless Medium Access Control(MAC) and Physical Layer(PHY) Specifications for Wireless Personal Area Networks(WPANs). IEEE Standard 8.02.15.1-2002
2. http://www.bluetooth.com/.
3. Massey and Rueppel, "Linear ciphers and random sequence generators with multiple clocks," Advances in Cryptology-Proceedings of EUROCRYPT 84 (LNCS 209), pages 74–87, 1985.
4. Creighton T. Hager and Scott F. Midkiff, "Demonstrating Vulnerabilities in Bluetooth Security," Proc. Of GLOBECOM 03.
5. Kim Zetter, "Security Cavities Ail Bluetooth," http://www.wired.com/news/privacy/0,1848,64463-2,00.html.
6. Bluetooth SIG Security Expert Group,"Bluetooth Security Whitepaper," Rev. 1, April 19, 2002.
7. Tom Karygiannis and Les Owens, "Wireless Network Security 802.11, Bluetooth, and Handheld devices," Special Publication 800-48, National Institute of Standards and Technology.

Part III

Security in Grid Computing

10

State-of-the-Art Security in Grid Computing

Giorgos Kostopoulos, Nicolas Sklavos and Odysseas Koufopavlou

CONTENTS

Abstract   In the last decade we have witnessed the dramatic increase of interest in grid computing as an innovative extension to distributed computing technology. This technology is achieving computing resource sharing among participants in a collection of virtual organizations. Grid computing is a computing model that provides the ability to perform higher throughput computing by taking advantage of many networked computers to model virtual computer architectures. This kind of architecture is able to distribute process execution across a parallel infrastructure.

This technology leverages a combination of hardware/software virtualization, and the distributed sharing of those virtualized resources. These resources can include all elements of computing, including: hardware, software, applications, networking services, pervasive devices, and complex footprints of computing power. Grid computing is one technology enabler for some of the most innovative and powerful emerging industrial solution approaches. The emergence of open standards has a great influence on this computing technology, especially in providing seamless grid interoperability and grid integration facilities. With the exception of financial firms, grid computing has not made inroads into the business community. Private industry has expressed concerns about the security of grid computing and various psychological barriers have prohibited it from being incorporated even in business LAN environments.

This chapter gives a fairly comprehensive security overview of grid computing. The main purpose is for the reader to obtain knowledge of security in high-performance computing. In grid technology, security tools are concerned with establishing the identity of users or services (authentication), protecting communications, and determining who is allowed to perform what actions (authorization), as well as with supporting functions such as managing user credentials and maintaining group membership information. The primary motivations behind privacy for grid computing are the need for secure communication (authenticated and also confidential) between elements and also the need to support security across organizational boundaries. Among them, it also requested to prohibit a centrally managed security system. The need to support "single sign-on" for users of the grid is also a proven crucial factor as is the delegation of credentials for computations that involve multiple resources and/or sites.

In this chapter, security mechanisms such as Message-Level Security, Transport-Level Security, and Authorization Frameworks will be described. These mechanisms are proven critical, since they support a variety of authorization schemes. The terms of Public Key Cryptography, Digital Signatures, Certificates, and Mutual Authentication from the aspect of Grid Computing are also examined and presented in detail.

10.1 Introduction

In the evolution of computational grids, security threats were overlooked in the desire to implement a high-performance distributed computational system. But now the growing size and profile of the grid require comprehensive security solutions as they are critical to the success of the endeavor. A comprehensive security system, capable of responding to any attack on grid resources, is indispensable to guarantee its anticipated adoption by both the users and the resource providers. Some security teams have started working on establishing in-depth security solutions. The evaluation of their grid security solutions requires excellent criteria to assure sufficient security to meet the needs of its users and resource providers.

Grid computing [4] is the aggregation of networked connected computers to form a large-scale distributed system used to tackle complex problems. By spreading the workload across a large number of computers, grid computing offers enormous computational, storage, and bandwidth resources that would otherwise be far too expensive to attain within traditional supercomputers. High-performance computational grids involve heterogeneous collections of computers that may reside in different administrative domains, run different software, be subject to different access control policies, and be connected by networks with widely varying performance characteristics. The security of these environments requires specialized grid-enabled tools that hide the mundane aspects of the heterogeneous grid environment without compromising performance. These tools may incorporate existing solutions or may implement completely new models. In either case, research is required to understand the utility of different approaches and the techniques that may be used to implement these approaches in different environments.

Grid computing is distinguished from conventional distributed computing by its focus on large-scale pervasive resource sharing, virtual and pluggable high-performance orientation. The electrical power grid's pervasiveness and reliability inspired computer scientists in the mid-1990s to explore the design and development of a new infrastructure, computational power grids for network computing. The real and specific problem that underlies the grid concept is coordinated resource sharing and problem solving in dynamic, multi-institutional virtual organizations. The sharing is not primarily file exchange but rather direct access to computers, software, data, and other resources, as is required by a range of collaborative problem-solving and resource brokering strategies emerging in industry, science, and engineering. This sharing is necessarily highly controlled, with resource providers and users defining clearly and carefully just what is shared, who is allowed to share, and the conditions under which sharing occurs. A set of individuals and/or institutions defined by such sharing rules form a virtual organization (VO).

The vast grid applications require a high degree of security. If an adversary can thwart the grid functioning by perturbing or pilfering the information, then the perceived usefulness of the grid endeavor will be drastically curtailed. Thus, security is a major issue that must be resolved in order for the potential of the grid to be fully exploited.

The heterogeneous nature of resources and their differing security policies are complicated and complex in the security schemes of a grid computing environment. These computing resources are hosted in different security domains and heterogeneous platforms. The major security requirement for the grid is centered on the dynamic configuration of its security services, such as data integrity, confidentiality, and information privacy in potentially volatile environments.

In general, the purpose of security mechanisms is to provide protection against malicious parties. Traditional security mechanisms typically protect resources from malicious users by restricting access to only authorized users. However, in many situations within distributed applications one has to protect oneself from those who offer resources so that the problem is in fact reversed. For instance, a resource providing information can act deceitfully by providing false or misleading information, and traditional security mechanisms are unable to protect against this type of threat.

10.2   Requirements for a Secure Grid Infrastructure

The security challenges faced in a grid environment can be grouped into three categories: integration with existing systems and technologies, interoperability with different hosting environments (e.g., J2EE servers, .NET servers, Linux systems), and trust relationships among interacting hosting environments. Relationships among these three categories of challenges are depicted in Figure 10.1.

The Virtual Organization (VO) is a key concept in the grid community. A VO can be seen as a temporary or permanent coalition of geographically dispersed individuals, groups, organizational units or entire organizations that pool resources, capabilities, and information to achieve common objectives. Depending on the context, dynamic ensembles of the resources, services, and people that comprise a scientific or business VO can be small or large, short- or long-lived, single- or multi-institutional, and homogeneous or heterogeneous. Trust and security challenges within the grid environment are driven by the need to support scalable, dynamic distributed VO [16].

The GGF has initiated the definition of the next generation of grid middleware by extending the emerging Web services technology that is currently being developed across the IT industry, under the umbrella of the Open Grid Services Architecture (OGSA). Trust and security requirements can be analyzed from different perspectives. This section analyzes requirements

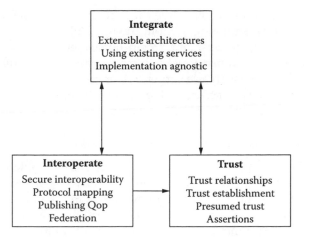

FIGURE 10.1
Categories of security challenges in a grid environment.

as defined by the GGF OGSA Security Workgroup, as well as through the different phases of a Virtual Organization.

10.2.1 Security Challenges According to GGF

The GGF OGSA Working Group has submitted a memo proposing a strategy for addressing security with OGSA [2]. According to the group, the security challenges faced in a grid environment can be grouped into three categories:

- integration solutions where existing services need to be used, and interfaces should be abstracted to provide an extensible architecture,

- interoperability solutions so that services hosted in different virtual organizations that have different security mechanisms and policies will be able to invoke each other, and

- solutions to define, manage, and enforce trust policies within a dynamic grid environment.

A solution within a given category will often depend on a solution in another category. For example, any solution for federating credentials to achieve interoperability will be dependent on the trust models defined within the participating domains and the level of integration of the services within a domain.

Defining a trust model is the basis for interoperability but a trust model is independent of interoperability characteristics. Similarly the level of integration implies a level of trust as well as a bearing on interoperability.

In a grid environment, where identities are organized in VOs that transcend normal organizational boundaries, security threats are not easily divided by such boundaries. Identities may act as members of the same VO at one moment and as members of different VOs the next, depending on the tasks they perform at a given time. Thus, while the security threats to OGSA fall into the usual categories (snooping, man-in-the-middle, intrusion, denial of service, theft of service, viruses, and Trojan horses, etc.) the malicious entity could be anyone. An additional risk is introduced, when multiple VOs share a virtualized resource (such as a server or storage system) where each of the participating VOs may not trust each other and therefore may not be able to validate the usage and integrity of the shared resource.

The Integration Challenge

For both technical and pragmatic reasons, it is unreasonable to expect that a single security technology can be defined that will both address all grid security challenges and be adopted in every hosting environment. Existing security infrastructures cannot be replaced overnight. For example, each domain in a grid environment is likely to have one or more registries in which user accounts are maintained (e.g., LDAP directories); such registries are unlikely to be shared with other organizations or domains.

Similarly, authentication mechanisms deployed in an existing environment that is reputed secure and reliable will continue to be used. Each domain typically has its own authorization infrastructure that is deployed, managed, and supported. It will not typically be acceptable to replace any of these technologies in favor of a single model or mechanism.

The Interoperability Challenge

Services that traverse multiple domains and hosting environments need to be able to interact with each other, thus introducing the need for interoperability at multiple levels. At the *protocol level*, it is required mechanisms that allow domains to exchange messages; this can be achieved, for instance, via SOAP/HTTP. At *the policy level*, secure interoperability requires that each party be able to specify any policy it may wish in order to engage in a secure conversation and that policies expressed by different parties can be made mutually comprehensible. Only then can the parties attempt to establish a secure communication channel and security context upon mutual authentication, trust relationships, and adherence to each other's policy. At the *identity level*, mechanisms for identifying a user from one domain in another domain are required.

The Trust Relationship Challenge

The VOs that underlie collaborative work within grids may form quickly, evolve over time and span organizations; as discussed before, their effective operation depends on trust. In the simple case, personal knowledge between parties in the VO allows policies to be derived from identifiable trust "anchors" (parties vouching for other parties).

An example in current grid systems is the use of certificate authorities to root certificate-based identity mechanisms. For these to work, one must "know" about the trustworthiness of the certificate authority used to establish the identity of a party in order to bind it to specific usage policies. However, personal knowledge does not scale for the case on nontrivial VOs, which are most of the VOs, and it is necessary that other technologies such as reputation management [15] are in place to create and monitor relationships.

10.2.2 Requirement Analysis through the VO Lifecycle

The VO Roadmap project [23] developed a VO lifecycle including phases such as identification, formation, operation/evolution, and dissolution. The identification phase deals with setting up the VO; this includes selection of potential business partners by using search engines or looking up registries. VO formation deals with partnership formation, including the VO configuration distributing information such as policies, agreements, etc., and the binding of the selected candidate partners into the actual VO. After the formation phase, the VO can be considered to be ready to enter the operation phase where the identified and properly configured VO members perform according to their role.

Membership and structure of VOs may evolve over time in response to changes of objectives or to adapt to new opportunities in the business environment. Finally, the dissolution phase is initiated when the objectives of the VO have been fulfilled. Here we summarize such requirements.

VO Identification

The identification phase addresses setting up the VO. This includes selection of potential business partners from the network of enterprises by using search engines or looking up registries. Generally, relevant identification information contains service descriptions, security grades, trust, and reputation ratings, etc. Depending on the resource types, the search process may consist of a simple matching (e.g., in the case of computational resources, processor type, available memory, and respective data may be considered search parameters with clear-cut matches) or in a more complex process, which involves adaptive, context-sensitive parameters. As an example, the availability of a simulation program may be restricted to specific user groups or only for certain data types, like less confidential data, etc. The process may also involve metadata such as security policies or Service Level Agreement (SLA) templates with ranges of possible values and/or dependencies between them, such as bandwidth depending on the applied encryption algorithm. The identification phase ends with a list of candidates that potentially could perform the roles needed for the current VO.

After this initial step from the potentially large list of candidates, the most suitable ones are selected and turned into VO members, depending on additional aspects that may further reduce the set of candidates. Such additional aspects cover negotiation of actual Quality-of-Service (QoS) parameters, availability of the service, "willingness" of the candidate to participate, etc. It should be noted that though an exhaustive list of candidates may have been gathered during the identification phase, this does not necessarily mean that a VO can be realized; consider the case where a service provider may not be able to keep the promised SLA at a specific date due to other obligations.

In principle, the intended formation may fail due to at least two reasons: (a) no provider (or not enough providers) is able to fulfill all given requirements for SLA, security, etc. or (b) providers are not (fully) available at the specified time. In order to circumvent these problems, either the requirements may be reduced ("choose the best available") or the actual formation may be delayed to be relaunched at a more suitable time. Obviously there may be the case where a general restructuring of the requirements leads to a repetition of the identification phase.

VO Formation

At the end of the (successful) identification phase the initial set of candidates will have been reduced to a set of VO members. In order to allow these members to perform accordingly their anticipated role in the VO they need to be configured appropriately. During the formation phase a central component, such as a VO manager, distributes the VO level configuration information, such as policies, SLAs, etc., to all identified members. These VO-level policies need to be mapped on local policies. This might include changes in the security settings (e.g., open access through a firewall for certain IP addresses, create users on machines on the fly, etc.) to allow secure communication or simple translation of XML documents expressing SLAs or obligations to a product-specific format used internally.

VO Operation

The operational phase could be considered the main lifecycle phase of a VO. During this phase the identified services and resources contribute to the actual execution of the VO's task(s) by executing predefined business processes (e.g., a workflow of simulation processes and pre- and postprocessing steps).

A lot of additional issues related to management and supervision are involved in this phase in order to ensure smooth operation of the actual task(s). Such issues cover carrying out financial arrangements (accounting, metering), recording of and reacting to participants' performance, updating and changing roles and therefore access rights of participants according to the current status of the executed workflow, etc. In certain environments persistent information of all operations performed may be required to allow for later examination, e.g., to identify fault sources.

Throughout the operation of the VO, service performance will be monitored. This will be used as evidence when constructing the reputation of the service providers. Any violation, e.g., an unauthorized access detected by the access control systems, and security threats, e.g., an event detected by an intrusion detection system, needs to be notified to other members in order to take appropriate actions. Unusual behaviors may lead to both a trust reassessment and a contract adaptation. VO members will also need to enforce security at their local site. For example, providing access to services and adapting to changes and the violations.

Evolution is actually part of the operational phase: As participants in every distributed application may fail completely or behave inappropriately, the need arises to dynamically change the VO structure and replace such partners. This involves identifying new, alternative business partner(s) and service(s), as well as renegotiating terms and providing configuration information during the identification and formation phases. Obviously one of the main problems involved with evolution consists in reconfiguring the existing VO structure so as to seamlessly integrate the new partner, possibly even unnoticed by other participants. Ideally, one would like the new service to take over the replaced partners' task at the point of its leaving without interruption and without having to reset the state of operation. There may be other reasons for participants joining or leaving the VO, mostly related to the overall business process, which might require specific services only for a limited period of time; since it is not sensible to provide an unused, yet particularly configured service to the VO for its whole lifetime, the partner may request entering or leaving the VO when not needed.

VO Dissolution

During the dissolution phase, the VO structure is dissolved and final operations are performed to annul all contractual binding of the partners. This involves the billing process for services used and an assessment of the performances of the respective participants (or more specifically their resources), such as the amount of SLA violations and the like. The latter may be of particular interest for further interactions for other potential customers. Additionally it is required that all security tokens, access rights, etc., be revoked in order to avoid a participant (mis)using its particular privileges. Generally the inverse actions of the formation phase have to be performed during termination. Obviously partial termination operations are performed during the evolution steps of the VO's operation phase.

10.3 Grid Security Model

Industry efforts have rallied around Web services (WS) as an emerging architecture that has the ability to deliver integrated, interoperable solutions. Ensuring the integrity, confidentiality, and security of Web services through the application of a comprehensive security model is critical, both for organizations and their customers, which is the fundamental starting point for constructing virtual organizations. The secure interoperability between virtual organizations demands interoperable solutions using heterogeneous systems. For instance, the secure messaging model proposed by the Web Services

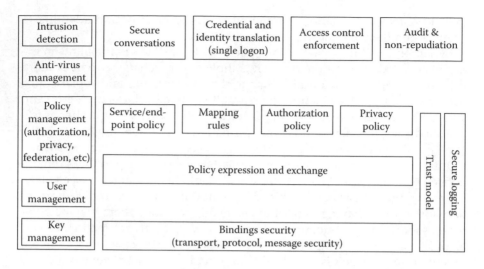

FIGURE 10.2
Components of grid security model.

Security roadmap [7] document supports both public key infrastructure (PKI) and Kerberos mechanisms as particular embodiments of a more general facility that can be extended to support additional security mechanisms.

The security of a grid environment must take into account the security of various aspects involved in a grid service invocation. This is depicted in Figure 10.2.

A Web service can be accessed over a variety of protocols and message formats it supports, as defined by its bindings [14]. Given that bindings deal with protocol and message formats, they should provide support for quality of service, including such security functions as confidentiality, integrity, and authentication.

Each participating end point can express the policy it wishes to see applied when engaging in a secure conversation with another end point. Policies can specify supported authentication mechanisms, required integrity and confidentiality, trust policies, privacy policies, and other security constraints. Given the dynamic nature of grid service invocations, end points will often discover the policies of a target service and establish trust relationships with it dynamically.

Once a service requestor and a service provider have determined the policies of each other, they can establish a secure channel over which subsequent operations can be invoked. Such a channel should enforce various qualities of service including identification, confidentiality, and integrity. The security model must provide a mechanism by which authentication credentials from the service requestors' domain can be translated into the service providers' domain and vice versa. This translation is required in order for both ends to evaluate their mutual access policies based on the established credentials and the quality of the established channel.

10.3.1 Binding Security

The set of bindings to be considered includes SOAP (SOAP/HTTP, SOAP over a message queue or SOAP over any other protocol) and IIOP bindings. The security of a binding is based on the security characteristics of the associated protocol and message format. If new protocols or message formats are introduced, care should be taken to address security requirements in those bindings so that, at a minimum, suitable authentication, integrity, and confidentiality can be achieved.

HTTP is an important protocol to consider because of its transparency to firewalls and wide adoption. In the case of bindings over HTTP, requests can be sent over SSL (i.e., https) and thus SSL can provide authentication, integrity, and confidentiality. However, SSL ensures these qualities of service only among participating SSL connection end points. If a request needs to traverse multiple intermediaries (firewalls, proxies, etc.), then end-to-end security needs to be enforced at a layer above the SSL protocol.

In the case of SOAP messages, security information can be carried in the SOAP message itself in the form of security tokens defined in the WS-Security specification [7]. SOAP messages can also be integrity and confidentiality protected using XML Digital Signature and XML Encryption support, respectively. Signature and encryption bindings defined in WS-Security can be used for this purpose.

Web services can be accessed over IIOP when the service implementation is based on CORBA [10]. In the case of IIOP, the security of the message exchange can be achieved by using the Common Secure Interoperability specification, version 2 (CSIv2) [11]. This specification is also adopted in J2EE [12].

In addition to, or in lieu of, binding-level security requirements, network security solutions (e.g., firewalls, IPSec, VPN, DNSSEC, etc.) remain useful components for securing a grid environment. Firewalls can continue to enforce boundary access rules between domains and other network-level security solutions can continue to be deployed in intradomain environments. Grid services deployment can take the topology into consideration when defining security policies. At the same time, deployment assumptions may be surfaced as policies attached to firewalls and network architecture.

The grid security model must be able to leverage security capabilities of any of these underlying protocols or message formats. For example, in the case of SOAP over HTTP requests, one can use WS-Security for end-to-end security functionality, HTTPs for point-to-point security, and SSL, TLS, or IPSec for other purposes. Security requirements for a given Web service access will be specified and honored based on the set of policies associated with the participating end points. For example, a policy associated with a Web service can specify that it expects SOAP messages to be signed and encrypted. Thus, service requestors accessing that service would be required to use WS-Security to secure their SOAP requests. Addressing the security of the service bindings will address the requirements related to integrity and confidentiality of messages, achieving delegation facilities, and facilitating firewall traversal.

10.3.2   Policy Expression and Exchange

Web services have certain requirements that must be met in order to interact with them. For example, a service may support specific message encoding formats or may require specific security credentials to perform a specific action. A hosting environment has access to policies associated with a hosted Web service so that it can enforce the invocation requirements when the service is accessed. It is important for service requestors to know about the policies associated with a target service. Once the service requestor knows the requirements and support capabilities of a target service, it can evaluate the capabilities and mechanisms that the service provider supports. At the end of the evaluation, both the service requestor and the service provider together select the optimal set of bindings to converse with one another. Note that the ability to acquire this knowledge is a privilege given by the hosting environment's policy.

In a dynamic environment like the grid, it is important for service requestors to discover these policies dynamically and make decisions at runtime. Such policies can be associated with the service definition (e.g., WSDL), service data (i.e., part of the grid service specification), or exchanged between service requestor and service provider (e.g., service provider can return a fault that contains information about the policy, or initiate negotiation). It should be noted that discovering and reacting to policies can be part of the bindings themselves. For example, in the case of IIOP bindings, service requirements, and capabilities are defined as part of the service reference (IOR) as a security-tagged component [11].

In addition to service provider policies that need to be exposed to a service requester (or similarly, service requester policies to the service provider), there may be other policies that a service requestor or a service provider needs to know about its environment but not necessarily expose in order to ensure a secure environment. For example, a service provider may have a set of authorization policies that indicate authorized requestors and this policy need not be (most likely will not be) exposed to service requestors. Similarly, service requestors may have policies specifying the identity of the service providers' hosting environments it may trust.

Based on the Web services roadmap document [7], WS-Policy will describe how both service providers and service requestors can specify their requirements and capabilities. WS-Policy will be fully extensible and will not place limits on the types of requirements and capabilities that may be described; however, the specification will likely identify several basic service attributes including privacy attributes, encoding formats, security token requirements, and supported algorithms. Grid service policies will also be specified and defined based on WS-Policy. In the case of grid services, these policies can be exchanged in a variety of ways including, but not limited to, SOAP messages, service data (part of grid service), part of bindings (e.g., CORBA security-tagged component) or by using a policy discovery service.

Policy expression and exchange facilities will address the grid security requirements to exchange policy between participating end points, securing the OGSI infrastructure and playing a critical part in achieving secure association between the end points.

The bindings and exchange layers discussed so far allow the service requestor and service provider to discover the policies of each other. The next layer of the model deals with the nature and enforcement of these policies: secure association between service end points, mapping of identities, and translation of credentials across domain boundaries between them, authorization policies and privacy policies, which together form the basis for enforcing control of access to protected services. These are reviewed in the following sections.

10.3.3 Secure Association

A service requester and a service provider are likely to exchange more messages and submit requests subsequent to an initial request. In order for messages to be securely exchanged, policy may require the service requester and service provider to authenticate each other. In that case, a mechanism is required so that they can perform authentication and establish a security context. This security context can be used to protect exchange of subsequent messages. As an added benefit, using the established security context will improve the performance of secure message exchanges. The period of time over which a context is reused is considered a session or association between the interacting end points. Security context establishment and maintenance should be based on a Web service context (to be) defined within Web or grid service specifications.

The notion of a context is tightly coupled with the bindings. Many existing protocols (e.g., IPSEC, SSL, IIOP) and mechanisms (e.g., Kerberos) already support secure association contexts. For example, in the case of IIOP, context establishment is based on the CSIv2 specification. In the case of SOAP, the context can be carried and secured as part of the SOAP messages. WS-SecureConversation will describe how a Web service can authenticate service requestor messages, how service requestors can authenticate service providers, and how to establish mutually authenticated security contexts. WS-SecureConversation will be designed to operate at the SOAP message layer so that the messages may traverse a variety of transports and intermediaries. Therefore, in the case of SOAP bindings, the grid security model should adopt WS-SecureConversation to establish security contexts and exchange messages securely. Alternatively, depending on the constraints of a VO's other technologies (e.g., SASL, BEEP, etc.) may be used. Therefore, the mechanism used to establish security contexts between end points will be based on the bindings used as well as the policy associated with the end points.

Facilitating secure association is required to establish the identity of a requestor to the service provider (and vice versa) so that the service provider (and service requestor) can satisfy the requirements to authenticate the identity on the other end and then enforce authorization and privacy policies based on the established identity.

The identities of the requestor and service provider are required for auditing purposes, so that audit logs will contain information about accessing identity.

10.3.4   Authorization Enforcement

Policies required in the grid security model also include authorization policies. Authorization is a key part of a security model and requires special mention. Each domain will typically have its own authorization service to make its own access decisions. In an Internet environment, authorization is typically associated with a service provider such that it controls access to a resource based on the identity of the service requestor. Clients, or service requestors, typically trust the server, or service provider. In case they do not, service provider authentication through SSL is one mechanism to establish service requestor trust in the service provider. In a grid environment, or even a B2B environment, more stringent rules apply from the service requestors' side. Service requestors evaluate their relationship with the service providers' environments prior to deciding whether to trust the service provider to handle the request.

The implementation of the authorization engine in each domain may also follow different models (e.g., role-based authorization, rule-based authorization, capabilities, access control lists, etc.). WS-Authorization will describe how access policies for a Web service are specified and managed. In particular it will describe how claims may be specified within security tokens and how these claims will be interpreted at the end points [7]. The grid authorization model should build on top of WS-Authorization. It should take into account that every domain is likely to have its own authorization model, authorization authority, and management facilities. Defining an authorization model will address that the requirement provide a secure grid environment by controlling access to grid services.

Grid computations may grow and shrink dynamically, acquiring resources when required to solve a problem and releasing them when they are no longer needed [6]. Each time a computation obtains a resource, it does so on behalf of a particular service requestor and based on a set of privileges associated with the requestor. Identity-based authorization is typical in most resource managers. It is necessary that any identity asserted by an end client (a service requestor) be recognizable and valid in the service provider's domain, facilitated by the identity and credential mapping functions. This is independent of whether the domain can associate the asserted identity with a real end user. There are circumstances where a user may want to remain anonymous, or use a different (possibly shared) identity. As long as an asserted identity can be associated with a set of privilege attributes or rights that can be evaluated and used to make access decisions, it does not matter if the identity is mapped to a real end user. Though a real user identity may not be required to perform authorization, it may be required to map the asserted identity to an end user for nonrepudiation purposes, by tracing through a set of mapping layers.

10.3.5 Privacy Enforcement

Maintaining anonymity or the ability to withhold private information is important in certain service environments. Organizations creating, managing, and using grid services will often need to state their privacy policies and require that incoming service requests make claims about the service provider's adherence to these policies. The WS-Privacy specification will describe a model for how a privacy language may be embedded into WS-Policy descriptions. The grid security model should adopt WS-Privacy in addition to WS-Policy to enforce privacy policies in a grid environment. The general practices and rules defined by the P3P effort [13] can prove useful in privacy policy enforcement. While the authorization and privacy functions in the grid security model build upon the WS-Policy, WS-Authorization, and WS-Privacy components, they do so by partitioning policy-related functions into specific functionality by abstracting the expression and exchange of policies from the actual policy itself. Mechanisms to express, expose, and exchange policies are covered by the policy expression and exchange layer in the proposed grid security model. Enforcement of policies pertaining to service end points, federation, authorization, and privacy should be built upon WSSecureConversation, WS-Federation, WS-Authorization, and WS-Privacy in the WS security architecture.

10.3.6 Trust

Each member of a VO is likely to have a security infrastructure that includes authentication service, user registry, authorization engine, network layer protection, and other security services. The security policies, authentication credentials and identities belonging to that member organization are likely to be managed, issued, and defined within the scope of the organization, i.e., a security domain. In order to securely process requests that traverse between members of a VO, it is necessary for the member organizations to have established a trust relationship. Such trust relationships are essential for services accessed between the members to traverse network checkpoints (e.g., firewalls) and satisfy authorization policies associated with a service achieved by translating credentials from one domain to another (e.g., Kerberos to PKI) and mapping identities across security domains. Therefore, defining and establishing these trust relationships in a grid environment, i.e., defining VO membership, is a necessary foundation of the security model. Such a model needs to define direct or mutual trust relationships between two domains, as well as indirect trust relationships brokered through intermediaries. These relationships will then often materialize as rules for mapping identities and credentials among the involved organization domains.

The grid trust model should be based on the Web services WS-Trust specification. Importantly, due to the dynamic nature of grids, trust relationships might also need to be established dynamically using trust proxies that act as intermediaries. Trust can be established and enforced based on trust policies defined either *a priori* or dynamically. Once such a model is defined, this will play a role in defining how trust assertions are to be consumed by a service provider or a requester as the case may be. The model will also form the basis to satisfy the requirements to achieve single logon based on trust of asserting authority or trust on requesting member of a VO.

10.4   Authentication in Grid Systems

Grid is a type of parallel and distributed system that enables the sharing, selection, and aggregation of geographically distributed "autonomous" resources dynamically at runtime depending on their availability, capability, performance, cost, and users' quality-of-service requirements.

A computational grid has been defined as "a hardware and software infrastructure that provides dependable, consistent, pervasive, and inexpensive access to high-end computational capabilities." Typically, grid resources are provided by various organizations and are used by people from diverse sets of organizations. A grid may support (or define) a single virtual organization or it may be used by more than one virtual organization. Individual pieces of hardware may be used in more than one grid, and people may be members of more than one virtual organization. The different resources in a grid may have different access policies, including how they authenticate and authorize users. If no common or overlapping authorizations exist among the resources, however, they do not form a usable grid.

Users, hosts, and services need to be able to authenticate themselves in the grid environment. Experience in using grids for remote computations has demonstrated the need for unattended user authentication in addition to interactive authentication. Unattended authentication of users is needed when a user is making frequent requests to remote servers and does not want to repeatedly type in a pass phrase and when a long-running job may need to authenticate itself after the user has left. Servers specific to a single host may need to be started at system boot time and run with their own or the host's identity. Some services may need to be started periodically on many different hosts and be able to authenticate themselves with a known identity.

Basically, authentication between two entities on remote grid nodes means that each party establishes a level of trust in the identity of the other party. In practical use an authentication protocol sets up a secure communication channel between the authenticated parties, so that subsequent messages can be sent without repeated authentication steps, although it is possible to authenticate every message. The identity of an entity is typically some token or name that uniquely identifies the entity.

10.4.1 Mutual Authentication

If two parties have certificates, and if both parties trust the CAs that signed each other's certificates, then the two parties can prove to each other that they are who they say they are. This is known as mutual authentication. The GSI uses the Secure Sockets Layer (SSL) for its mutual authentication protocol, which is described below. (SSL is also known by a new, IETF standard name: Transport Layer Security, or TLS.) Before mutual authentication can occur, the parties involved must first trust the CAs that signed each other's certificates. In practice, this means that they must have copies of the CAs' certificates, which contain the CAs' public keys, and that they must trust that these certificates really belong to the CAs.

To mutually authenticate, the first person (A) establishes a connection to the second person (B). To start the authentication process, A gives B his certificate. The certificate tells B who A is claiming to be (the identity), what A's public key is, and what CA is being used to certify the certificate. B will first make sure that the certificate is valid by checking the CA's digital signature to make sure that the CA actually signed the certificate and that the certificate has not been tampered with. (This is where B must trust the CA that signed A's certificate.)

Once B has checked out A's certificate, B must make sure that A really is the person identified in the certificate. B generates a random message and sends it to A, asking A to encrypt it. A encrypts the message using his private key, and sends it back to B. B decrypts the message using A's public key. If this results in the original random message, then B knows that A is who he says he is. Now that B trusts A's identity, the same operation must happen in reverse. B sends A her certificate, A validates the certificate and sends a challenge message to be encrypted. B encrypts the message and sends it back to A, and A decrypts it and compares it with the original. If it matches, then A knows that B is who she says she is. At this point, A and B have established a connection to each other and are certain that they know each others' identities.

10.4.2 Grid Certification Authorities

A Grid Certification Authority is defined as a CA that is independent of any single organization and whose purpose is to sign certificates for individuals who may be allowed access to the grid resources, hosts, or services running on a single host. Typically, a Grid CA will only sign certificates for these end entities and not for subordinate CAs. A Grid CA is substantially different from a traditional organizational CA, which signs certificates only for members of its organization and is closely linked with the authority that defines who those members are. Those certificates are then used to access resources within the organization. There are two implications of this difference: one in the format of the Distinguished Names and the other in the methods of vetting user identification.

Elements of Distinguished Names (DN)

In identity certificates issued by an organizational CA, the Distinguished Name often contains a number of attributes taken from the organization's X.500 or LDAP directory (e.g., organizational unit, location, and email). Often, an underlying assumption is that the X.509 certificate is stored in the directory entry for the user. An organizational CA is in the position to find existing LDAP entries, verify the correctness of the name elements, issue certificates for such a user, and store the certificate back in the LDAP entry. As a result of this paradigm a Distinguished Name could have several vetted components. A Grid CA breaks this paradigm by being independent of its subscribers. Even in the organizational environment there are often problems related to putting too much information in a Distinguished Name, since whenever any part of the information changes (e.g., an employee changes departments or gets a new email address), the certificate must be reissued.

Since a Grid CA is independent of the organizations to which its subscribers belong, it does not have a way to verify much information about a subscriber or to know when such information changes. The prudent approach for a Grid CA is to put as little information in the certificate as possible. A minimal set that is used to be chosen by several grid projects is:

- an organization element that identifies the grid to which the CA belongs
- a class designator that identifies the certificate as representing a person, host, or service, which is intended to be used when storing and retrieving certificates in the Grid CA's publishing directory
- a common name that reasonably identifies the entity for which the certificate is issued.

An email address can be added as an alternative name for the sake of convenience, but not for identification.

Since the operator of Grid CA does not personally know the persons who are requesting certificates and does not have access to a trusted directory of such users, it must rely on registration agents (RAs). These are individuals who are likely to know a subset of subscribers firsthand or secondhand. If the users of a grid can be grouped by actual or virtual organizations, an RA may be chosen for each such organization and given the responsibility to approve requests from members of that organization only. The rules for establishing member identities should be published by each RA, and the procedures for verifying the identities and certificate requests should be consistent among all the RAs and approved by the CA.

A topic of much discussion is the meaning ascribed to the subject name. On the one hand, most CAs specify that the common name component of the subject name should be an official and recognized name for the person who requested the certificate and the identity vetting process should assure this. On the other hand, the name should be treated by a relying party simply as an identity token that can be used to assure that the entity making the current connection is the same entity that has used this token before. In either case, before the name is added to any lists that authorize access to resources, the name must be checked by the authorizing party against some other database or virtual organization authority to see what rights should be allowed for the holder of this certificate. Using only the subject name for authorization is not safe, because subject names are guaranteed to be unique only within the domain of a single CA. Hence, either both the subject name and the issuer (CA) name must be used, or some other means must be used to limit what name spaces may be signed for by which CAs.

Virtual Organization Authority vs. Grid CA

Another subject of discussion is the role of a virtual organization (VO) as a trusted third party. If an entity is authorized to use resources because it is a member of a VO, the relying party needs to verify that a token belongs to a member of the VO. In this case, it might be more efficient to have the VO issue the certificates in the first place. In this case any entity that holds a certificate from the VO could be assumed to be a member of that VO. This approach has two drawbacks, however.

First, since not every user of grid resources is a member of an accepted VO, having VO CAs does not eliminate the need for a broader Grid CA. Second, some persons are members of more than one VO, and they would end up with a certificate from each VO. This has the advantage that it would allow a user to act in different capacities (or roles) by using different identity certificates.

On the other hand, managing different certificates is also a burden on the user, especially in view of the primitive tools available for certificate handling. A more serious objection is the merging of authorization into the concept of identity and authentication. The consensus is that X.509 certificates should be used purely for authentication of identity and that authorization should be handled as a separate issue.

Offline CA

A certificate authority can be configured such that the signing engine is available only to the CA administrator, on a host that is never available on any network or by any other means except personal use by the administrator. The host should be kept in a locked, secured facility, and the administrator's access and use carefully logged and controlled. Signing requests are conveyed to the offline CA through removable media, and signed certificates and revocation lists published by writing on removable media and transferring this data back to the "public" part of the CA. While wholly dependent on the administrator's behavior and subject to a variety of theoretical attacks, in practice this is a very good security solution for private key protection for a small PKI.

10.5   Authorization and Confidentiality in Grid Systems

Authorization deals with the verification of an action that an entity can perform after authentication was performed successfully. In a grid, resource owners will require the ability to grant or deny access based on identity, membership of groups or virtual organizations, and other dynamic considerations. Thus policies must be established that determine the capabilities of allowed actions. Authorization is closely related to access control trust.

There are several architectural proposals for handling authorization in grids. One of the earliest attempts at providing authorization in VOs was in the form of the Globus Toolkit Gridmap file. This file simply holds a list of the authenticated distinguished names of the grid users and the equivalent local user account names that they are to be mapped into. Access control to a resource is then left up to the local operating system and application access control mechanisms. As can be seen, this neither allows the local resource administrator to set a policy for who is allowed to do what, nor does it minimize his/her workload. The Community Authorization Service (CAS) [30] was the next attempt by the Globus team to improve upon the manageability of user authorization. CAS allows a resource owner to grant access to a portion of his/her resource to a VO (or community, hence the name CAS), and then let the community determine who can use this allocation. The resource owner thus partially delegates the allocation of authorization rights to the community. This is achieved by having a CAS server, which acts as a trusted intermediary between VO users and resources. Users first contact the CAS asking for permission to use a grid resource. The CAS consults its policy (which specifies who has permission to do what on which resources) and if granted, returns a digitally self-signed capability to the user optionally containing policy details about what the user is allowed to do. The user then contacts the resource and presents this capability. The resource checks that the capability is signed by a known and trusted CAS and if so maps the CAS's distinguished name into a local user account name via the Gridmap file.

The data being processed in a grid may be subject to considerable confidentiality constraints, either due to privacy concerns or issues of intellectual property. For instance, grid applications may involve medical data, bioinformatics and genomic databases, and industrial design information.

As mentioned, confidentiality is usually associated with the encryption of data only; however there are other aspects to be considered for the case of grids. The use of grids implies that confidential data is stored in online accessible databases. Access to their interfaces must be carefully controlled, both to allow access only to appropriate users, and also to allow queries and simulations to run over these highly confidential data without that data being compromised or revealed. If the database is to be shared in a grid, it might need to be operated by a trusted third party. A further novelty of grid applications is that they may entail running confidential code or using confidential data on a remote resource; running a job on a dynamically selected cluster according to load may be good resource management, but the data owner may know nothing about the trust status of the cluster selected by the grid software. Confidentiality also extends to the privacy requirements of the actual users and resources. Users are protected under privacy laws and these must be adhered to by all components of proposed grid technology.

10.6 Relationship to Security Standards

The grid environment and technologies address seamless integration of services with existing resources and core application assets. As discussed in the Grid Security Model section, the grid security model is a framework that is extensible, flexible, and maximizes existing investments in security infrastructure. It allows the use of existing technologies such as X.509 public key certificates, Kerberos shared-secret tickets, and even password digests. Therefore, it is important for the security architecture to adopt, embrace, and support existing standards where relevant. Given grid services are based on Web services, grid security model will embrace and extend the Web services security standards proposed under the WS Security roadmap [7].

Specifically, given that OGSA is a service-oriented architecture based on Web services (i.e., WSDL-based service definitions), the OGSA security model needs to be consistent with Web services security model. The Web services security roadmap [wssecurity-roadmap] provides a layered approach to address Web services, and also defines SOAP security bindings.

Figure 10.3 illustrates the layering of security technology and standards that exist today and how they fit into the grid security model.

10.7 Restricted Delegation

Grid experience has shown the need for unattended authorization. It has also demonstrated the need for both local and remote processes to run on behalf of a user and with his authorization rights. These issues have been addressed by the Globus Toolkit Grid Security Infrastructure (GSI) by allowing for short-term proxy certificates, stored with unencrypted private keys, to which a user has delegated his identity. These certificates are correctly formatted X.509 certificates, except that they are marked as proxy certificates and are signed by an end entity rather than a CA. The choice of the lifetime of proxy certificates requires a compromise between allowing long-term jobs to continue to run

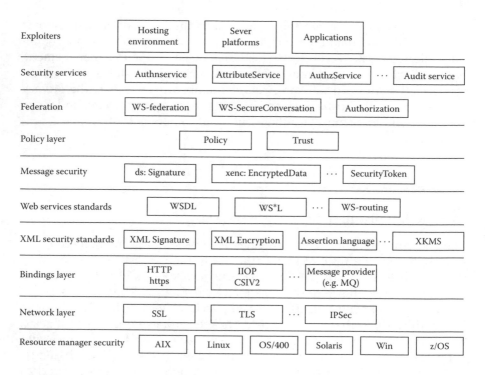

Exploiters	Hosting environment	Sever platforms	Applications		
Security services	Authnservice	AttributeService	AuthzService ⋯	Audit service	
Federation	WS-federation	WS-SecureConversation	Authorization		
Policy layer		Policy	Trust		
Message security	ds: Signature	xenc: EncryptedData ⋯	SecurityToken		
Web services standards	WSDL	WS*L ⋯	WS-routing		
XML security standards	XML Signature	XML Encryption	Assertion language ⋯	XKMS	
Bindings layer	HTTP https	IIOP CSIV2 ⋯	Message provider (e.g. MQ)		
Network layer	SSL	TLS ⋯	IPSec		
Resource manager security	AIX	Linux	OS/400	Solaris	Win	z/OS

FIGURE 10.3
Building blocks for grid security architecture.

as authenticated entities and the need to limit the damage that might be done in the event that a proxy is stolen. Proxy certificates with restricted rights are another way of limiting the damage done by a stolen proxy. Authorization software run by relying parties must be able to recognize proxy certificates and search the certificate chain until the end-entity certificate is found in order to do the authorization based on that identity token. Such software may also want to enforce policy decisions based on the lifetime of the proxy or on the number of levels of delegation that have been done. While restriction of proxy rights may make a site more secure, it will likely break some grid software attempting to run at that site. The delegation of credentials may take place on the machine on which the original credential resides, or may take place between two machines in different administrative domains. In the latter case, the delegation expands the trust relationships to include an additional domain (and the delegation software that runs there). The relying party should be aware of this situation; but in the absence of secure DNS, it is difficult to include trusted domain name information in a certificate chain. Although several schemes for including trace delegation information in the proxy delegation chain were discussed, no standard was agreed upon. At the current time, only the number of times a proxy has been delegated can be deduced from the chain of delegated proxies.

10.8   Firewalls and VPNs

Firewalls or VPNs between the user's host and the server host, or between different server hosts present a serious challenge to grid security measures. Grids that span administrative sites and encourage the dynamic addition of resources are not likely to benefit from the security that static, centrally administered commercial firewalls or VPNs provide. On the contrary, grids need to enforce their own security and a firewall is likely to prevent grid-authorized accesses. Typically firewalls only allow access from or to specific hosts and to specific ports. The grid infrastructure servers can be configured to run on known ports that can be allowed by the firewalls. User-provided servers and code tend to be more unpredictable in their port usage and it may not be possible to run them on hosts that are behind firewalls.

Also jobs that are scheduled to run on the "best" set of hosts may break if the request does not arrive from an allowed host. VPNs usually require some specific authentication and authorization in order to make a connection. Some VPNs support x509 identity certificates for authorization and might be able to use grid IDs. Such a VPN might present a way to get through firewalls and allow the standard grid access control to work.

10.9   OGSA Security

To address the grid-specific security requirements of OGSA, the OGSA Security Group has proposed an architecture leveraging as much as possible from the Web Services Security specifications [2].

As we mentioned previously, secure operation in a grid environment requires that applications and services be able to support a variety of security functionalities, such as authentication, authorization, credential conversion, auditing, and delegation. These functionalities are based on mechanisms that may evolve over time as new devices are developed or policies change. As suggested in [2], grid applications must avoid embedding security mechanisms statically. Exposing security functionalities as services (i.e., with a WSDL definition) achieves a level of abstraction that helps provide an integrated, secure grid environment. An OGSA infrastructure may use a set of primitive security functions in the form of services themselves. Reference [2] suggests the following security services:

- An authentication service: An authentication service is concerned with verifying proof of an asserted identity. One example is the evaluation of a user ID and password combination, in which a service requestor supplies the appropriate password for an asserted user ID. Another example involves a service requestor authenticating through a Kerberos mechanism, and a ticket being passed to the service provider's hosting environment, which determines the authenticity of the ticket before the service is instantiated.

- Identity mapping service: The identity mapping service provides the capability of transforming an identity that exists in one identity domain into an identity within another identity domain. The identity mapping service is not concerned with the authentication of the service requestor; rather it is strictly a policy-driven name mapping service.

- Authorization service: The authorization service is concerned with resolving a policy-based access control decision. The authorization service consumes as input a credential that embodies the identity of an authenticated service requestor and for the resource that the service requestor requests, resolves based on policy, whether the service requestor is authorized to access the resource. It is expected that the hosting environment for OGSA-compliant services will provide access control functions, and it is appropriate to further expose an abstract authorization service depending on the granularity of the access control policy that is being enforced.

- VO policy service: The VO policy service is concerned with the management of policies. The aggregation of the policies contained within and managed by the policy service comprises a VO's policy set. The policy service may be thought of as another primitive service, which is used by the authorization, audit, identity mapping, and other services as needed.

- Credential conversion service: The credential conversion service provides credential conversion between one type of credential to another type or form of credential. This may include such tasks as reconciling group membership, privileges, attributes, and assertions associated with entities (service requestors and service providers). For example, the credential conversion service may convert a Kerberos credential to a form that is required by the authorization service. The policy-driven credential conversion service facilitates the interoperability of differing credential types, which may be consumed by services. It is expected that the credential conversion service would use the identity mapping service. WS-Trust defines such a service.

- Audit service: The audit service, similar to the identity mapping and authorization services, is policy driven. The audit service is responsible for producing records, which track security-relevant events. The resulting audit records may be reduced and examined to determine if the desired security policy is being enforced. Auditing and subsequently reduction tooling are used by the security administrators within a VO to determine the VO's adherence to the stated access control and authentication policies.

- Profile Service: The profile service is concerned with managing a service requester's preferences and data that may not be directly consumed by the authorization service. This may be service requester specific personalization data, which, for example, can be used to tailor or customize the service requester's experience (if incorporated into an application that interfaces with end-users.) It is expected that primarily this data will be used by applications that interface with a person.

- Privacy Service: The privacy service is primarily concerned with the policy-driven classification of personally identifiable information (PII). Service providers and service requestors may store personally identifiable information using the privacy service. Such a service can be used to articulate and enforce a VO's privacy policy.

10.10 Web Services Security

Web services offer an interoperable framework for stateless, message-based, and loosely coupled interaction between software entities. These entities can be spread across different companies and organizations, can be implemented on different platforms, and can reside in different computing infrastructures. Web services expose functionality via XML messages, which are exchanged through the SOAP protocol. The interface of a Web service is described in detail in an XML document using the "Web Service Description Language" (WSDL).

In order to provide security, reliability, transaction abilities, and other features, additional specifications exist on top of the XML/SOAP stack. The creation of the specifications is a cross-industry effort, with the participation of standardization bodies such as W3C and OASIS. A key element in the Web services specifications is the so-called combinability. Web services specifications are being created in such a way that they are mostly independent of each other; however, they can be combined to achieve more powerful and complex solutions. In this section we describe some individual specifications, specifically focusing on those dealing with secure and reliable transactions.

10.10.1 Reliability

The WS-ReliableMessaging specification describes a protocol for reliable delivery of SOAP messages in the presence of system or network failures. To do so, the initial sender retrieves a unique sequence identifier from the ultimate receiver of the sequence to be sent. Each message in the sequence is uniquely bound to that identifier, together with a sequence number. The receiver of the sequence acknowledges to the sender what messages have already been received, thus enabling the sender to determine which messages have to be re-transmitted based on the sequence number. WSReliableMessaging should be used in conjunction with WS-Security, WS-Secure-Conversation, and WSTrust in order to provide security against attackers at the network layer.

10.10.2 Policies

The Web Services Policy Framework, WS-Policy, provides a general-purpose model to describe Web service-related policies. WS-Policy by itself only provides a framework to describe logical relationships between policy assertions, without specifying any assertion. WS-PolicyAttachment attaches policies to different subjects. A policy can be attached to an XML element by embedding the policy itself or a link to the policy inside the element or by linking from the policy to the subject that is described by the policy. WS-PolicyAttachment also defines how policies can be referenced from WSDL documents and how policies can be attached to UDDI entities and stored inside a UDDI repository. WSMetadataExchange defines protocols to retrieve metadata associated with a particular Web services endpoint. For example, a WS-Policy document can be retrieved from a SOAP node using WS-Metadata. WS-PolicyAssertions specifies some common WS-Policy assertions, related to text encoding, required SOAP protocol version and so-called "MessagePredicate" assertions that can be used to enforce that a particular header combination exists in a given SOAP message.

10.10.3 Security

WS-SecurityPolicy defines certain security-related assertions that fit into the WS-Policy framework. These assertions are utilized by WS-Security, WS-Trust, and WS-SecureConversation. Integrity and confidentiality assertions identify the message parts that have to be protected and it defines what algorithms are permitted. For instance, the "SecurityToken" assertion tells a requestor what security tokens are required to call a given Web service. Visibility assertions identify what particular message parts have to remain unencrypted in order to let SOAP nodes along the message path be able to operate on these parts. The "MessageAge" assertion enables entities to constrain after what time a message is to be treated as expired. The WS-Security specification defines mechanisms for integrity and confidentiality protection, and data origin authentication for SOAP messages and selected parts thereof. The cryptographic mechanisms are utilized by describing how XML Signature and XML Encryption are applied to parts of a SOAP message. That includes processing rules so that a SOAP node (intermediaries and ultimate receivers) can determine the order in which parts of the message have to be validated or decrypted. These cryptographic properties are described using a specific header field, the <wsse:Security> header. This header provides a mechanism for attaching security-related information to a SOAP message, whereas multiple <wsse:Security> headers may exist inside a single message. Each of these headers is intended for consumption by a different SOAP intermediary. This property enables intermediaries to encrypt or decrypt specific parts of a message before forwarding it or enforces that certain parts of the message must be validated before the message is processed further. Besides the cryptographic processing rules for handling a message, WS-Security defines a generic mechanism for associating security tokens with the message. Tokens generally are either identification or cryptographic material or may be expressions of capabilities (e.g., signed authorization statements).

The WS-Trust specification introduces the concept of "security token services" (STS). A security token service is a Web service that can issue and validate security tokens. For instance, a Kerberos ticket granting server would be an STS in the non-XML world. A security token service offers functionality to issue new security tokens, to renew existing tokens that are expiring and to check the validity of existing tokens. Additionally, a security token service can convert one security token into a different security token, thus brokering trust between two trust domains. WS-Trust defines protocols including challenge-and-response protocols to obtain the requested security tokens, thus enabling the mitigation of man-in-the-middle and message replay attacks. The WS-Trust specification also permits that a requestor may need a security token to implement some delegation of rights to a third party. For instance, a requestor could request an authorization token for a colleague that may be valid for a given time interval.

WS-Trust utilizes WS-Security for signing and encrypting parts of SOAP messages as well as WSPolicy/SecurityPolicy to express and determine what particular security tokens may be consumed by a given Web service. WS-Trust is a basic building block that can be used to rebuild many of the already existing security protocols and make them fit directly in the Web services world by using Web service protocols and data structures.

WS-Federation introduces mechanisms to manage and broker trust relationships in a heterogeneous and federated environment. This includes support for federated identities, attributes, and pseudonyms.

"Federation" refers to the concept that two or more security domains agree to interact with each other, specifically letting users of the other security domain access services in their own security domain.

For instance, two companies that have a collaboration agreement may decide that employees from the other company may invoke specific Web services. These scenarios with access across security boundaries are called "federated environments" or "federations." Each security domain has its own security token service(s), and each service inside these domains may have individual security policies. WS-Federation uses the WS-Security,

WS-SecurityPolicy, and WS-Trust specifications to specify scenarios to allow requesters from the one domain to obtain security tokens in the other domain, thus subsequently getting access to the services in the other domain.

10.10.4 Web Services Specification in Implementing the VO Life Cycle

Some of the requirements presented in the analysis of requirement through the VO lifecycle can be met by application of Web services specification, as shown in [17].

The identification phase includes defining VO-wide policies as well as selecting potential business partners who are both capable of providing the required services and of fulfilling the trustworthiness requirements of the VO. The selection of potential business partners involves looking at repositories. The usual Web service technology to be applied is WSDL/UDDI. WSDL describes messages and operations while UDDI offers a discovery mechanism. To include the provision of SLA, "Web Service Level Agreements" (WSLA) has been developed, a XML language for specifying and monitoring SLA for Web services, which is complementary to WSDL. Determining the required service providers and a proper negotiation requires secure communication. The WS-Security specification and data origin authentication for SOAP messages can be used between the entities to secure the communication.

The realization of the VO requires the creation of federations, where two or more security domains agree to interact with each other, specifically letting users of the other security domain access services in their own security domain. The WS-Federation specification deals with federations by providing mechanism to manage and broker trust relationships in a heterogeneous and federated environment. This includes making use of WS-Trust to support federated identities, attributes, and pseudonyms. The dissemination of configuration information requires secure communication as provided by the WS-Security specification.

Throughout the operation of the VO, service performance will be monitored. This will be used as evidence when constructing the reputation of the service providers. Any violation, e.g., an unauthorized access detected by the access control systems — and security threats, e.g., an event detected by an intrusion detection system — need to be made known to other members in order to take appropriate actions.

VO members will also need to enforce security at their local site. For example, providing access to services and adapting to changes and the violations. Monitoring can be supported by event management and notification mechanisms using the WS-Eventing and WS-Notification specifications. This allows the monitoring service to partner to receive messages when events occur in other partners. A mechanism for registering interest is needed because the set of Web services interested in receiving such messages is often unknown in advance or will change over time.

10.11 Grid Security Infrastructure

In grid computing environments, the mutual authentication and information service are serious issues. Before their applications are running, the users need to choose hosts based on security, availability, and many other aspects. The GSI (Grid Security Infrastructure) is designed as one very important part of the Globus grid toolkit. The GSI uses public key cryptography (also known as asymmetric cryptography) as the basis for its functionality.

The primary motivations behind the GSI are the need for secure communication (authenticated and perhaps confidential) between elements of a computational grid and the need to support security across organizational boundaries, thus prohibiting a centrally managed security system. Finally, a fundamental issue is to support "single sign-on" for users of the grid, including delegation of credentials for computations that involve multiple resources and/or sites.

GSI, which is designed to solve security in the Globus system, is based on RSA encystations algorithm and employs a standard (X.509v3) for encoding credentials for security principals, and thus enables secure authentication and communication over open network. At the same time GSI enables interoperability with local security solutions without changing anything.

The GSI is a specific implementation of an OGSA-based grid security architecture that is included as part of the Globus Toolkit Version 4 (GT4). Given the prominent use of Globus within the grid community, let us briefly revise such implementation.

- Authentication. GSI defines a credential format based on X.509 identity certification. An X.509 certificate, in conjunction with an associated private key, forms a unique credential set that a grid entity (requestor or service provider) uses to authenticate itself to other grid entities (e.g., through a challenge-response protocol such as TLS).
- Identity Federation. GSI uses gateways to translate between X.509-based identity credential and other mechanisms. For example, the Kerberos Certificate Authority (CKA) and SSLK5/PKNIT provide translation from Kerberos to GSI and vice versa, respectively. These mechanisms allow a site with an existing Kerberos infrastructure to convert credentials between Kerberos and GSI as needed.
- Dynamic Entities and Delegation. GSI introduces X.509 proxy certificates, an extension to X.509 identity certificates that allows a user to assign dynamically a new X.509 identity to an entity and then delegate some subset of its rights to that identity.
- Message-Level Security. The Globus Toolkit Version 4 (GT4) uses the Web services security specifications to allow security messages and secured messages to be transported, understood, and manipulated by standard Web services tools and software.

In relation to stateful and secured communication, GSI supports the establishment of a security context that authenticates two parties to each other and allows for the exchange of secured messages between the two parties. GT4 achieves security context establishment by implementing preliminary versions of WS-SecurityConversation and WS-Trust specifications. Once the security context is established, GIS implements message protection using the Web services standards for secured messages XML-Signature and XML-Encryption. To allow for communication without the initial establishment of a security context, GT4 offers the ability to sign messages independent of any established security context, by using XML-Signature specification.

- Trust Domains. The requirement for overlaid trust domains to establish VOs is satisfied by using both proxy certificates and security services such as CAS. GSI has an implicit policy that any two entities bearing proxy certificates issued by the same user will inherently trust each other. This policy allows users to create trust domains dynamically by issuing proxy certificates to any services that they want to interoperate.

UP (User Proxy) and RP (Resource Proxy) are involved in GSI. A user proxy is a session manager process given permission to act on behalf of a user for a limited period of time. User proxy acts as a stand-in for the user. It has its own credentials, eliminating the need to have the user repeat to type his password to offer his credentials to the resource.

GSI supports multitrusted CA domains to access each other. To research the access policy in multitrust domains, we set up two CA centers and enable users in the two trusted CA domains to identify each other. [1–2, 5, 7] For reducing the delay in mutual authentication, the root CA has been located in each node. In the same time we should create a signing-policy file, which is an EACL (Extend Access Control List) file. A different CA domain has a different signing policy. In our condition, we should create two signing-policy files and put them on each node in the testbed. We implement the mutual authentication among multi-CA domains by changing the EACL policies to allow the one who has the certificate published by the trusted CA access.

As shown in Figure 10.4, GSI may be thought of as being composed of four distinct functions: message protection, authentication, delegation, and authorization. Implementations of different standards are used to provide each of these functions:

- TLS (transport-level) or WS-Security and WS-SecureConversation (message level) are used as message protection mechanisms in combination with SOAP.
- X.509 End Entity Certificates or Username and Password are used as authentication credentials
- X.509 Proxy Certificates and WS-Trust are used for delegation
- SAML assertions are used for authorization

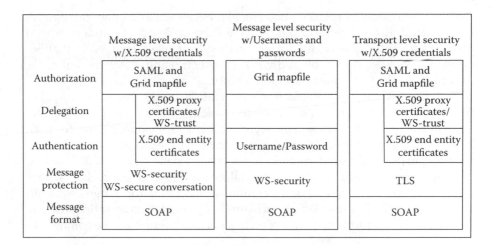

	Message level security w/X.509 credentials	Message level security w/Usernames and passwords	Transport level security w/X.509 credentials
Authorization	SAML and Grid mapfile	Grid mapfile	SAML and Grid mapfile
Delegation	X.509 proxy certificates/ WS-trust		X.509 proxy certificates/ WS-trust
Authentication	X.509 end entity certificates	Username/Password	X.509 end entity certificates
Message protection	WS-security WS-secure conversation	WS-security	TLS
Message format	SOAP	SOAP	SOAP

FIGURE 10.4

Overview of the GT4 grid security infrastructure and standards used for different functions. The two left figures show message-level security, with X.509 credentials and username/password authentication. The figure on the right shows transport-level security with X.509 credentials.

10.11.1 Message Protection

The Web services portions of GT4 use SOAP as their message protocol for communication. Message protection can be provided either by transporting SOAP messages over TLS, known as Transport-Level security, or by signing and/or encrypting portions of the SOAP message using the WS-Security standard, known as Message-Level Security. In this section we describe these two methods.

Message-Level Security

The SOAP specification allows for the abstraction of the application-specific portion of the payload from any security (e.g., digital signature, integrity protection, or encryption) applied to that payload, allowing GSI security to be applied in a consistent manner across SOAP messages for any GT4 Web service-based application or component.

GSI implements the WS-Security standard and the WS-SecureConversation specification to provide message protection for SOAP messages. (We use the term specification to denote a scheme that has been well documented but has not passed through a public-standards body.) The WS-Security standard from OASIS defines a framework for applying security to individual SOAP messages; GSI conforms to this standard. GSI uses these mechanisms to provide security on a per-message basis, i.e., to an individual message without any pre-existing context between the sender and receiver (outside sharing some set of trust roots).

WS-SecureConversation is a proposed standard from IBM and Microsoft that allows for an initial exchange of a message to establish a security context that can then be used to protect subsequent messages in a manner that requires less computational overhead (i.e., it allows the trade-off of initial overhead for setting up the session for lower overhead for messages). Note that SecureConversation is only offered with GSI when using X.509 credentials as described in the subsequent section on authentication.

Both WS-Security and WS-SecureConversation are intentionally neutral to the specific types of credentials used to implement this security. GSI, as described further in the subsequent section on authentication, allows for both X.509 public key credentials and the combination of username and password for this purpose. GSI used with either username/password or X.509 credentials uses the WS-Security standard to allow for authentication; that is, a receiver can verify the identity of the communication initiator. When used with X.509 credentials GSI uses WS-Security and WS-SecureConversation to allow for the following additional protection mechanisms (which can be combined):

- Integrity protection: a receiver can verify that messages were not altered in transit from the sender.
- Encryption: messages can be protected to provide confidentiality.
- Replay prevention: a receiver can verify that it has not received the same message previously.

The specific manner in which these protections are provided varies between WS-Security and WS-SecureConversation. In the case of WS-Security, the keys associated with the sender's and receiver's X.509 credentials are used. In the case of WS-SecureConversation, the X.509 credentials are used to establish a session key that is used to provide the message protection.

Bibliography

1. N. Nagaratnam, P. Janson, J. Dayka, A. Nadalin, F. Siebenlist, V. Welch, S. Tuecke, and I. Foster. *Security Architecture for Open Grid Services.*
2. I. Foster, C. Kesselman, G. Tsudik, and S. Tuecke. Proc. *A Security Architecture for Computational Grids.* 5th ACM Conference on Computer and Communications Security Conference, pp. 83–92, 1998.
3. I. Foster, C. Kesselman, J. Nick, and S. Tuecke. *The Physiology of the Grid: An Open Grid Services Architecture for Distributed Systems Integration.* January, 2002.
4. C. Lai, G. Medvinsky, and B.C. Neuman, *Endorsements, Licensing, and Insurance for Distributed System Services.* In Proc. 2nd ACM Conference on Computer and Communication Security, 1994.
5. N. Sklavos and P. Souras, *Economic Models and Approaches in Information Security for Computer Networks,* International Journal of Network Security (IJNS), Science Publications, Vol. 2, No. 1, January, pp. 14–20, 2006.
6. N. Sklavos and O. Koufopavlou, *Access Control in Networks Hierarchy: Implementation of Key Management Protocol,* International Journal of Network Security (IJNS), Science Publications, Vol. 1, No. 2, September, pp. 103–109, 2005.

7. I. Foster, C. Kesselman, and S. Tuecke. International J. Supercomputer Applications, *The Anatomy of the Grid: Enabling Scalable Virtual Organizations*, 2001.
8. *Security in a Web Services World: A Proposed Architecture and Roadmap*, http://www-106.ibm.com/developerworks/library/ws-secmap/.
9. *The SSL Protocol Version 3.0.* http://home.netscape.com/eng/ssl3/draft302.txt.
10. *RFC 2246: The TLS Protocol. ftp://ftp.isi.edu/in-notes/rfc2246.txt.*
11. *The Common Object Request Broker: Architecture and Specification, Version 2.3.1. The Object Management Group (OMG),* http://www.omg.org/cgi-bin/doc?formal/99-10-07.
12. *Common Secure Interoperability Version 2 Final Available Specification. The Object Management Group (OMG),* http://www.omg.org/cgi-bin/doc?ptc/2001-06-17.
13. *Java 2 Platform, Enterprise Edition, v1.5 (J2EE).* http://java.sun.com/j2ee.
14. *The Platform for Privacy Preferences 1.0 (P3P1.0) Specification,* W3C Recommendation 16 April 2002, http://www.w3.org/TR/P3P/.
15. S. Tuecke, K. Czajkowski, I. Foster, J. Frey, S. Graham, and C. Kesselman. *Grid Service Specification.* Draft 2, 6/13/2002, http://www.globus.org.
16. M.D. Abrams, M.V. Joyce. *Trusted Computing Update. Computers and Security,* 14(1):57–68, 1995.
17. R. Alfieri et al. *VOMS: An Authorization System for Virtual Organizations.* In Proceedings of 1st European Across Grids Conference, Santiago de Compostela, 2003. Available from: http://gridauth.infn.it/docs/VOMS-Santiago.pdf.
18. A.E. Arenas, I. Djordjevic, T. Dimitrakos, L. Titkov, J. Claessens, C. Geuer-Pollman, E.C. Lupu, N. Tuptuk, S. Wesner, and L. Schubert. *Towards Web Services Profiles for Trust and Security in Virtual Organisations.* IFIP Working Conference on Virtual Enterprises PRO-VE05, Valencia, Spain, 2005.
19. S. Boeyen et al. *Liberty Trust Models Guidelines.* In J. Linn (editor), Liberty Alliance Project. Liberty Alliance, draft version 1.0, 2003.
20. M. Brady, D. Gavaghan et al. *eDiamond: A Grid-Enabled Federated Database for Annotated Mammograms.* In F. Berman, G. Fox, T. Hey (editors), Grid Computing: Making the Global Infrastructure a Reality, Wiley, 2003.
21. J. Bradshaw, A. Uszok, R. Jeffers et al. *Representation and Reasoning about DAML-based policy and domain services in KAoS.* In Proc. of the 2nd Int. Joint Conf. on Autonomous Agents and Multi Agent Systems (AAMAS2003). 2003.
22. P.J. Broadfoot and G. Lowe. *Architectures for secure delegation within grids.* Oxford University Computing Laboratory Technical Report, PRG-RR-03-19, 2003.
23. P.J. Broadfoot, A.P. Martin. *A Critical Survey of Grid Security Requirements and Technologies.* Oxford University Computing Laboratory Technical Report, PRG-RR-03-15, 2003.
24. L.M. Camarinha-Matos and H. Afsarmanesh. *A Roadmap for Strategic Research on Virtual Organizations.* Proceedings of IFIP Working Conference on Virtual Enterprises, PRO-VE03, Lugano, Switzerland, pages 33–46, 2003.
25. C. Castelfranchi and R. Falcone. *Principles of Trust for MAS: Cognitive Anatomy, Social Importance, and Quantification.* In Y. Demazeau (editor), Proceedings of the Third International Conference on Multi-Agent Systems. IEEE C.S., Los Alamitos, 1998.
26. C. Castelfranchi, R. Falcone, B. Sadighi, and Y-H Tain. Guest Editorial. *Applied Artificial Intelligence,* 14(9), Taylor and Francis, 2000.
27. D. Chadwick. *Authorisation in Grid Computing.* Information Security Technical Report, Elsevier, 10(1):33–40, 2005.

28. S. Crompton, B. Matthews, A. Gray, A. Jones, and R. White. *Data Integration in Bioinformatics Using OGSA-DAI.* In Proceedings of Fourth All Hands Meeting, AHM2005, U.K. 2005.

29. G. Denker, L. Kagal, T. Finin, M. Paolucci, and K. Sycara. *Security for DAML Web Services: Annotation and Matchmaking.* In D. Fensel, K. Sycara, and J. Mylopoulos (editors), The Semantic WebISWC 2003. Proceedings of the 2nd International Semantic Web Conference, Sanibel Island, FL, October 2003, LNCS 2870, 2003.

30. T. Dimitrakos. System Models, *e-Risk and e-Trust. Towards Bridging the Gap? In Towards the ESociety: E-Business, E-Commerce, and E-Government, B. Schmid, K. Stanoevska-Slabeva, and V. Tschammer* (editors). Kluwer Academic Publishers, 2001.

31. T. Dimitrakos, D. Golby and P. Kearney. *Towards a Trust and Contract Management Framework for Dynamic Virtual Organisations.* In eAdoption and the Knowledge Economy: eChallenges 2004. Vienna, Austria, 2004.

32. L. Pearlman, V. Welch, I. Foster, C. Kesselman, and S. Tuecke. *A Community Authorization Service for Group Collaboration.* In Proceedings of the IEEE 3rd International Workshop on Policies for Distributed Systems and Networks, 2002.

33. P. Sussner, P.M. Pardalos, and G.X. Ritter. On integer programming approaches for morphological template decomposition, *Journal of Combinatorial Optimization*, Vol. 1, No. 2 (1997), pp. 177–188.

11

Unifying Grid and Organizational Security Mechanisms

David W. Chadwick

CONTENTS

Abstract   This chapter describes the authentication and authorization security mechanisms that protect grid-enabled resources. These are implemented in the grid middleware software package, the Globus Toolkit (GT). Unfortunately these mechanisms are mostly grid specific, which means that most GT users need different credentials for accessing grid-enabled resources than for accessing Web-based and organizational resources. This is not optimum from either a usability or administrative perspective. Two new technologies, Shibboleth and PERMIS, will allow harmonization of user accesses to organizational, Web-based and grid-enabled resources. Shibboleth is a protocol suite that is providing users with single sign-on access to distributed federated resources, using their normal organizational credentials. PERMIS is a policy-based authorization infrastructure that says if a user is granted or denied access to any type of resource, based on the user's credentials and the authorization policy for the resource. Several recent research projects are enabling these three technologies to be combined together so that users can use the same set of credentials to access grid, organizational, and Web-based resources. Furthermore, administrators are empowered to write the same authorization policies for protecting their resources, regardless of whether they are being accessed locally or via the Web or the grid.

11.1   Introduction

The primary grid security mechanisms today are based on public key cryptography for authentication, proxy certificates [1] for delegation and the grid-mapfile for authorization. These are described in section 2 below. While public key authentication provides strong security for grid jobs, one of its main drawbacks is that it is not the usual authentication mechanism used today in most organizations and university campuses. Public key cryptography is used in some high-security domains such as banking and defense, but ubiquitous rollout has been prevented since it is notoriously difficult for the average user to comprehend public key technology and use the associated keys effectively [3, 10]. The most common authentication mechanism today remains the username and password, with more sophisticated mechanisms such as Kerberos [2], one-time passwords and biometrics (such as fingerprint readers) being used sporadically. The net result is that users need to have different authentication tokens for running grid jobs than for their everyday computing tasks. This is not helpful to users, and may be an inhibitor to the wider take-up of grid computing. This chapter will show how the latest research is enabling a user's normal organizational authentication token to be used to authenticate grid jobs.

Proxy certificates on the other hand provide a good mechanism for delegation of authority from the user to his primary grid job, and from the primary job to spawned jobs on the grid. Their benefits are several. They allow one job to autonomously spawn countless other subtasks that can run on different machines on the grid, while still authenticating as belonging to the original user. They do not need the user to be present at his terminal for the duration of the job, which is an important consideration when some jobs can run for many hours. Furthermore, with the addition of MyProxy [8], the user gains the mobility to be able to launch a grid job from any location. However, proxy certificates rely on public key cryptography. If we are to introduce alternative authentication mechanisms for users that are based on their existing organizational credentials, and which do not rely on the users being public key infrastructure (PKI) enabled, then we either need to introduce an alternative delegation mechanism that is as effective as proxy certificates but does not rely on PKI, or we need to allow the new authentication mechanisms to work effectively with proxy certificates. We shall see that the latest research is proposing to adopt the latter technique.

The grid-mapfile is a very simplistic mechanism for authorization, and merely maps a user's Distinguished Name[1] (DN) into a local user name. The grid-mapfile lacks the ability to specify fine-grained access control, or the ability to specify conditional rules for granting access, such as *only between*

---

[1] Each public key certificate contains the globally unique Distinguished Name of the user to whom the certificate has been issued. A DN typically looks like CN=David Chadwick, O=University of Kent, C=GB.

9 am and 5 pm or *only if requested resource usage is less than a certain predefined limit.* Furthermore, it either requires new local usernames to be created for all the different remote users who might run their grid jobs locally, or it is unable to effectively differentiate between different remote users. Finally, any access control mechanism that relies solely on listing the names of authorized users is not scalable to the proportions needed for ubiquitous rollout of the grid to existing computer users, since there are millions of them. We need a common authorization mechanism that can be used for controlling access to computing resources regardless of whether they are being accessed via a grid, via a remote login, or by a local user. Authorization ought to be based primarily on the attributes of the user requesting access. This set of attributes might include the role of the user, the organization affiliation of the user, and might also include his name, but the latter should not be mandatory. In this chapter we describe a policy-based authorization mechanism that allows a set of resources to be accessed via a common policy regardless of whether the user is accessing the resource via the Web, the grid or locally.

The rest of this chapter is structured as follows. Section 2 describes the Globus Toolkit v4 (GT4) conventional authentication and authorization mechanisms for running grid jobs, which rely on public key cryptography, proxy certificates, and the grid-mapfile. The MyProxy facility that provides global roaming access to grids is also described. Section 3 describes the Shibboleth architecture, which provides Web-based users with a single sign-on capability to multiple distributed resources, as well as authorization based on their attributes. Shibboleth is currently being rolled out at many campuses worldwide, and it allows a user to authenticate to distributed resources using his local organization-allocated security tokens. Section 4 describes the PERMIS role-based access control authorization infrastructure that allows access to resources to be controlled via a policy written by the owner of the resource(s). PERMIS can be used for controlling access to grid-based, Web-based and organizational-based resources, and does not differentiate between them. Section 5 describes the latest research in which Globus, Shibboleth and PERMIS are being integrated together to allow users to access grid and organizational-based computing resources in a consistent manner, using the same authentication tokens and attributes throughout. It also allows resource administrators to control access to resources regardless of whether the users are locally or remotely based, or are accessing the resource via the grid or the Web.

11.2 Globus Toolkit Security Mechanisms

Globus Toolkit uses the Grid Security Infrastructure (GSI) [4] to provide security to grid jobs. This provides mutual authentication, transport, and message-level security and user authorization. GSI uses the SSL protocol [5] to provide transport-level security and mutual authentication between the grid user and the grid infrastructure. Message-level security is provided to GT4's SOAP messages by implementing the WS-Security [6] and WS-SecureConversation [7] specifications. Authorization is provided by callouts to pluggable authorization components, with the grid-mapfile being the built-in default authorization component. A full description of GT4 security can be found in [9].

The first step a grid user has to take before he can run a grid job is to obtain a (long-lived) public key certificate from a recognized Certification Authority (CA). While this is no mean task, we do not propose to describe this procedure here. These long-lived certificates are typically valid for a year or more, but may be revoked by the CA if something goes wrong, e.g., the user loses his private key, or is dismissed from his job. Once the user has obtained his public key certificate, he must extract his Distinguished Name (DN) from it, and get the DN registered in the grid-mapfiles of every grid resource that he needs to access. This registration will provide him with authorization to run his jobs at each resource site by mapping his DN into an appropriate local user name at each resource site. Some grid resource sites may provide users with tools to do this automatically, other sites may require the user to go via the site or resource administrator.

Now that the user is able to authenticate and authorize his grid jobs, when he is ready to run his first grid job, he needs to create a proxy certificate at his local site.[2] This creates a temporary asymmetric key pair and a short-lived proxy certificate for the public key. The proxy certificate is signed by the private key corresponding to the user's long-lived public key certificate, and contains his name plus an indication that this is a proxy certificate. The short-lived private key and proxy certificate are needed so that the user can use them during their short lifetime for mutual authentication of his grid jobs without him needing to enter his private key password again, since the temporary private key will be automatically used to sign messages as required. Since the lifetime of proxy certificates is relatively short (the default is 12 hours) their temporary private keys do not need to be protected as securely as the users' long-lived private keys, which are usually stored encrypted with a password known only to their respective owners. Furthermore if the user submits a job of long duration and leaves it running, if the job subsequently needs to spawn a subjob to run at another grid site, it will be able to do so by automatically using the temporary private key to sign the proxy certificate for the spawned job. The process of authenticating and authorizing a grid job, along with spawning a subjob, is shown in Figure 11.1.

Global roaming is provided to grid users through the MyProxy server [8]. A user initializes his entry in the MyProxy credential store by storing a medium-lived proxy certificate and corresponding private key there, protected by a username and password of his choice. The proxy certificate in the store is signed by the user's long-lived private key corresponding to his CA-issued public key certificate. When the user wishes to run a grid job while away from his home site, he no longer needs to take his CA-issued certificate and private key with him. Instead, he uses the Web to log into the grid portal that

---

[2] This is usually achieved by calling the grid-proxy-init command provided by GT4.

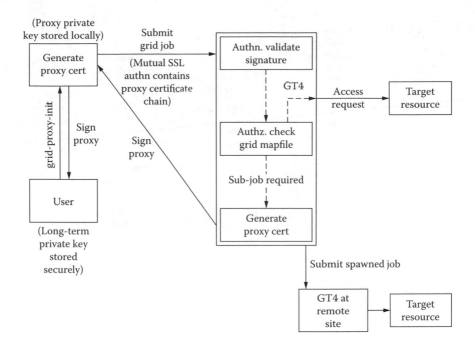

FIGURE 11.1
Grid security: Authentication, proxy certificate delegation, and authorization.

front ends the MyProxy server, enters his username and password, unlocks the proxy certificate and private key that are stored therein, and uses these to create a new short-lived proxy certificate with which to run his grid jobs. The process of retrieving a new short-lived proxy certificate from the MyProxy server is shown in Figure 11.2. Once he has this short-lived proxy certificate he can submit grid jobs in the same way that he can from his home site. One drawback of the original MyProxy design was that the user needed yet another username/password pair in order to login to the MyProxy server. More recent developments allow the user to login to the MyProxy server using his existing organizational-based authentication credentials, through the use of Pluggable Authentication Modules (PAM) and/or the Simple Authentication and Security Layer (SASL) [20].

11.3 Shibboleth

Shibboleth [11] is an Internet2/MACE-developed protocol that provides cross-domain single sign-on and attribute-based authorization for users that require interinstitutional sharing of Web-based resources. The main idea behind Shibboleth is that instead of having to login and be authorized at each resource site,

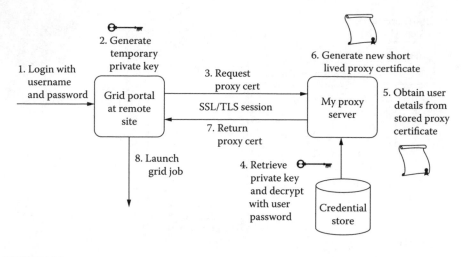

FIGURE 11.2
Generating a new proxy certificate using the MyProxy server.

users authenticate just once to their home site, which then passes the user's attributes to the various resource sites. A key feature of Shibboleth is privacy protection of the user's identity. This is enabled because each resource site (called the Service Provider (SP) in Shibboleth terminology) trusts the user's home site (called the Identity Provider (IdP) in Shibboleth terminology) to authenticate the user correctly using its existing authentication mechanism. The IdP then passes the user's attributes to the SP, which grants access based upon these attributes. The Shibboleth software is already in use today and is being widely experimented with as an interrealm access control solution for research and education applications.

The Shibboleth security protocol is based on OASIS Security Assertion Markup Language (SAML) [12] assertions, which are transacted between the two main participants: the Identity Provider (IdP) and the Service Provider (SP). Shibboleth conformant software is deployed separately at each site. The Shibboleth IdP software comprises three main functional components:

- the Authentication Authority that is responsible for authenticating users and issuing authentication statements about them,
- the Attribute Authority (AA) that is responsible for issuing attribute assertions about authenticated users; and
- the Single Sign On (SSO) service that is the first point of contact for the SP.

The SSO receives requests (directly or indirectly) from the SP, and initiates an authentication dialogue between the user and the backend Authentication Authority (step 4 in Figure 11.3). After the user has been successfully authenticated, the SSO generates a random temporary handle that is used to

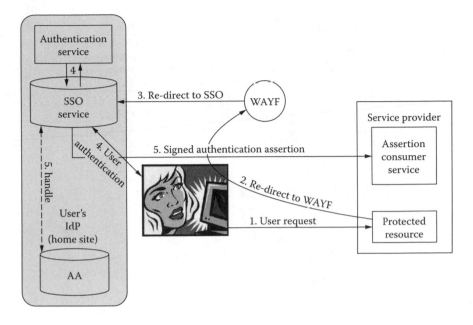

FIGURE 11.3
User authentication in Shibboleth.

identify the user, and gives this to the local AA and also includes it in the authentication assertion that is sent to the SP (step 5). The use of this one-time random handle in communication with the SP preserves the user's privacy. The SP subsequently contacts the AA at the IdP and requests the attributes for the user who is identified by this handle (step 7 in Figure 11.4). The IdP AA can use the existing institutional identity management infrastructure, such as an LDAP directory, for storing the user's attributes. Typical attributes will include the user's role, organizational affiliation, project memberships, etc.

The Shibboleth SP software controls access to the protected resources by requesting and consuming the necessary attributes from the IdP. The SP software comprises three functional components:

- access control software that protects the resource
- the Assertion Consumer Service that receives the authentication assertion from the IdP (step 5). This validates the authentication assertion and if it is acceptable, establishes a security context, passes the user's handle to the Attribute Requester, and redirects the user back to the resource (step 6).
- the Attribute Requester that establishes a back channel to the IdP and requests the attributes that belong to the user identified by a particular handle (step 7). Once the attributes have been returned to the Attribute Requester they are given to the access control software (step 9) so that it can make appropriate access control decisions about user's request.

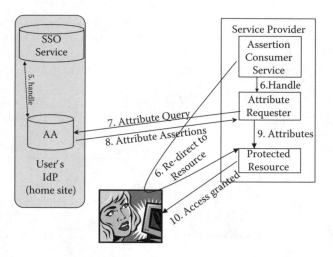

FIGURE 11.4
User authorization in Shibboleth.

If a user does not have an established security context when he first contacts the resource (step 1 in Figure 11.3) then the user is redirected to a Where Are You From Service (WAYF) that is responsible for directing the user to his home site for authentication. The WAYF service typically presents the user with a picking list of sites that are members of the federation, and when the user picks one of them he is redirected to the SSO service of that site (step 3).

If the user does have a security context already established when he contacts the resource (step 1), then the user will be granted or denied access straight away (Step 10). The access control decision will depend upon the user's request and the attributes that have been asserted on his behalf.

The Shibboleth security context is recorded in a cookie, and this is passed between the various Web resources via the user's browser.

11.4 PERMIS

PERMIS [14] is a policy-based authorization system that uses policies written in XML to support the role-based access control (RBAC) paradigm [13]. Given a user's DN, a resource and an action, the PERMIS decision engine says whether the user is granted or denied access based on the RBAC policy for the resource and the user's validated roles and attributes. The PERMIS decision engine comprises two components:

- an Attribute Assertion Validation Service (AAVS), which validates a user's attribute assertions according to the configured Role Assignment Policy (RAP);
- a Policy Decision Point, which says if a user is granted or denied access to a resource based on his validated attributes and the Target Access Policy (TAP).

PERMIS is based on the X.509 Privilege Management Infrastructure (PMI) model [21], which uses the X.509 attribute certificate (AC) as the primary assertion syntax for binding a user's unique DN to one or more of his privilege attributes, in much the same way that a public key certificate binds the user's DN to his public key. An AC is signed by the Attribute Authority (AA) that issued it and AAs are subordinate to the root of trust for a PMI target resource, which is called the Source of Authority (SOA). The SOA is similar to the root CA in a PKI, and AAs are similar to subordinate CAs. An X.509 PMI can easily support the RBAC model by storing users' roles and attributes in ACs and allowing managers to act as AAs to issue the attributes and roles they control. Resource owners may then write policies that specify the access rights and privileges that they have granted to each role or attribute.

PERMIS implements the hierarchical RBAC model, which means that user roles (attributes) are organized hierarchically with superior roles inheriting the privileges of the subordinate ones. Each user is assigned one or more roles (attributes) as ACs, and each role or attribute is given a set of permissions in the PERMIS authorization policy that is written by the resource owner and stored in his LDAP entry as a policy AC (see Figure 11.5).

In order to gain access to a protected target resource a user has to either present his attribute assertions to the PERMIS decision engine (the push mode), or the PERMIS decision engine will pull them itself from the user's attribute assertion store (typically an LDAP directory) (see Figure 11.5). The attribute assertion validation service (AAVS) of PERMIS evaluates all the pushed or pulled attribute assertions against the RAP, discards untrusted ones, and passes all validated attributes to the policy enforcement point (PEP). The PEP in turn passes these to the PERMIS policy decision point (PDP), along with the user's access request, and any environmental parameters, such as the current date and time. The PDP makes an access control decision based on the TAP, and passes its granted or denied response back to the PEP. The PEP then either allows or forbids the user from accessing the resource.

The PERMIS toolkit provides an easy-to-use graphical user interface (the Policy Editor) for creating its policies. Once created, the policies are converted into XML for subsequent evaluation by the AAVS and PDP. Each policy may be digitally signed by its author (the SOA) in order to stop it from being tampered with. It can then be stored in the LDAP directory entry of the author for subsequent retrieval by the PERMIS decision engine.

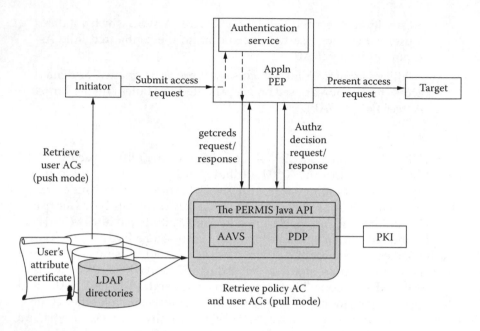

FIGURE 11.5
The PERMIS decision engine.

PERMIS provides an API in the Java language for accessing the AAVS and PDP, through calls to *getCreds* and *decision*, respectively. The API caller provides the authenticated name of the user, who is identified by either his LDAP DN or his public key (or proxy) certificate, to *getCreds*. A fuller description of PERMIS can be found in [14].

11.5   GT4, Shibboleth, and PERMIS Integration

Several different research projects have recently been carried out to integrate these three different security technologies together. The first project, carried out jointly between the University of Kent and the Globus Team, integrated Globus Toolkit and PERMIS, so that GT4 could replace grid-mapfile authorization with PERMIS policy-based authorization. This was enabled by specifying an open protocol, under the auspices of the Global Grid Forum (GGF), which allows a grid application to make a call to any external PDP. The protocol is a profile of the OASIS SAML Authorization Decision Request-Response protocol [12] and is specified in [15]. A full description of the integration can be found in [16]. In essence, the Globus Toolkit has been modified so that it can be configured to call multiple PDPs and Policy Information Points (PIPs) in parallel, including external ones (see Figure 11.6). PIPs are used to collect information, such as user attributes, that are needed by the PDPs. GT4 combines all the access control decisions of the multiple configured PDPs with a

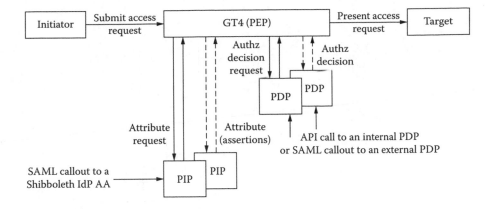

FIGURE 11.6
Globus Toolkit callouts to PIPs and PDPs.

"deny overrides" policy, meaning that if any single PDP denies access, or if no PDP grants access, then the user's request will be rejected. PERMIS was also modified so that it can now run as a standalone external decision engine as well as an integrated internal one.

The next project, SIPS, at the University of Kent, integrated PERMIS with Shibboleth, so that a Shibboleth resource can utilize the PERMIS policy-based authorization infrastructure. After the Attribute Requester in the Shibboleth SP has received the user's attributes from the AA in the IdP, these attribute assertions are forwarded to the external PERMIS decision engine in order for it to make an access control decision based on the policy set by the re-source owner. This is shown schematically in Figure 11.7. Conceptually, this entailed building a Shibboleth PEP that will call an external PERMIS PDP. The mod_permis module shown in Figure 11.7 fulfills this function. A full description of this integration can be found in [17].

The next project, GridShib, run by the Globus team, has integrated the Shibboleth IdP's AA with GT4, so that grid users can leverage the attributes they have been assigned for accessing Shibboleth-protected Web resources to access grid resources as well. It also means that the same administrative AA function can be used to assign attributes to users regardless of whether they are subsequently needed to access grid or Web-based applications. This should significantly reduce the administrative burden on organizations. There was however a major conceptual hurdle to overcome with this integration, due to the inherently different naming philosophies of GT4 and Shibboleth. Grid jobs are always run by users who have globally unique and static DNs. Shibboleth resource accesses are always based on dynamically assigned ran-dom user handles that change with every Shibboleth session. So how can GT4, which has successfully authenticated the user's DN using his current proxy certificate, know which dynamic handle to use in order to retrieve the user's attributes from the Shibboleth AA? Clearly it cannot. The GridShib project has temporarily sidestepped this problem by providing a DN mapping plug-in

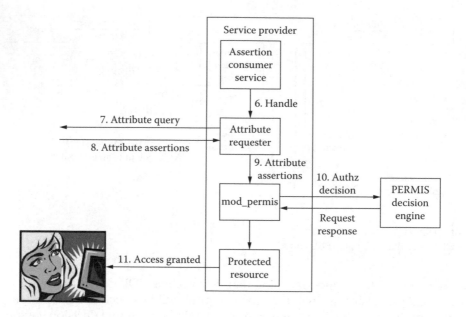

FIGURE 11.7
Shibboleth and PERMIS integration.

for Shibboleth IdPs (called the Distinguished Name Binder in Figure 11.8) that allows the IdP to map the user's DN into its local (static) name for the user. This temporary fix is not ideal, since user privacy is lost — the user is always known by his permanent name — and furthermore there is an unnecessary management overhead since a separate name mapping has to be carried out

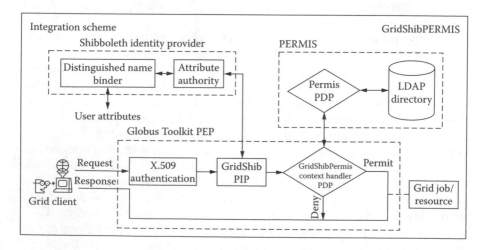

FIGURE 11.8
Globus Toolkit callouts to a Shibboleth PIP and a PERMIS PDPn.

for every user. The Globus team is currently working on a more flexible and permanent solution to this problem, but at the time of this writing it is not clear what this will be. Another problem this project faced was how to tell GT4 the location of the user's IdP AA, i.e., what component will provide the equivalent functionality of the WAYF service of Shibboleth? The initial solution chosen by the Globus team is to require the user to pass the provider ID parameter to GT4 in his initial grid access request. Again, this solution is not ideal, as it puts too much burden on the user, and so the Globus team is currently working on a more flexible solution to this. One idea is to include the provider ID parameter in the proxy certificate generated by the MyProxy server, but this will require modifications to MyProxy. A full description of the current design of GridShib can be found in [18].

In parallel with the GridShib project, the team at Kent integrated PERMIS with GridShib, so that GT4 can now call PERMIS as an internal PDP via its API function, after it has called the GridShib PIP. This is shown in Figure 11.8. This integration was relatively straightforward, since once GT4 has a clean API call to a PDP, then it should be relatively easy to plug in any PDP such as PERMIS. A full description of this can be found in [19].

The final piece in the integration jigsaw puzzle is to allow a user who does not have a public key certificate to run a grid job using his normal organizational authentication credentials. The University of Manchester is performing this integration in the SHEBANGS project [22]. A key component of their design is a new component called the Credential Translation Service (CTS), which behaves like a standard Shibboleth SP in the Shibboleth world, and an online CA and myProxy client in the grid world. The sequence of events is shown in Figure 11.9. The user accesses the CTS Web service using his normal Web browser (step 1) in the same way that he would access any other Shibboleth-protected resource. Following the standard Shibboleth procedure, the user is redirected to the WAYF service (step 2), chooses his home

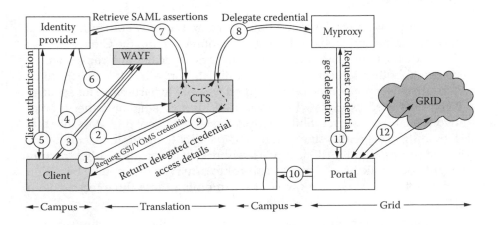

FIGURE 11.9
Integrating Shibboleth authentication with the Grid University of Manchester.

organization (step 3), is redirected there (step 4) and authenticates using his normal authentication method (step 5). The user is then redirected back to the Shibboleth SP (step 6) and the CTS contacts the IdP to pick up the user's attributes (step 7). Up to this point in time the system has followed the standard Shibboleth procedure. The CTS now behaves like an online CA and generates a new key pair and public key certificate for the user. The certificate differs from that of a standard CA certificate in that it is valid for only a short period of time, typically the same as that of a proxy certificate, and it contains the attributes of the user returned from the IdP AA. The short lifetime is so that the CA does not need to revoke any of its issued certificates. The name inserted into the certificate is based on the IdP's returned attributes, which should contain the local username of the user so that the user's DN can be (almost) the same for every grid session. This is important if the user needs to retrieve the output of a grid job in a different session to the one in which it was started. Of course, user privacy is lost in this case since the user's identity in the grid world remains fixed throughout all the sessions.

The final CTS component is a MyProxy client, whose functionality is to contact an external MyProxy server, register the new user in it, and deposit a new proxy certificate in there (step 8). The private key that is generated by the online CA component never needs to leave the CTS since it is only needed in order to interact with the external MyProxy server in order to sign the proxy certificate that is created there. The private key can be destroyed once the CTS has finished its work. The CTS finally responds to the user with the username, password, and url of the MyProxy server that contains his new proxy certificate (step 9). The user can now launch his grid job in the normal way using a grid portal front end that connects to the MyProxy server (steps 10–12)

11.6   Conclusions

In this chapter we have described how users are typically authenticated and authorized to run grid jobs and use grid resources, using the Globus Toolkit (GT). Since these mechanisms are usually grid specific, this causes inconvenience to both grid users and grid administrators alike. If grids are to be rolled out to user populations at large, then we need to harmonize the security mechanisms used by grids with those used for accessing any other type of network-enabled resource. Shibboleth is a relatively new protocol suite that is providing users with a single sign-on capability to multiple federated distributed resources. PERMIS is a relatively new authorization infrastructure that provides policy-based access control to any type of resource based on a policy set by the resource owner. By combining these three technologies together, we simultaneously enable

- users to access any type of network resource, whether via the grid or the Web, in a consistent manner using the same set of authentication and authorization credentials,

- administrators to assign attributes to users in a common way, in order to allow users to be granted access to both grid or Web-based resources
- resource owners to write common policies for who can access their resources, under what conditions, regardless of whether they are connecting to the resource via a grid, a Web service, a remote login service or locally.

We have described several recent research projects that are working toward this common objective. While not all the issues in this problem space have been solved entirely, nevertheless we now have workable solutions that are being piloted in various applications around the world. We expect the mechanisms described here to be rolled out to much larger sets of users in the coming years.

Acknowledgments

The author would like to thank the U.K. JISC for funding part of this work under the Grid API, SIPS, and GridShibPERMIS projects.

Bibliography

1. S. Tuecke, V. Welch, D. Engert, L. Pearlman, and M. Thompson. "Internet X.509 Public Key Infrastructure (PKI) Proxy Certificate Profile." RFC3820, June 2004.
2. C. Neuman, T. Yu, S. Hartman, and K. Raeburn. "The Kerberos Network Authentication Service (V5)." RFC 4120, July 2005.
3. A. Whitten, and J.D. Tygar, "Why Johnny Can't Encrypt. A Usability Evaluation of PGP 5.0." Proceedings of the Eighth USENIX Security Symposium, August 23–26, 1999, pp. 169–184.
4. "Overview of the Grid Security Infrastructure." Available from http://www.globus.org/security/overview.html.
5. P. Karlton Frier, and P. Kocher, "The SSL 3.0 Protocol," Netscape Communications Corp., Nov. 18, 1996.
6. OASIS, "Web Services Security: SOAP Message Security 1.1 (WS-Security 2004)," OASIS Standard Specification, 1 February 2006.
7. Steve Anderson et al. "Web Services Secure Conversation Language (WS-SecureConversation)," February 2005. Available from ftp://www6.software.ibm.com/software/developer/library/ws-secureconversation.pdf.
8. J. Basney, M. Humphrey, and V. Welch, "The MyProxy Online Credential Repository," Software: Practice and Experience, Volume 35, Issue 9, July 2005, pages 801–816.

9. GT4 Security at http://www.globus.org/toolkit/docs/4.0/security/key-index.html.

10. D. Balfanz, G. Durfee, D.K. Smetters, and R.E. Grinter, "In Search of Usable Security: Five Lessons from the Field." IEEE Security and Privacy, Sept/Oct 2004, pp. 19–24.

11. Tom Scavo and Scott Cantor. "Shibboleth Architecture: Technical Overview." Working Draft 02, 8 June 2005. Available from http://shibboleth.internet2.edu/shibboleth-documents.htm.

12. OASIS. "Assertions and Protocol for the OASIS Security Assertion Markup Language (SAML) v1.1." 2 September 2003. See http://www.oasis-open.org/committees/security/.

13. ANSI. "Information technology — Role Based Access Control." ANSI INCITS 359, 2004.

14. D.W. Chadwick and A. Otenko "The PERMIS X.509 Role Based Privilege Management Infrastructure." Future Generation Computer Systems, 936 (2002), December 1–13, 2002. Elsevier Science BV.

15. Von Welch, Rachana Ananthakrishnan, Frank Siebenlist, David Chadwick, Sam Meder, and Laura Pearlman. "Use of SAML for OGSI Authorization," Aug 2005, Available from https://forge.gridforum.org/projects/ogsa-authz.

16. David W Chadwick, Sassa Otenko, Von Welch. "Using SAML to link the GLOBUS toolkit to the PERMIS authorisation infrastructure." Proceedings of Eighth Annual IFIP TC-6 TC-11 Conference on Communications and Multimedia Security, Windermere, U.K., 15–18 September 2004, pp. 251–261.

17. Wensheng Xu, David Chadwick, and Sassa Otenko. "Development of a Flexible PERMIS Authorisation Module for Shibboleth and Apache Server." Proceedings of 2nd EuroPKI Workshop, University of Kent, July 2005.

18. Tom Barton, Jim Basney, Tim Freeman, Tom Scavo, Frank Siebenlist, Von Welch, Rachana Ananthakrishnan, Bill Baker, and Kate Keahey. "Identity Federation and Attribute-Based Authorization Through the Globus Toolkit, Shibboleth, GridShib, and MyProxy." NIST PKI Workshop, April 2006.

19. D.W. Chadwick, A. Novikov, and O. Otenko, "GridShib and PERMIS Integration." *Campus-Wide Information Systems.* Vol. 23, No. 4. 2006, pp. 297–308.

20. J. Myers. "Simple Authentication and Security Layer (SASL)." RFC 2222. October 1997.

21. ISO 9594-8/ITU-T Rec. X.509. (2001). The Directory: Public-key and attribute certificate framework.

22. Shibboleth Enabled Bridge to Access the National Grid Service, http://www.mc.manchester.ac.uk/research/projects/shebangs.

12

Grid Security Architecture: Requirements, Fundamentals, Standards and Models

Jose L. Vivas, Javier Lopez and Jose A. Montenegro

CONTENTS

Abstract   Grid computing is concerned with the creation of distributed virtual organizations across multiple control domains to enable the sharing of diverse remote resources. Due to its multi-institutional nature, securing the grid is one of the main challenges in grid computing. In this chapter we provide an overview of the grid security fundamentals, standards, requirements, models, architecture, and use patterns. We survey the major security challenges and requirements for grids, the main grid security models that address these requirements, current grid security architectures, emerging grid security standards and standard bodies, the convergence of grid and Web services, and the emerging enterprise grids.

12.1　Introduction

A grid may be defined as a collection of computing resources distributed over a local or wide area network, and available to an end user as a single large computing system. Originally, the grid focused on the areas of computing power, data access, and storage resources. It was intended for large-scale and distributed scientific computing that requires efficient and dynamically determined access to large amounts of data and computational resources that are distributed along several independently administered networks. However, the use of grid computing has been expanding lately to include deployment of grid technologies within the context of business [46], which significantly widens the range of applicability of grid technologies. Standard interfaces for business services have also been leveraged by grid computing. grid computing has been targeting such differing areas as finance, medicine, decision-making, collaborative design, and utility computing. The focus today is on coordinated resource sharing distributed across virtual organizations. However, shareable on-demand resources in commercial applications greatly complicate resource sharing and introduce new challenges related to federated security and integration.

Fundamental to grid computing is the notion of scalable virtual organization (VO) [1], which may be defined as a dynamic set of individuals and/or institutions that share resources and services according to a set of well-defined rules and policies. The grid vision is to provide unlimited power and information access to end users through the creation of dynamic VOs for secure and agile resource sharing among individuals and organizations. VOs may span several administrative domains, each one with its own security requirements and policies. Hence, interoperability among the multiple domains involved in a VO requires that VO-defined policies comply with domain-level policies, while at the same time maintaining a clear separation among virtual and real protection domains in a context in which they may superpose and intersect each other in a variety of ways.

Security has been a central issue in grid computing from the outset, and has been regarded as the most significant challenge for grid computing [6]. This is particularly true for enterprise grids. Significant compromises in security might be the result of an inadequate understanding of the security implications of a grid. The security requirements and policies are determined largely by the architectures developed for these types of applications, which are distinguished from client-server architectures by the fact that grid environments assume a dynamic and simultaneous use of a large number of resources from a number of administrative domains. Although the intention has been from the outset to use available security mechanisms as much as possible, this requirement could not be met by mechanisms that were devised largely for insulating and protecting networks from their environment, as in intranets and virtual private networks. As a result, novel security technologies have been evolving all the time within the grid community, including solutions for the management of credentials and policies, new resource management protocols for coallocation of multiple resources and for secure remote access to data and computing resources, and new information query protocols and data management services [7].

The requirements of grid computing are to a great extent in contradiction with the security policies and mechanisms related to administrative domains, since the objective of a grid is basically to circumvent the barriers imposed by these mechanisms by the establishment, in an ad hoc manner and for any desired period of time, of a virtual domain emulating the behavior of real domains. In consequence, grid computing has given rise to new security challenges, both for providers and for users, that could not be immediately met by available security technologies as the latter were intended to meet a set of requirements that was in many cases in contradiction to the requirements associated with grid computing. In order to provide resources to nonlocal members, system administrators must accommodate mechanisms and policies that are not completely under their control and that force them to open some previously closed access points. Therefore, the task of grid security engineering has been largely to reconciliate these antagonistic set of requirements, thus enabling components to be administered independently, according to local policies, and allowing users to achieve the desire level of quality of service regarding confidentiality, integrity, and availability requirements.

A key challenge here is the assignment of users, resources, and organizations to a VO. Security issues related to this task include the specification of federation, delegation, and access control among the participants. A further requirement, which gives rise to new security problems for current administrative domains, is the need to have hundreds of processes in different domains collaborating with each other in order to carry out a particular computing task. This kind of computation requires the dynamic establishment of multiple trust and security relationships among processes, turning authentication, delegation, and authorization into major challenges.

The presence of multiple administrators raises also many issues concerning accountability and responsibility. Other key issues concern interaction with firewalls and the process of creation and destruction of VOs. The requirements associated with grid computing can therefore not be met within the framework of client-server relationships with tight access control by individual domains.

Recently we have seen an evolution toward a grid system architecture based on Web services concepts and technologies and message-level security [2]. Grid computing is rapidly turning into a multifaceted discipline driven by international bodies and research projects, also attracting the interest of commerce and industry. This development, together with the adoption of Web services, has exerted a great impact on its architecture, infrastructure, standards and protocols, and also produced a much more fragmented landscape

[20]. As a result there is presently no broad consensus on which standards to follow and on the implementation of the architecture.

Enterprise grid computing poses also new challenges and unique requirements. In enterprise grids typically a single organization is responsible for managing a shareable set of resources and composing higher-order services with value for the business. Those resources may be owned by several businesses, e.g., independent service providers or outsourcing services firms with no geographic limitations. Unique security requirements are associated with enterprise grids because of the needs of organizational security, privacy, and regulatory compliance goals.

In this text we present the security requirements and challenges encountered in grid environments. We provide an overview of the grid security fundamentals, standards, requirements, models, architecture and use patterns, survey the major security challenges and requirements, the grid security models addressing these requirements, current grid security architectures, emerging grid security standards and standard bodies, the current convergence of grid and Web services, and the emerging enterprise grids. We focus mainly on the security model associated with the service-oriented Open Grid Services Architecture (OGSA) [24], and the OGSA suite of security services and components. It is our hope that this chapter will give those with a background in computer security, but otherwise unacquainted with grid computing, a good introduction to the field. We concentrate on the high-level aspects of grid computing, and omit questions about mechanisms, technologies, and implementation.

The rest of the chapter is organized as follows. In Section 12.2 we give an overview of grid security standards and corresponding standard bodies. Section 12.3 is dedicated to a description of use patterns, general requirements, assumptions, and challenges concerning grid computing. Finally, in Section 12.4 we concentrate on the presentation of the most important grid security models and architectures.

12.2 Grid Security Standards

The requirements concerning integration and interoperability in grids call for an extensive use of standard interfaces. Standardization is a key to the realization of the grid vision, enabling the portability, interoperability, and reusability of components and systems, as well as discovery, access, allocation, and monitoring of services and resources in grid environments. By facilitating the adoption of good practices, it is also important for security in general.

In this chapter we present the most relevant standards bodies (Section 12.2.1) and standards (Section 12.2.2) related to grids and grid security. Since Web service security standards are now an integral part of grid computing, we dedicate Section 12.2.3 to a presentation of WS-Security standards. For more details see also [20].

12.2.1 Standard Bodies

The main standard body for the grid is the Global Grid Forum [40], which works together with industrial organizations and has a decisive impact over the definition of security requirements and the adoption of infrastructures. Other important standard-setting bodies in grid computing are the World Wide Web Consortium, the Web Services Interoperability Organization, the Advancement of Structured Information Standards, and the Distributed Management Task Force. These and other relevant standard bodies are presented below.

Global Grid Forum. The GGF [40] was formed in 1998 and consists of community-initiated working groups developing best practices and specifications for grid computing. GGF creates four types of documents: informational, experimental, community practice, and recommendations. Work is divided into seven areas, one of which is concerned with technical and operational security issues in grid environments, including authentication, authorization, privacy, confidentiality, auditing, firewalls, trust establishment, policy establishment, scalability, and management. GGF drafts define the delegation protocol for remote creation of *X.509 Proxy Certificates and* GSS-API [22] extensions for grid computing. There are currently three groups working on security:

- *Open Grid Service Architecture Authorization* (OGSA AUTHZ-WG), whose objective is to define specifications to facilitate interoperability and plugability of authorization components in the OGSA framework.
- *Firewall Issues* (FI-RG).
- *Trusted Computing* (TC-RG), whose purpose is to evaluate how the capabilities of TC can be used in a grid context.

OASIS. Founded in 1993, OASIS [41] is a not-for-profit global consortium that promotes standards for e-business, focusing primarily on higher-level functionality, including security, authentication, and reliable messaging. There is a committee dedicated to the development of security standards for e-business and Web services applications. OASIS is responsible for the WS-Security standard, recognized as the foundation for securing distributed applications and Web services.

World Wide Web Consortium. The W3C [42] is an international organization initiated in 1994 to develop Web standards and guidelines, and promote common and interoperable protocols. It created the first Web services specification in 2003, focusing on SOAP and the Web Services Description Language (WSDL).

Distributed Management Task Force. The DTMF [43] is an industry-based organization founded in 1992 to develop management standards and interoperability for Enterprise and Internet environments. It formed an alliance with the GGF in 2003 in order to build a unified approach to the provisioning and sharing and management of grid resources and technologies. Two working groups are dedicated to security issues.

- *Security Protection and Management (SPAM) Working Group.* The goal of this working group is to ease the manageability of heterogeneous security systems within an enterprise or service provider environment.

- *User and Security Working Group.* The objective of this working group is to provide a set of relationships between the representations of users, their credentials, privileges and permissions, and the resources and resource managers involved in security management.

Internet2. Internet2 [44] is a consortium of groups from academia, industry, and government, formed in 1996 to develop and deploy advanced network applications and technologies for research and higher education. Several Internet2 working groups target grid standards, e.g., the *Higher Education PKI Technical Activities Group, the Peer-to-Peer Working Group*, and the *Shibboleth* project. Internet2 is part of the *EDUCAUSE/Internet2 Computer and Network Security Task Force*, which promotes practices and solutions for the protection of information assets and critical infrastructures for higher education, and is advised by *SALSA*, an oversight group consisting of technical representatives from the higher education community. The *SALSA-NetAuth Working Group* deals with the data requirements and implementation, integration, and automation technologies associated with understanding and extending network security management.

Liberty Alliance. The Liberty Alliance [45] is an international alliance of companies, nonprofits, and government organizations formed in 2001 to develop an open standard for federated network identity that supports network devices and addresses technical, business, and policy challenges concerning identity and Web services. It has developed the Identity Federation Framework, which enables identity federation and management.

Web Services Interoperability Organization. The WS-I [47] is an open industry organization that promotes Web services interoperability across platforms, operating systems, and programming languages. WS-I provides guidance, recommended practices, and supporting resources. WS-I creates and supports generic protocols for the interoperable exchange of messages between Web services. There are currently six working groups, one of them dedicated to security, the Basic Security Profile Working Group, which is developing an interoperability profile dealing with transport security, SOAP messaging security, and other security issues. A set of usage scenarios and related message exchange patterns is being developed by the Working Group. A working draft with interoperability and security recommendations was released in March 2006 [33].

Enterprise Grid Alliance. The EGA [46] consortium is an open, vendor-neutral organization formed to develop enterprise grid solutions and accelerate the deployment of grid computing in enterprises. EGA promotes open, interoperable solutions, and best practices focusing exclusively on the needs of enterprise users. The EGA is addressing requirements for deploying commercial applications in a grid environment. Initial focus areas include reference models, provisioning, security, and accounting. The EGA's Grid Security Working Group (EGA-GSWG) is dedicated to the identification of the unique security threats, issues, and requirements associated with enterprise grid architectures and computing.

12.2.2 Grid Security Standards

The de facto standard middleware for grid computing is the Globus Toolkit (GT) [48], and for grid security the GT's *Grid Security Infrastructure* (GSI) [49]. The Globus Toolkit is an open-source software that provides a set of services supporting collaboration across dynamic, multi-institutional virtual organizations. GSI was implemented by the Globus Toolkit, and uses X.509 identity and proxy certificates. GSI is based on standard technologies, such as TLS and secure Web services specifications.

The most important grid standard today is the *Open Grid Service Architectures* (OGSA) [50] presented in Section 12.4.1. OGSA is promoted by the OGSA Working Group of the Global Grid Forum, created in September 2002 to draft specifications. The Globus Toolkit has adopted this standard in the latest versions.

The first introduction of OGSA was the Open Grid Services Infrastructure, OGSI v1.0 [23] released in June 2003. OGSI is based on the concept of *grid service*. Dissatisfaction with OGSI, which required modifications to standard WSDL, led to an effort to define an alternative infrastructure based on pure Web services specifications. On January 2004 the WS-Resource Framework (WSRF) [51] was announced. WSRF contains specifications for expressing the relationship between stateful resources and Web services. After revision, the final result was submitted to two OASIS technical committees, the WS-Resource Framework (WSRF) TC and the WS-Notification (WSN) TC. Several specifications were standardized by both committees.

Alternatives to the WSRF include:

- **Basic Profile from the WS-I**: The Basic Profile [33] contains guidelines for using Web service standards SOAP, WSDL, and UDDI.
- **Web Services Grid Application Framework**: The WS-GAP [15] proposes to extend basic Web services functionality in order to meet the

needs of Grid applications; it uses the Web services standard WS-Context to make services stateful.

- **WS-I+**: created by the Open Middleware Infrastructure Institute (OMII), WS-I+ [19] is a set of Web services specifications that can be used to build interoperable Web service grids. Basically, WS-I+ extends WS-I profiles in order to provide access to core functionalities required by many e-Science applications. OMII [39] is an institute established by the U.K. e-Science Programme to act as a center for expertise in grid middleware. The OMII specified a roadmap to allow the capture of generic middleware components from multiple projects in a way that facilitates interoperability with grid services standards and OGSA developments.

12.2.3   WS-Security

WS-Security [26] is a Web service standard initially released by Microsoft in October 2001. In April 2002 IBM and Microsoft released a joint "Security in a Web Services World" document [27]. This defined a security framework for Web services, the first of which is WS-Security. Later specifications for Web services security include *WS-Trust* [28], *WS-Policy* [29], *WS-SecureConversation* [31], *WS-Federation* [32], *WS-Privacy* (unpublished), and *WS-Authorization* (unpublished). In 2002 WS-Security specification was submitted to the OASIS standards body. A Web services security group was formed in OASIS in order to develop WS-Security as an OASIS standard. WS-security standards are now an integral part of grid computing.

WS-Security is primarily for securing SOAP messages. It defines security tokens in SOAP messages and how they and other parts of a SOAP message can be encrypted and signed by XML Security specifications, i.e., *XML Signature* and *XML Encryption*. WS-Security includes specifications such as WS-Trust, WS-Policy, and WS-SecureConversation.

WS-security defines element names in order to package security tokens into SOAP messages. On top of it there is a conceptual model that abstracts different security technologies into "claims" and "tokens." A claim is a statement relating a subject with a property, e.g., an identity, and may be used for access control. A token is an XML representation of security information, e.g., a password, X.509 digital certificates, or a Kerberos ticket. Further specifications build on these concepts and show how to apply for security tokens, how tokens are related to identity, and how to associate security information with a Web service.

Interoperability across domains with different security technologies are as important for Web services as for grids, and similar solutions apply. WS-Security provides a level of abstraction for companies using different security technologies to communicate securely using SOAP. In this way, existing or new security technologies and infrastructures can be used for both Web and grid services security.

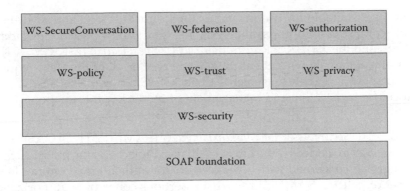

FIGURE 12.1
WS-Security model.

The WS-Security model defines scenarios where the integrity and the confidentiality of SOAP messages are ensured while traversing intermediaries, which may themselves perform security functionality. Additional specifications, such as WS-Trust and WS-Policy, define how security tokens are issued.

The model for WS-Security is shown in Figure 12.1. Each specification depends on its predecessors. SOAP, which is transport-independent, is at the base of the diagram. WS-Security is on top of SOAP. It provides a means for encrypting and signing portions of a SOAP message using XML Signature and XML Encryption, and for enclosing security tokens in a SOAP message. We will describe the XML and Web service standards that are relevant for grid security.

XML Signature. *XML Signature* (XML-SIG) [16] was the first XML security standard to reach recommendation status. XML Signature is a building block for WS-Security. It provides integrity for data, and is used also for authentication and nonrepudiation. WS-Security provides a SOAP binding for XML signature by defining how an XML signature can be placed in a SOAP message. XML Signature makes it possible to express a signature in a standardized XML format, and to sign only part of an XML document. It contains a KeyInfo element that can be used to reference the public key of the signer.

XML Encryption. *XML Encryption* (XML-ENC) [17] allows confidentiality to be satisfied on an end-to-end basis. Portions of an XML document can be selectively encrypted, and encrypted data can be expressed using XML. XML may also express an encrypted key, information about how an agreement was reached on the encrypted key, reference to the encrypted data, information about the data type of the encrypted document, and the encryption method used. XML-ENC uses the *KeyInfo* block from XML Signature.

XKMS. The *XML Key Management Specification v2.0* (XKMS 2.0) [18] is a W3C recommendation that specifies protocols for distributing and registering public keys, suitable for use in conjunction with XML-SIG and XML-ENC. XKMS comprises two parts, the *XML Key Information Service Specification* (X-KISS) and the *XML Key Registration Service Specification* (X-KRSS). X-KISS is a protocol that allows a client to delegate part or all of the tasks required to process XML Signature elements to an XKMS service. X-KRSS defines a protocol for registration and management of public key information.

SAML. The *Secure Assertion Markup Language* (SAML) [37] is an OASIS specification, later extended by the Liberty Alliance Project and the Internet2 Shibboleth group, concerned with access control for authenticated entities based on a set of policies. SAML allows trust assertions concerning authorization, authentication, and attributes of specific entities, to be specified using XML. An assertion is either a claim, a statement, or a declaration, and can be accepted as true to the extent that the certification authority that issued the claim can be trusted. Thus, authorities for attributes and authentication are crucial elements in the SAML model. SAML also defines a client/server protocol for exchanging XML messages. Typically, the underlying transport protocol is SOAP running over HTTP. SAML also enables portable trust by supporting authentication assertions between multiple administrative domains, a capability that is very important for grid services. Furthermore, it allows the mapping of access control elements between different systems. SAML has been proposed as a message format for expressing and requesting authorization assertions from an OGSA authorization service [11]. SAML 2.0, which became an OASIS standard in March 2005, added features to enable communication between SAML authorities, to enhance authentication methods, and to protect privacy.

XACML. The *Extensible Access Control Markup Language* (XACML) [38], developed at OASIS, is a language for expressing access control policies. XACML has the ability to express the complex policies that are not embedded into application code, and can also associate actions, called obligations, with access control decisions. Important for grid services and context-based authorization is the ability endowed by XACML to base decisions on a resource's properties, or on environmental factors such as date, time, and location. It may also take into account properties such as role or group membership of all the entities involved in a request, including intermediaries to the request. Of fundamental importance for grids is the ability of XACML to operate in large-scale environments with multiple administrators that create policies. Specific features have been defined to enable XACML and SAML to work together. XACML 2.0 was approved as an OASIS Standard in February 2005.

WSS: SOAP Message Security. The *Web Services Security (WSS): SOAP Message Security v1.0* specification [21] defines the use of *security tokens* and *digital signatures* to protect and authenticate SOAP messages. Three main mechanisms are provided: message integrity, message confidentiality, and the ability to send security tokens as part of a message, which can be associated with message contents. Message integrity and confidentiality are provided by the encryption and digital signature of XML elements in the message. WSS 1.0 became an OASIS Standard in 2004, and WSS 1.1 in February 2006.

WS-Policy. The *WS-Policy* specification [29] defines a policy data model and an extensible grammar for expressing the capabilities, requirements, and general characteristics of a Web service. It is used to convey the conditions for an interaction between a Web service requestor and a Web service provider. WS-Policy defines fundamentals used for creating security policies, such as the type of security tokens a service will accept, supported algorithms for encryption and data signatures, and privacy attributes. WSDL bindings are typically used in order to attach policy information to a Web service.

WS-SecurityPolicy. The *WS-SecurityPolicy* [30] defines a set of security policy assertions for use with the WS-Policy framework with respect to security features provided in WSS: SOAP Message Security, WS-Trust, and WS-SecureConversation.

WS-Trust. *WS-Trust* [28] is thought to enable applications to construct trusted SOAP message exchanges. It defines extensions that build on WS-Security to broker trust relationships and to provide a framework for requesting and using security tokens, managing trusts, and establishing and assessing trust relationships,. The extensions also provide methods for issuing, renewing, and validating security tokens. Trust relationships can be direct or brokered. In the latter case a *trust proxy* is used to read the WS-Policy information and request security tokens from an issuer. WS-Security is able to transfer security tokens using XML Signature and XML Encryption for integrity and confidentiality. The trust model also allows delegation and impersonation.

WS-Privacy. *WS-Privacy* (unpublished) uses a combination of WS-Policy, WS-Security, and WS-Trust to communicate privacy policies. For privacy, incoming SOAP requests are required to contain claims that the sender conforms to desired privacy policies. These claims are encapsulated into verifiable security tokens with the help of the WS-Security specification. WS-Privacy defines also how to express privacy requirements in WS-Policy descriptions, and WS-Trust is used to evaluate the privacy claims included in SOAP messages.

WS-SecureConversation. *WS-SecureConversation* [31] defines extensions that build on WS-Security and WS-Trust to provide secure communication across messages. WS-SecureConversation is designed for the SOAP message layer, and has been described as "SSL at the SOAP level." Since there is no concept of a session for a group of SOAP messages, WS-SecureConversation allows a requestor and a Web service to mutually authenticate using SOAP messages, and also to establish a mutually authenticated security context that uses session keys, derived keys, and per-message keys. Asymmetric encryption is used to negotiate a symmetric key that can be used for a series of SOAP messages, thus avoiding message-level authentication. WS-SecureConversation builds upon WS-Security and WS-Trust to securely exchange contexts in order to negotiate and issue keys.

WS-Federation.   *WS-Federation* [32] acts at a layer above WS-Policy and WS-Trust, and explains how federated trust scenarios and relationships may be constructed and managed using WS-Security, WS-Policy, WS-Trust, and WS-SecureConversation. It describes how to manage and broker the trust relationships in a heterogeneous federated environment, including support for federated identities and management of pseudonyms. WS-Policy and WS-Trust are used to determine which tokens are consumed, and how to apply for tokens from a security token issuance service.

WS-Authorization.   *WS-Authorization* (unpublished) describes how access control policies for a Web service may be specified and managed. The specification is extensible with respect to both authorization format and authorization language, and supports both ACL-based and RBAC-based authorization.

12.3   Grid Security Requirements, Challenges and Use Patterns

The special security requirements of grid applications derive mainly from the dynamic nature of grid applications and the notion of virtual organization (VO), which requires the establishment of trust across organizational boundaries. In this kind of environment, security relationships can be dynamically established among hundreds of processes spanning several administrative domains, each one with its own security policies. Important requirements in this context are heterogeneity and site autonomy: a site must keep control over its resources and usage policies. As a result, the grid security requirements are complex and pose significant new challenges. Existing intradomain security solutions and infrastructures must be integrated into the overall security architecture and interoperate with interdomain security solutions, since organizations are not as a rule prone to change their internal security requirements and policies in order to become a part of a wider organization. Complex patterns of trust between the different organizations within a VO must thus be established by entities that must be able to determine the identities and rights of other entities to ensure that only legitimate ones may access the required resources. Grids focus on the users and their needs, allowing them to take advantage of multiple distributed resources located in several administrative domains. *Authentication* and *authorization,* and in general *trust policies and management*, have thus been major challenges in grid security. Moreover, two general nonfunctional requirements for grids have deep implications for security: integration and interoperability. In Section 12.3.3 we present these requirements, but to make the presentation more concrete we first show some typical usage scenarios in grid computing highlighting the corresponding security concerns involved, and introduce the underlying assumptions about grids as well as some terminology. Finally, Section 12.3.4 is dedicated to a presentation of the security requirements of the emerging Enterprise Grid Computing.

12.3.1 Underlying Assumptions and Terminology

We give here a short account of the terminology and underlying assumptions related to a grid system, its participants, entities, components, and specific policies [3].

The participants involved in a grid computation include *subjects* or *users*, *user proxies* operating on behalf of the user, *resources*, and *resource proxies* that are agents or processes operating on behalf of resources. *Credentials* are pieces of information about the identity of a user, such as passwords and certificates. Trust domains are administrative units with a local security policy, and consist of users and resources.

A grid environment consists of a *virtual organization*, multiple trust domains with local security policies that cannot be overridden by the grid security policy. Operations confined to a trust domain are subject solely to local policy, and can be implemented by a variety of mechanisms.

Subjects must have globally defined names besides local names, and there may exist partial mappings from global to local subject names. If there is a mapping in a trust domain for a determined global name that has been globally authenticated, then the subject is also assumed to be locally authenticated. The identity of the user needs to be passed transparently between sites during the execution of a job. This is the basis of *single sign-on*, which is made possible by the existence of a global identity. Access control decisions are always made locally on the basis of the local name.

Mutual authentication is required when an operation involves entities located at different trust domains. It is possible for a user to delegate a subset of his rights to a process to act on his behalf, thus enabling the execution of long-lived processes without user interaction, as well as the creation of new processes. Moreover, processes running on behalf of a single subject may share the same set of credentials, thus enhancing the scalability of the security architecture by avoiding the need to issue a unique credential for each process.

12.3.2 Typical Usage Scenarios

A variety of scenarios are typical of grid environments [5]. We briefly present some typical ones, together with some security issues each one brings forth.

A job execution request. A user submits a request to initiate a job, accompanied by a description of the job and the user's grid credentials, either personal credentials or VO-issued credentials. The request is thereafter evaluated by different policy evaluation points (PEPs) against both local and VO policies. If the request is authorized, it is mapped to a set of local credentials and enforced by local enforcement mechanisms. During job execution the user may make management requests to the job [12].

Resource allocation.   Resource allocation can be initiated by a user proxy or a process. The first step is to identify the resource proxy. Mutual authentications are then executed, upon which a request, possibly signed, will be sent to the resource. The resources check the requester's credentials, and if authorized the resource is allocated and a process is created on that resource if needed. The request can fail either because of an allocation, authentication, or authorization failure. It is the responsibility of the resource to enforce local authorization policies.

A job execution on a specified grid computer with local I/O.   Here the user designates the execution host and submits a job, possibly together with the code, to a grid gateway, i.e., a process that accepts remote resource requests. The job uses only remote computation cycles and possibly temporary file storage, input data is uploaded at job submission, the output is returned along the connection for job submission. The security requirements in this case include: (i) mutual authentication of user and grid gateway on the host; (ii) grid gateway on the host must map grid ID to a local one; (iii) a request must be submitted by the grid gateway to the resource gateway in a manner that enables the job to run as the authorized local user. Authorization to use the resource is performed here by the grid gateway.

A job execution on a specified grid computer with nonlocal I/O.   In this case the remote job must access nonlocal files, and therefore delegation in some form becomes necessary. Additional security requirements are as follows: (i) if file transfer must occur before execution, authorization must be given to transfer these files on behalf of the user, and delegation thus becomes necessary; (ii) otherwise, credentials must be obtained upon startup to obtain the data; (iii) a Kerberos ticket from the user may be needed since the remote job writes output to a local file server of type AFS or DFS; (iv) if the output is in the form of files that must be copied back to the user's machine submitting the job, credentials to be authorized with the grid gateway on the local machine are needed, as well as some form of delegation.

A job execution requiring a combination of resources from multiple sites. In this case, a user starts a coordinated job that needs to combine resources from multiple sites. Specific resources may be selected by a third-party service such as a scheduler, eventually following some explicit QoS or other kinds of user requirements. Remote execution at multiple sites may thus be required, together with the corresponding data manipulation. Possible security requirements associated with this scenario might involve (i) authorization to execute the required jobs or access data in each of the target grid machines according to the user's credentials, and, recursively, to access any resources that might be requested by any of the started processes; (ii) authentication, single or mutual, for any agents involved during job execution, starting with the user; (iii) mapping of Grid IDs to local IDs; (iv) possibly some kind of credential or privilege delegation, since the scheduler or any remote job might be required to act on the user's behalf.

A job execution requiring advanced scheduling. In some jobs, advance reservation of data storage, network bandwidth or compute cycles may be required. Possible security requirements associated with this scenario are: (i) delegation of a user's rights to a scheduler to make reservations; (ii) bandwidth reservations may require that a bandwidth broker knows at reservation time that the user's connection will come from an authorized site; (iii) the user should be able to authenticate itself as the entity that made the reservation; in the context of group membership and reservation made on behalf of a group, the user should be able to prove group membership; (iv) nonrepudiation: the resource proxy should not be able to falsely deny granting of reservation.

Job control. A job might be disconnected by the user and reattached later, possibly from another location, or a user might want to monitor a job's progress or enter steering information. Another user or collaborator may be allowed to monitor the job at some specific time. Possible security requirements associated with this scenario are: (i) access policy for a job may be required; (ii) authentication by the collaborator; (iii) auditing may be required since the grid software must provide a means of identifying which grid user started a local job.

Accessing Grid Information Services. Information Services are present in most Grid architectures for helping in the location of services and determining their status and availability. Typically, users will be able to read from the Directory Service, and entities such as processes will be able to enter information and set access policies for their information. Possible security requirements associated with this scenario are: (i) authentication between users and the Information Services; (ii) implementation of required access control policy by the Information Service; (iii) confidentiality or message integrity on the communication from the publisher to the Information Service; (iv) the Information Service must be trusted by the publisher.

Setting or querying security parameters. Entities in a grid environment may want to have the capability to constrain the manner in which they interact with each other. For instance, a user or resource provider may want to define message integrity and confidentiality parameters, stakeholders may want to

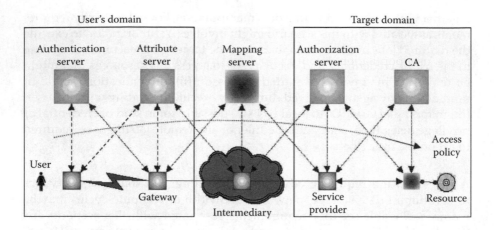

FIGURE 12.2
A service request scenario involving intermediaries.

set authorization policies or to revoke access, principals may want to specify trust grid hosts, require confidentiality on stored data, etc. All these scenarios are complex and meeting the requirements is in general difficult. For further details consult [5].

Auditing use of grid resources. A typical scenario of this kind is when a grid administrator may want to check a list of past requests and allow or deny accesses. This implies that (i) the resource gateway must keep an unforgeable log of all access including time of access and user identity; (ii) access to the log should be carefully restricted; (iii) a mechanism must exist to signal troublesome access requests. The usefulness of such a log file depends on how trusted a server is. Restricted access to the log may also be desirable. In this case there should be mechanisms to restrict access to the logs.

A typical service request scenario. In this scenario, drawn from [10] and illustrated in Figure 12.2, we show an example involving grid services, explained in Section 12.4.

A user in his own domain wishes to invoke a grid service in the target domain. The user first authenticates to an authentication server local to his domain, and obtains an identity credential. Thereafter the request is routed through a gateway, which may consult an attribute server to obtain the user's privilege attributes and rights. The assertions are then sent together with the service request. The request may be routed through an intermediary that is able to translate the assertions into a form that is understandable by the target domain and forwards the request according to a set of policies. Thereafter the target may receive the request and validate the certificate, and if successful, it can map the user's identity to a local one and make the appropriate authorization decisions using the locally defined policies.

The scenarios shown above illustrate many features associated with grid environments [3]:

- user population, resources pool, and the group of processes running on different sites are potentially large and dynamic.
- processes may communicate by a variety of mechanisms such as unicast or multicast.
- different authentication and authorization mechanisms can be present in a single job computation, according to the local security policies of the sites involved.
- a user may be associated with different local name spaces or credentials.
- local authentication, authorization and access control may apply at different sites.
- individual users may be associated with different local name spaces, credentials and accounts at different sites.

12.3.3 Security Requirements

We present below the most common general security requirements and challenges associated with grids.

Authentication.   Authentication mechanisms and policies are supposed to constitute the basis on which local security policies can be integrated within a VO [3]. Because of its complexity and heterogeneity, in a grid environment it is desirable to separate authentication from authorization. Difficult issues with respect to authentication in grids are scalability, trust across different certification authorities, revocation, key management, and delegation. Since processes with delegated authority act on behalf of their owner, there is a question of authentication in delegation, which becomes even more complex when delegation is chained. Key management is also an issue for several reasons. Due to user mobility users may require a portable medium for media storage. Furthermore, users may have different credentials, for instance, to cover different roles, which in practice means that numerous key pairs might become necessary.

Confidentiality.   Both privacy and intellectual property concerns require confidentiality in the use of data. Encryption is one of the mechanisms used to enforce confidentiality. The nature of grids forces data to be stored in accessible online databases. A confidential code may be requested to execute on a remote host, and confidential data may need to be used at remote locations. Data may also need to be replicated at multiple sites, and thus should be stored in an encrypted form and remain consistent throughout. Furthermore, not only data but also users and resources may have privacy requirements, and users may be protected under privacy laws to which all components must adhere. Mechanisms for protection of confidentiality should also protect against the deducibility of data. Finally, laws regarding privacy rights and encryption vary among countries and must be taken into account when deploying grid technologies across international borders.

Integrity.   Many applications have strong code or data integrity concerns. The trust status of remote resources is important when data arises from remote processing as the accuracy of results can be trusted only to the extent that the remote host generating the data is trusted. Integrity is also an issue with regard to delegation, since the set of rights that has been delegated must not be modified maliciously.

Authorization and access control.   Authorization is the process by which a subject is eventually allowed to access some resource. In grids local access mechanisms should be applied whenever possible, and the owner of a resource should be able to enforce local user authorization. Users also need a consistent way to get authorization to access grid resources across organizations. The first condition a user must meet in order to access the grid is that he is a member of the VO, but eventual roles played by the user or other attributes may also be taken into consideration. Authorization by identity is very common, but in a grid context resource owners may want to grant access based on, e.g., roles, group membership, credit worthiness, static or dynamic and context-based attributes. Confirmation that a user has the VO membership and the required roles and attributes must be possible to obtain.

   A resource provider in a grid environment must have reached some form of agreement with the VO to allow the use of the resource. The VO may wish to specify a portion of the resource usage policies, to manage jobs running on VO resources, or to give some group of users the ability to manage those jobs. The authorization policy system must thus be able to combine policies from the resource owner and the VO, express policies about resource usage, manage VO-wide jobs and resource allocations, and dynamically enforce fine-grained policies about resource usage [12].

Revocation.   Revocation is crucial for authentication in case of a compromised key, and for authorization when a VO is terminated or a user proves untrustworthy.

Distributed trust.   Trust is a complex theoretical issue. A grid must be constructed in a dynamic fashion from components whose trust status is hard to determine. For instance, a user that trusts R may not necessarily trust R to delegate the user's rights further. Determining trust relations between participant entities in the presence of delegation is important, and delegation mechanisms must rely upon stringent trust requirements.

Freshness. Freshness is related to authentication and authorization and is important in many grid applications. Validity of a user's proof of authentication and authorization is an issue when user rights are delegated and the duration of a job may span several weeks. Furthermore, some applications may want to state the number of times a given user may access a resource, a nontrivial problem when one user's rights are delegated to another user that may thereafter wish to access the resource.

Scalability. A grid must be easy to extend and capable of progressive replacement. Fault recovery and dynamic optimization should be usually possible, and degradation should happen gracefully.

Trust. Trust refers to the assured reliance on someone or something. Since VOs can span multiple security domains, trust relationships between domains are of paramount importance. Sites in a grid must be able to enter into trust relationships with grid users and maybe other grid sites as well. In a grid environment trust is usually established through exchange of credentials, either on a session or a request basis. Due to the dynamic nature of grid environments, trust can scarcely be established prior to session execution.

Single sign-on. A user should be able to authenticate only once, whereupon he may acquire, use, and release resources without further authentication. This is required since a user may want to access a large number of resources with different patterns of availability, access control policies, etc., that cannot be determined statically. Moreover, users may want to initiate computations running for long periods of time without needing to remain logged on all the time.

Delegation. Privilege delegation for operations executed by a proxy is a basic requirement for grid environments, among other reasons in order to satisfy the single sign-on requirement. Delegation of user rights depends upon the security requirements of the application. Delegation is hard to achieve securely in practice, since enabling the delegation of a user's rights gives rise to many unresolved subtle issues and has a great impact on the overall security of a system [13].

Privacy. Privacy is the ability to keep information from being disclosed to determined actors. Privacy can be important in many grid applications, for instance, in medical and health grids [14].

Nonrepudiation. Nonrepudiation refers to the inability to falsely deny the performance of some action. It is especially important in e-commerce involving money transactions. With the advent of enterprise grid this requirement becomes very important.

Credentials.   A credential is a piece of information that may be used to prove the identity of a subject, e.g., a password or a private key. Interdomain access requires a uniform way of expressing the identities of users or resources, and must thus employ a standard for the encoding of credentials. Furthermore, user credentials must be protected.

Exportability.   Code is required to be exportable and executable in multinational testbeds. As a result, bulk encryption cannot be required.

Secure group communication.   Authenticated communications for dynamic groups is required since the composition of a process group may change dynamically during execution.

Multiple implementations.   It should be possible to enforce security requirements with distinct security technologies and mechanisms.

Interoperability.   In the context of grids, interoperability means that services within a single VO must be able to communicate across heterogeneous domains. Interoperability guarantees that services located in different administrative domains are able to interact at multiple levels. This gives rise to many serious security concerns related to authentication, privacy, authorization, and policy enforcement. Services may be hosted in domains with different security mechanisms and policies, and interoperability between these services will depend on the trust models adopted.

With regard to policy management, security interoperability means that the security policies established by different parties in a VO can be made compatible, thus allowing the establishment of secure communications channels and security contexts following mutual authentication. This requires that users in different domains be able to identify each other. As a result, mechanisms for identity federation, mapping of identities, and credentials, must be made available, since global identities would be very impractical.

Interoperability with local security solutions.   Access to local resources is normally enforced by local security policies and mechanisms. Interoperability between sites and domains with differing local policies is necessary in a grid environment. In order to accommodate interdomain access, one or several entities in a domain may act as agents of external entities for local resources.

Integration.   In order to allow the use of existing services and resources, integration requirements call for the establishment of an extensible architecture with standard interfaces. Security integration is facilitated by the use of existing security mechanisms. The latter is also in part a consequence of the requirement for site autonomy with regard to security policies, and also of the fact that no single security technology would be able to address the inherent complexity of grid computing.

Uniform credentials and certification infrastructure. A common way of expressing identity, e.g., by a standard such as X.509, is necessary for interdomain access.

12.3.4 Enterprise Grid Computing

Enterprise grid computing [34] is the use of grid computing in the context of a business or enterprise. There are many requirements and challenges that are unique for enterprise grid architectures, managed by a single enterprise or business. Resources consists basically of computing, network, storage, and service capabilities. Resources and services need not necessarily be owned by an organization, they may also be available through service providers or outsourcing firms. The boundaries of the enterprise grid are defined by its sphere of management responsibility and control. An enterprise grid may extend across several data centers, and no geographical limitations exist.

Based upon the assessment of threats and risks, many security requirements have been highlighted that are specific for enterprise grids, which we show below, together with more general requirements, following [34]. We follow the terminology used in this document; for details see Section 12.3.4. In this model, a grid consists of entities called components, and the Grid Management Entity (GME) is a logical entity that manages those components and their mutual relationships.

Confidentiality. Communication must be secure between grid components for confidentiality, and the confidentiality of sensitive data must be preserved through the life cycle of grid components.

Integrity. Grid components must be validated for security and integrity in accordance with the grid security policy; integrity checks must be executed to guard against tampering with the wire; images used to provision grid components and settings during configuration processes, as well as information preserved from provisioning resources, must be validated for integrity.

Availability. Availability must be enforced often since it is obviously very important in many enterprise grids.

Identification. All components and user communities must be uniquely identifiable, and identities must be preserved.

Authentication. Communicating entities must be able to authenticate to each other; the GME must provide a functionality equivalent to an ordinary AAA (Authentication, Authorization, Auditing) server, including support for policy-based, extendible, and strong authentication mechanisms, and for role-based resource access control.

Authorization.   Grid components must be authorized to communicate with each other; authorization can be strict or loose depending on the nature of the organization.

Auditing.   It must be possible to track and resolve the dynamic binding of grid components; audit data must also be meaningful after reprovisioning or decommission of audited components.

Separation of Duties and Least Privilege.   The standards of access control policy, separation of duties, and least privilege, apply to enterprise grids.

Defense in Depth.   Traditional defense in depth measures such as DMZs (demilitarized zones) should be preserved in enterprise grids; additional security measures can be taken by utilizing security measures to reinforce systemic security at every layer of the DAGs (directed acyclic graphs) provided by the EGA reference model.

Secure failures.   The GME and the enterprise grid as a whole must be designed to fail securely, i.e., grid components must not be able to enter a vulnerable state.

Grid life cycle security.   A number of security requirements associated with the life cycle and reuse of grid components are unique for enterprise grids. These include the following.

- *Secure packaging*: Grid components must be logically packaged for provisioning from resources. This allows components to be logically isolated from each other, packages to be easily modifiable, revised and managed for integrity. Packages should be also digitally signed or encrypted according to the security policy of the site.
- *Secure update of deployed components*: secure communication with components to query state, update, and check pointing changes should be provided.
- *Secure archival*: it should be easy to extract needed imformation from a provisioned resource.
- *Secure reuse of grid components*.

Interoperable security.   Support for interoperable security across heterogeneous grid components must de provided since a homogeneous environment cannot be assumed in enterprise grids.

Secure isolation.   Since shareable pools of grid components may be used, the same secure isolation requirements associated with physically or logically siloed environments apply for enterprise grids.

Trust relationships. Trust relationships in Enterprise Grids include relationships between users, administrators, applications, and services to the GME and Grid components; important questions here include how trust is established, maintained, and terminated, and how trust violations are detected and addressed.

12.4   Grid Security Architectures and Models

In the early days of grid computing, the definition of grid was centered on computational aspects. A computational grid was defined as "a hardware and software structure that provides dependable, consistent, pervasive, and inexpensive access to high-end computational capabilities" [4]. Several custom middleware solutions were created, but interoperability was hard to achieve. Later, the focus changed to coordinated resource sharing according to well-defined policies, easier integration, security, and QoS aspects. With the advent of Web services, in recent years we have seen a merging of Web and grid services technologies. Today, the Open Grid Services Architecture (OGSA), announced in February 2002 by the Global Grid Forum, has become the standard architectural model for grid systems. Section 12.4.1 is entirely dedicated to OGSA and OGSA security, and in Section 12.4.2 we give a brief presentation of the Enterprise Grid Alliance reference model security.

12.4.1   OGSA

The Open Grid Services Architecture (OGSA) is a service-oriented architecture (SOA) that represents an evolution toward a grid system architecture based on Web services concepts and technologies, autonomic computing principles, and open standards for integration and interoperability.

Components in Web services are typically defined in terms of access methods, bindings of access methods to chosen communication mechanisms, and service discovery mechanisms. Some mechanisms are becoming de facto standards in Web services, such as the *Simple Object Access Protocol* (SOAP) [35], which uses XML technologies for messaging with HTTP as the underlying transport protocol, and the *Web Services Description Language* (WSDL) [36], in which signatures and bindings to protocols may be expressed in an XML document.

OGSA builds on concepts and technologies from both grid computing and Web services. OGSA introduces the notion of *grid service*, a potentially transient kind of Web service that conforms to a set of conventions for grid interaction expressed in terms of WSDL interfaces, extensions, and behaviors. OGSA also extends Web services with the important notion of *stateful service*, a service that may keep state information by retaining data between invocations.

OGSA was created to meet the challenges related to the integration of services across VOs running on top of different native platforms [2]. In the context of grid services, for instance, access control to resources amounts to controlling access to services through security protocols and policies. OGSA defines a set of core capabilities and behaviors addressing several aspects of grid computing and the need for standardization. It specifies a set of characteristics describing how service requestors should interact with OGSA service providers. An important concept related to grid services is the notion of *service virtualization*, which enables mapping of service semantics onto native platform facilities. OGSA also envisages mapping of security parameters between domains.

OGSA Security

Web services standards did not meet all grid security requirements from the beginning and were thus expanded with new service definitions. Grid requirements played a central role in the definition of WSDL 2.0 and in the review of *WS-Security*, a standard for creating secure message exchanges that provides mechanisms for authentication, confidentiality, encryption, and message integrity. OGSA introduces new challenges for security.

Web services security specifications include the Web Services Security Policy (*WS-Policy*) [30], XACML [38], SAML [37], WS-Security [26] for security token exchange, as well as the standards *WS-SecureConversation* [31] and *WS-Trust* [28] for authentication, establishment of security contexts, and trust relationships. However, OGSA introduces new challenges for security, and the specifications above have to be extended to address specific grid security requirements.

The *OGSA security model* builds on Web services security with specific extensions to cope with the challenges posed by virtual organizations. Security arises at various levels of the OGSA architecture. WS-Security is used to allow service requests to provide suitable tokens, for purposes of, e.g., authentication and authorization. For user authentication, delegation, and single sign-on, the OGSA uses the Grid Security Infrastructure (GSI) [49] protocol. End-to-end message protection is also required by the OGSA architecture, and provided by mechanisms such as XML encryption and digital signatures. Security components are also rendered as services, e.g., the OGSA authorization service that uses *WS-Agreement* with SAML and XACML (unpublished to date).

In the context of grid services, some security challenges gain a new dimension:

- **Integration**: reuse of existing services and interface abstraction for extensibility
- **Interoperability**: services located in different VOs and with different mechanisms and policies should be able to invoke each other
- **Trust relationships**: services should make access requirements available in order to enable access to them, and trust policies should be

specified and enforced, e.g., through exchange of credentials. Moreover, heterogeneity calls for some form of federation among security mechanisms

Special security challenges related to trust relationships are associated with the notion of *transient services*, a class of grid services that implements an interface that creates new grid service instances [2]. Transient services are created by end-users to perform some request-specific task that may involve execution of user code. Those challenges include the following:

- the requirement that it must be possible to control the *authorization status* under which transient services execute;
- *policy enforcement* by service providers even when users want to establish policies for the transient services they create;
- availability of the *assurance level* of a hosting environment for the benefit of the end user, including privacy, virus protection, firewall and VPN usage;
- *security policy composition* in the case that several policies are generated from different sources;
- *authority delegation* to enable transient services to perform actions on behalf of a user.

The OGSA *security model* stipulates that security mechanisms should be *pluggable* and *discoverable* by service requestors from a service description, enabling service providers to select their preferred mechanisms.

The Global Grid Forum's OGSA 1.0 [24] document targets security requirements, including authentication and authorization, security infrastructures, perimeter security solutions, isolation, delegation, policy exchange, intrusion detection and protection, and secure logging. It also specifies security services associated with message integrity, confidentiality and privacy, auditing, intrusion prevention, and access control. We show these requirements below.

- **authentication:** plug points for multiple authentication mechanisms should be provided.
- **delegation:** support should be provided for enabling delegation of access rights from requestors to services, and for the specification of delegation policies.
- **single sign-on:** authentication to a VO should happen only once per session for the end user.
- **credential renewal:** the user should be notified whenever the expiration time of a credential is approaching.
- **authorization:** various access control models should be allowed, and access to OGSA services based on the authorization policies of each service should be possible, as well as the specification of invocation policies by requestors.

- **privacy:** both service requestors and providers should be able to specify and enforce privacy policies.
- **confidentiality:** confidentiality should be possible to maintain both in point-to-point transport and store-and-forward mechanisms.
- **message integrity:** unauthorized changes to a message should be detectable.
- **policy exchange:** service requestors and providers should be able to exchange policy information in order to establish a security context.
- **secure logging:** facilities for time-stamp and reliable logging are required, and are the basis for other important security requirements such as notarization, nonrepudiation, and auditing.
- **assurance** means should be provided to qualify the security assurance level of a hosting environment, for instance, with regard to virus protection or firewall usage.
- **manageability**: security functionality should be manageable, e.g., identity, policy or key management.
- **firewall traversal** mechanisms should be provided for cleanly traversing firewalls without compromising local control of firewall policy.
- **securing the OGSA infrastructure:** security of the components of the OGSA infrastructure must be provided.

OGSA Security Services

OGSA security services are intended to support the enforcement of security policies. The architecture is assumed to be implementation-agnostic, extensible, and easy to integrate with existing security services. OGSA components must enable systems to interoperate securely since services may traverse multiple domains. Also, due to heterogeneity of security infrastructures, required trust relationships are supposed to be established through some form of federation among the security mechanisms.

The model for security services in OGSA v1.0 [24] proposes a language to understand and describe *security policies*, which are defined as statements about *entities, interaction mechanisms*, and *contexts*. The statements specify restrictions on associated *attribute values* and *properties*, and their *relationships*. Entities refer to *users, subjects*, or *services*, and interact through mechanisms within a context. *Interaction mechanisms* refer to the different communication protocols, such as HTTP, SOAP, or SSL/TSL. A *context* is related to interactions, and is a way of putting them in perspective, for instance, by the establishment of a secure association. The *policy statements* are thus expressed in terms of entities, resources, and environment characteristics, and involve aspects such as authorization, authentication, trust, identity mapping, delegation, and assurance levels.

Security services are designed to support security policy enforcement, and are defined as "entities with interaction patterns that facilitate the administration, expression, publishing, discovery, communication, verification, enforcement and reconciliation of the security policy" [24].

With regard to security, grid applications differ from Web services by focusing on security services that enable cross-organizational interactions among entities. These entities have specific attributes and properties within a virtual community that differ from those in their home domain. Hence, the OGSA security services model has to support the concurrent enforcement of multiple policies that have to be evaluated each one within its own context.

Delegation of rights is needed in order to let services work on behalf of other entities. Since those services may become compromised, the delegated rights are limited to those rights that are truly needed by the service according to the *least-privilege delegation model*. This model requires the nontrivial calculation of the adequate number of rights required by the invoked service operations. The idea is to use the job directives expressed in a suitable language to specify the job requirements, which are matched against the capabilities of resources according to a language used to express resource capabilities. The latter should thus be able to match up with the language used to express job directives.

Security services should provide the required security functional capabilities. Figure 12.3, extracted from the OGSA 1.0 document [25], shows key

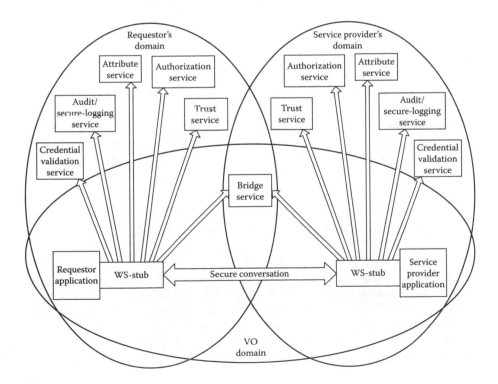

FIGURE 12.3
Security services in a virtual organization setting.

relationships among service requestors, providers, and security services. It illustrates how different security services are invoked by the service requestor or the service provider. It can be seen that call-outs are made from within the stubs and thus are transparent to applications. Policy enforcement should in part be established in this way, thus keeping security-specific code at a minimum for application developers. The figure also shows that call-outs are made to different security service instances managed in different organizations, allowing compliance through configuration with the services and security policies of the requestor, the provider, and the VO. It can be also seen that the service requestor and the service provider are within the same VO but each is subject to their respective domain's policies. Requestor and provider are federated by the Bridge/Translation service that has credentials in both domains and may thus issue identity and capability assertions that can be validated in both domains. Outgoing arrows represent the interfaces to the security services from the requestor and provider, which must be specified in terms of OGSA interfaces.

Many of the following capabilities are considered in the OGSA 1.0 document.

Authentication.   This capability is part of the Credential Validation Service and Trust Service shown in Figure 12.3. Examples of authentication services are a combination of user ID and password or Kerberos authentication.

Identity mapping.   This capability is provided by the Trust, Attribute, and Bridge/Translation services. Identity mapping provides the possibility of associating identities existing in different identity domains.

Authorization.   This service provides the means to make policy-based access-control decisions. Resource access is typically authorized or denied according to the resource access policy and the requestor's credentials. It is expected that the hosting environment provides access control functions.

VO policy.   This service is concerned with the policy management. The policy service may be requested by services such as the authorization, audit, and identity mapping services.

Credential conversion.   The capability of converting a credential from one type to another is provided by the Trust, Attribute and Bridge/Translation Services. Credential conversion may enable the reconciliation of group membership, privileges, attributes and assertions associated with service requestors and providers, and facilitates also the interoperability of differing credential types. Credential conversion may require the service of identity mapping.

Audit and secure logging.   The audit service is policy-driven and responsible for recording security-relevant events. This service is typically used by security administrators within a VO to check adherence to access-control and authentication policies. Auditing requires that events are logged in a secure fashion. Logging services and secure access to logs in a distributed setting is a complex problem since logs may reside in different administrative domains. Logs should be secured and tamper-proof, and capable of ensuring message integrity. Among the events that requires auditing are security events, e.g., an intrusion, which should be dealt with by the security services.

Profile. This service concerns the management of the preferences and personalized data of the service requestor that may not be directly consumed by the authorization service. This data may be used by applications that interface with a person.

Privacy. This service is concerned with the classification of personally identifiable information (PII) that may be stored by provider or requestors.

Figure 12.4, from [10], provides a view of the relationships between the components of the grid security model as a layered stack of related services. The layering shows that application-specific components such as Secure Conversations depend on policies and rules for the components at the layer below, e.g., *Service/End-Point Policy* or *Authorization Policy*. Further, the figure also shows that in order to apply and manage the policies and rules of a layer, e.g., the one in which the Authorization Policy resides, languages for *Policy Expression and Exchange* are required, as well as secure communication mechanisms through bindings to transport protocols or message security. Management components such as *Intrusion Detection* or *Policy Management* are shown in the left box in the picture.

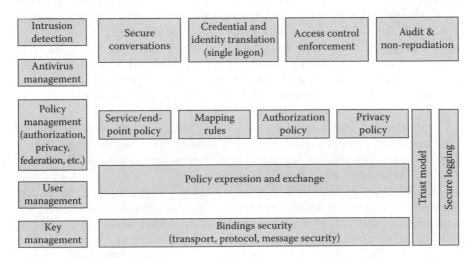

FIGURE 12.4
Components of grid security model.

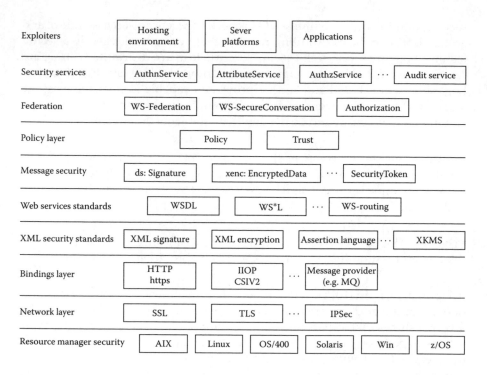

FIGURE 12.5
Security specifications "stack."

Figure 12.5, extracted from [8], shows how the layering of existing security technologies and standards fit into the grid security model. A determined security function can be implemented at different levels, for instance, at the network layer via IPSec or SSL/TLS, which provide only point-to-point security. SOAP and WS-Security, on the other hand, provide message-level mechanisms at higher levels that can be used to achieve end-to-end security.

All OGSA security interfaces need to be standardized. Compliant implementations are supposed to be able to make use of existing services and policies through configuration, and to provide the associated and possibly alternative security services.

Invocations of OGSA services are usually subject to the enforcement of relevant security policies. OGSA security services may be closely connected to other services of a higher level, and one security service may be a consumer of other OGSA services.

The Global Grid Forum produced a roadmap [7] leveraging existing and emerging Web services security specifications and enumerating a set of proposed specifications to ensure interoperable implementations of the OGSA security architecture. The proposal builds on the framework described by the WS Security Architecture [26], which consists of layered modules, including WS-Security, WS-Policy, WS-Federation, WS-SecureConversation, WS-Privacy, WS-Trust, and WS-Authorization. These modules are proposed to become building blocks for OGSA security. A set of profiles for WS security specifications has been proposed. It is recommended that WS security specifications be modified in Global Grid Forum specifications when they do not meet OGSA security requirements. The OGSA security specifications proposed include services such as naming, delegation, audit and secure logging, translation between security realms, and authorization, trust, privacy, and VO policy management. Other proposed specifications concern support for multiple authentication mechanisms, authorization services plugability, security policy expression and exchange, interoperability through firewalls, and secure service operation.

12.4.2 Enterprise Grid Alliance Reference Model Security

The EGA reference model defines an *Enterprise Grid* as a collection of interconnected (networked) *grid components* under the control of a *grid management entity* [34]. A *grid component* is defined as a superclass of object from which all of the components that are managed within an enterprise grid are descended or derived. Components include servers, network components, ERP services, online bookstores, etc. Grid components can as a rule be combined together into more sophisticated components.

Components have security properties and attributes, and may define specific dependencies that can be used to support enforcement of security policies and to ensure minimal exposure. The concept of enterprise grid wide dependencies and constraints supports the secure provision, configuration and enabling of entire services or business functions. To help minimize risk, these attributes, dependencies, and constraints should be enforced.

The Grid Management Entity (GME) is defined by the EGA reference model as the logical entity that manages the grid components, the relationships among them, and their entire life cycles. The GME should support the definition and enforcement of the security policy of the enterprise grid. The security functions that the GME should manage include: the user identities and administrative roles; authentication of identities; authorization of actions taken by principals; access restrictions to the grid components; capture, storage, analysis, and reporting of audit-related events; key management; enforcement of secure communications across the grid and of secure isolation of shared grid components and services; validation of individuals or groups with regard to their expected security states; and ensuring that local and remote management and troubleshooting operations are secured in accordance with the organization's security policy.

The EGA reference model defines three life cycle states of a grid component: provision, ongoing management, and decommission/repurposing.

Provisioning. Provisioning involves adding, creating, configuring, and starting a grid component. The security attributes and properties associated with provisioning include questions such as the identity of the provisioner of the grid component, the provisioning history, component verification and validation, satisfaction of required dependencies, and eventual constraints on the use of the component.

Ongoing management.   Management of a grid component involves any management-related activities when the component is in an active state. The security issues related to ongoing management include questions such as who is authorized to create or modify components or administrative roles, the location for performing management functions, restrictions concerning the authentication of the administrator, management of administrative roles, management of grid components and security attributes, distribution and updating of security policies, validation of security configuration, failure detection and repercussions of failures, detection of unauthorized changes, notification of security events, access control, and user authentication.

Decommissioning and repurposing.   Decommissioning involves the retirement or repurposing of a service or grid component. Relevant security issues here include authorization to decommission/repurpose a component's security attributes, the history and other details of a resource's provisioning/decommission/repurpose, and conditions under which a resource can be decommissioned/repurposed.

Bibliography

1. I. Foster, C. Kesselman, and S. Tuecke. The Anatomy of the Grid: Enabling Scalable Virtual Organizations. *International Journal of High Performance Computing Applications,* 15(3). 200–222. 2001.
2. I. Foster, C. Kesselman, J. Nick, and S. Tuecke. The Physiology of the Grid. *Global Grid Forum,* June 2002.
3. Ian T. Foster, Carl Kesselman, Gene Tsudik, and Steven Tuecke. A Security Architecture for Computational Grids. *ACM Conference on Computer and Communications Security,* 1998.
4. I. Foster and C. Kesselman. *The Grid: Blueprint for a New Computing Infrastructure.* Morgan Kaufmann: San Francisco, 1999.
5. M. Humphrey and M. Thompson. Security implications of typical grid computing usage scenarios, HPDC 10, August 2001.
6. M. Humphrey, M.R. Thompson, and K.R. Jackson. *Security for Grids* (August 14, 2005). Lawrence Berkeley National Laboratory, Paper LBNL-54853.
7. F. Siebenlist, V. Welch, S. Tuecke, I. Foster, N. Nagaratnam, P. Janson, J. Dayke, and A. Nadalin. *OGSA Security Roadmap, Global Grid Forum,* 5 Document, Open Grid Security Architecture Security Working Group, July 2003.
8. F. Siebenlist, V. Welch, S. Tuecke, I. Foster, N. Nagaratnam, P. Janson, J. Dayke, and A. Nadalin. *OGSA Security Roadmap, Global Grid Forum,* 5 Document, Open Grid Security Architecture Security Working Group, July 2003, page 6, ©Global Grid Forum (2003), all rights reserved.

9. G. Della-Libera et. al. *Security in a web services world: A proposed architecture and roadmap.* White paper, IBM Corporation and Microsoft Corporation, April 2002.
10. N. Nagaratnam, P. Janson, J. Dayka, A. Nadalin, F. Siebenlist, V. Welch, I. Foster, and S. Tuecke. *The Security Architecture for Open Grid Services.* Open Grid Services Security Architecture WG, Global Grid Forum, 2.9 (Draft Version 1), July 2002.
11. V. Welch, F. Siebenlist, D. Chadwick, S. Meder, and L. Pearlman. *Use of SAML for OGSA Authorization,* Global Grid Forum, 2004.
12. K. Keahey, V. Welch, S. Lang, B. Liu, and S. Meder. Fine-Grained Authorization for Job Execution, in *The Grid: Design and Implementation. Concurrency and Computation: Practice and Experience,* April 2004, 16(5): 477–488.
13. P.J. Broadfoot and G. Lowe. *Architectures for Secure Delegation Within Grids.* Technical Report PRG-RR-03-19, Oxford University Computing Laboratory, September 2003.
14. J. Herveg, F. Crazzolara, S.E. Middleton, D.J. Marvin, and Y. Poullet. *GEMSS: Privacy and Security for a Medical Grid.* In *Proceedings of HealthGRID* 2004, Clermont-Ferrand, France.
15. S. Parastatidis, J. Webber, P. Watson, and T. Rischbeck. *A Grid Application Framework Based on Web Services Specifications and Practices,* Document version: 1.0, 12 August 2003.
16. D. Eastlake, J. Reagle, and D. Solo. *XML-Signature Syntax and Processing.* World Wide Web Consortium, Recommendation REC-xmldsig-core-20020212, February 2002.
17. D. Eastlake, J. Reagle, and D. Solo. *XML-Signature Syntax and Processing.* World Wide Web Consortium, Recommendation REC-xmlenc-core-20021210, December 2002.
18. P. M. Hallam-Baker, S.H. Mysore. *XML Key Management Specification (XKMS),* Version 2.0, World Wide Web Consortium, Recommendation REC-xkms2-20050628, June 2005.
19. M. Atkinson, D. DeRoure, A. Dunlop, G. Fox, P. Henderson, T. Hey, N. Paton, S. Newhouse, S. Parastatidis, A. Trefethen, and P. Watson. *Web Service Grids: An Evolutionary Approach,* Report UKeS-2004-05, UK e-Science Technical Report Series, 2004.
20. M. Baker, A. Apon, C. Ferner, and J. Brown. Emerging Grid Standards, *IEEE Computer,* Volume 38, Number 4 (April 2005), pages 43–50.
21. *Web Services Security: SOAP Message Security 1.0* (WS-Security 2004), OASIS Standard 200401, March 2004.
22. J. Linn. *Generic Security Service Application Program Interface,* Version 2, RFC 2078.
23. *Open Grid Services Infrastructure* (OGSI), Version 1.0, Global Grid Forum, 27 June 2003.
24. *The Open Grid Services Architecture,* Version 1.0, Global Grid Forum, 29 January 2005.
25. *The Open Grid Services Architecture,* Version 1.0, Global Grid Forum, 29 January 2005, p. 43. Copyright ©Global Grid Forum (2002–2005), all rights reserved.
26. *Web Services Security: SOAP Message Security 1.1,* OASIS Standard Specification, 1 February 2006.
27. www-128.ibm.com/developerworks/library/specification/ws-secmap.
28. *Web Services Trust Language* (WS-TrusT), February 2005.
29. *Web Services Policy Framework* (WS-Policy), March 2006, Version 1.2.
30. *Web Services Security Policy Language* (WS-SecurityPolicy), Version 1.1., July 2005.
31. *Web Services Secure Conversation Language,* February 2005.

32. *Web Services Federation Language* (WS Federation), Version 1.0, July 8, 2003.
33. *Basic Profile Version 1.0. Web Services Interoperability Organization*, 16 April 2004.
34. *Enterprise Grid Security Requirements*, Version 1.0, Enterprise Grid Alliance Security Working Group, 8 July 2005.
35. www.w3.org/TR/soap.
36. www.w3.org/TR/wsdl.
37. www.oasis-open.org/committees/security.
38. www.oasis-open.org/committees/xacml.
39. www.omii.ac.uk.
40. www.ggf.org.
41. www.oasis-open.org.
42. www.w3.org.
43. www.dtmf.org.
44. www.internet2.edu.
45. www.projectliberty.org.
46. www.gridalliance.org.
47. www.ws-i.org.
48. www.globus.org/toolkit.
49. www.globus.org/security/overview.html.
50. www.globus.org/ogsa.
51. www.globus.org/wsrf.

13

A Trust-Based Access Control Management Framework for a Secure Grid Environment

James B. D. Joshi, Siqing Du and Saubhagya R. Joshi

CONTENTS

Abstract   Grid computing represents a significant new paradigm aimed toward harnessing the collective computational power of distributed computing resources. Such an aggregation of computational resources provides tremendous opportunities for enabling support for applications that are highly compute and resource intensive. Besides, the newly emerging service-oriented architecture and the peer-to-peer computing paradigms are naturally being integrated with grid computing to address significant scalability and manageability issues related to the deployment and efficient management of a grid computing infrastructure. At the same time, the security issues for such an emerging grid environment are becoming increasingly complex and hence they pose a significant bottleneck to its successful deployment. In this chapter, we focus on the key issue of access control specification and enforcement for the protection of resources and shared information in a grid, and address the problem of ensuring secure interoperation among independent grid components that may be unknown to each other but have to engage in transient interactions by establishing trust in an ad hoc manner.

13.1   Introduction

Recent advances in computing and networking technologies have enabled the development of large-scale application infrastructures that allow unprecedented levels of sharing of information and resources. Grid computing epitomizes such an infrastructure that is posed to harness the collective power of globally distributed computing facilities. In simple terms, grid computing can be defined as *coordinated resource sharing and problem solving in dynamic, multi-institutional virtual organizations* [15]. A key goal of a grid is to enable its users to seamlessly process resource-intensive jobs on aggregated remote resources owned by other members of the grid community. Grid systems are traditionally also viewed as a trust community of virtual organizations (VOs) with a persistent service infrastructure that is centralized (*tightly coupled*) or distributed but achieves hierarchical coordination [5, 14, 19, 38]. At the same time, recently emerging paradigms of service-oriented architecture (SOA) and peer-to-peer (P2P) computing are significantly influencing the design of a grid infrastructure, providing means to achieving scalability, modularity, reusability, and service orientation, among others [5, 36]. In particular, increased manageability of a large number of transient, *loosely coupled* interactions among peers in an unstable environment without any central coordination is a key characteristic of the P2P computing approach [12]. The SOA paradigm, on the other hand, provides a distributed architecture designed for service interoperability, easy integration, and simple, extensible, and secure access [36]. In an SOA-enabled grid environment, the relationship between service consumers and providers is, instead of being established *a priori*, formed dynamically in an ad hoc manner based on service requirements [8].

The inevitable convergence of SOA and P2P technologies with grid computing has been seen as the new direction for the successful design and deployment of a large, scalable grid infrastructure that can help unleash the potential to truly support a wide range of high-end computing and resource-intensive applications [23]. While such a convergent grid environment[1] (Figure 13.1) benefits from the various advantages provided by the SOA and P2P technologies, they significantly exacerbate security problems [12, 35]. Such a service-based grid needs to facilitate seamless, dynamic, trust-based, loosely coupled, interdomain interactions among grid components, potentially unknown to each other *a priori*, so that information and resource sharing can be done in a secure manner. The key challenges include the development of the following capabilities:

- *A comprehensive access control model to address the dynamic, context-based access requirements of the grid and its components, and*

- *An elaborate trust framework to facilitate secure loosely coupled interactions among constituent domains in a grid environment.*

---

[1] In the remainder of this chapter grid refers to this convergent grid environment.

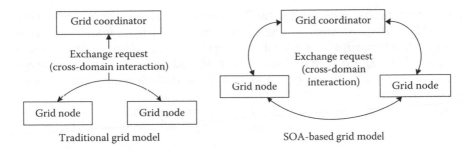

FIGURE 13.1
Traditional and SOA-based grid models.

A grid environment is highly dynamic in nature and hence access to its resources may need to be restricted based on the context information [8]. For instance, different resources may need to be made available at different times and under different system loads. Grid nodes should be able to express such fine-grained context-based access policies that restrict use of their resources. In a grid environment, the interactions between entities, previously known or unknown to each other, typically between the VO layer and the grid nodes as well as between different nodes, may be very dynamic or transient, and driven by service requirements. A robust security solution should allow for efficiently capturing a generic set of parameters required for establishing trust between such interacting partners. Furthermore, the behavior of the participating grid components, context information, and resource availability can evolve rapidly, thereby affecting established trust values during cross-domain interactions. Efficient techniques are needed to manage and sustain such evolving trust. Another key issue is that of the integration or mapping of security policies of partner domains contributing to the grid, which entails various challenges such as managing semantic heterogeneity, ensuring secure interoperability, and policy evolution management [20]. In particular, ensuring secure interoperation involves making sure the *principles of security* and *autonomy* are preserved [20, 30]. The *principle of security* states that the accesses that are originally allowed within a domain *must* also be allowed when the domain interoperates with other domains, while the *principle of autonomy* states that an access that is not allowed originally within a domain *must not* be allowed when the domain interoperates with other domains. It has been shown that even when two domains are perfectly secure, interactions between them can make them insecure; furthermore, ensuring secure interoperability has been shown to be *undecidable* [16].

In this chapter, we present our work related to a trust-based access management framework for such a loosely coupled multidomain grid environment. Our work is based on the Generalized Temporal Role-Based Access Control (GTRBAC) model and its extensions [3, 22]; a key reason for this is that RBAC has been shown to be the most promising approach for addressing fine-grained access control requirements in such large, distributed systems [30].

It not only encompasses traditional discretionary and mandatory access control (DAC/MAC) models, but also offers many beneficial features, such as policy neutrality, support for least privilege and efficient access control management [21, 22]. RBAC has also been considered as the most promising approach for integrating heterogeneous policies in multidomain environments, such as the grid [18].

The key components of the framework proposed in this chapter include: *trust negotiation, policy negotiation* and *mapping, trust sustenance* and *evolution*. The *policy negotiation* process interleaves with the *trust negotiation* process and together they enable the requirements-driven secure interoperation. Once negotiations are done, *trust tickets* are generated to support fast accesses for the agreed-upon services under the given context. We assume that the grid domains involved in interoperation employ policies expressed using X-RBAC, an XML-based language for expressing context-based policies, and policy mappings in multidomain environment. To the best of our knowledge, little or no work exists in the literature that provides integrated trust-based secure interoperation framework for a multidomain environment such as a grid. The existing access control approaches for grids lack adequate expressiveness and flexibility [18, 23]. Further, policy integration and evolution issues are in the very early stages of research [13].

The remainder of the chapter is organized as follows. In Section 13.2, we present an overview of the GTRBAC model and the X-RBAC language. We present the trust-based negotiation framework in Section 13.3, policy mapping in Section 13.4, related work in Section 13.5, and finally, the conclusion in Section 13.6.

13.2   The GTRBAC Framework

In this section, we present an overview of the GTRBAC model and the X-RBAC language that can be used to specify GTRBAC as well as generic context-based access control policies, and multidomain policy mappings.

13.2.1   Overview

The GTRBAC model introduces the separate notion of *role enabling* and *role activation*, and provides constraints and event expressions associated with both [22]. An *enabled* role indicates that a valid user can activate it, whereas an *activated* role indicates that at least one user has activated the role. The basic GTRBAC model allows specification of the following set of constraints [22]: (*i*) temporal constraints on role enabling/disabling that allow specification of intervals and durations in which a role is enabled; (*ii*) temporal constraints on user role and role-permission assignments that allow specifying intervals and durations in which a user or permission is assigned to a

TABLE 13.1

GTRBAC Constraint Expressions

Constraint Categories	Constraints		Expression
Periodicity Constraint	User-role assignment		$(I, P, pr : assign\mathcal{U}/deassign_{U}\ r\ to\ u)$
	Role enabling		$(I, P, pr : enable/disable\ r)$
	Role-permission assignment		$(I, P, pr : assignP/deassign_{P}\ p\ to\ r)$
Duration Constraints	User-role assignment		$([(I, P)|D], D_{U}, pr : assign_{U}/deassign_{U}\ r\ to\ u)$
	Role enabling		$([(I, P)|D], D_{R}, pr : enable/disable\ r)$
	Role-permission assignment		$([(I, P)|D], D_{P}, pr : assign_{P}/deassign_{P}\ p\ to\ r)$
Duration Constraints on Role Activation	Total active role duration	Per-role	$([(I, P)|D], D_{active}, [D_{default}], pr : active_{R\_total}\ r)$
		Per-user-role	$([(I, P)|D], D_{uactive}, u, pr : active_{UR\_total}\ r)$
	Max role duration per activation	Per-role	$([(I, P)|D], D_{max}, pr : active_{R\_max}\ r)$
		Per-user-role	$([(I, P)|D], D_{umax}, u, pr : active_{UR\_max}\ r)$
Cardinality Constraint on Role Activation	Total no. of activations	Per-role	$([(I, P)|D], N_{active}, [N_{default}], pr : active_{R\_n}\ r)$
		Per-user-role	$([(I, P)|D], N_{uactive}, u, pr : active_{UR\_n}\ r)$
	Max. no. of concurrent activations	Per-role	$([(I, P)|D], N_{max}, [N_{default}], pr : active_{R\_con}\ r)$
		Per-user-role	$([(I, P)|D], N_{umax}, u, pr : active_{UR\_con}\ r)$
Trigger			$E_{1}, \ldots, E_{n}, C_{1}, \ldots, C_{k} \rightarrow pr : E\ after\ \Delta t$
Constraint Enabling			$pr:enable/disable\ c\ where\ c\ (\{(D, D_{x}, pr : E), (C), (D, C)\})$
	Users' activation request		$(s : (de)activate\ r\ for\ u\ after\ \Delta t))$
			$(pr : assign_{U}/de-assign_{U}\ r\ to\ u\ after\ \Delta t)$
Run-time Requests	Administrator's run-time request		$(pr : enable/disable\ r\ after\ \Delta t)$
			$(pr : assign_{P}/de-assign_{P}\ p\ to\ r\ after\ \Delta t)$
			$(pr : enable/disable\ c\ after\ \Delta t)$

role; (*iii*) activation constraints that allow specification of restrictions on the activation of a role, such as specifying the total duration for which a user may activate a role, or the number of concurrent activations of the role at a particular time; (*iv*) run-time events allow an administrator and users to dynamically initiate the various role events, or enable the duration or activation constraints; (*v*) constraint-enabling events that enable or disable duration and role-activation constraints mentioned earlier; and (*vi*) triggers that allow expressing dependencies among events and conditions.

Table 13.1 summarizes the constraint types and expressions of the GTRBAC model. The periodic expression used in the constraint expressions is of the form (I, P), where P is an *expression* denoting an infinite set of periodic time instants, and $I = $ [begin, end] is a time interval denoting the lower and upper bounds that are imposed on instants in P [22]. D expresses the duration specified for a constraint. In the duration and role activation constraint expressions, D_{x} and N_{x} indicate the constrained durations and cardinalities. If the subscript x starts with u then it is a *per-user-role* constraint; otherwise it is a *per-role* constraint. For instance, D_{active} indicates how long the specified role can be active, whereas, $D_{uactive}$ indicates how long the specified user may

TABLE 13.2

Example GTRBAC Access Policy for Grid Environments

	a.	(*OffPeakTime*, *enable* edu Grid)
1	b.	((M, W, F), *assign$_U$ users@nasa.gov to* gov Grid)
	c.	([10am, 3pm], *assign$_U$ joe@pitt.edu to* media Grid)
2	a.	([7/1/06, 12/31/06], *9pm-6am, disable* gov Grid)
	b.	c_1 = (9 *hours*, 4 *hours, enable* edu Grid)
3	a.	(*enable* PSC Grid → *enable* c_1)
	b.	(*enable* gov Grid → *disable* media Grid *after* 10 *min*)

activate the specified role. The following example illustrates the specification of a GTRBAC policy. For more details, we refer the reader to [11].

Example 13.1

Table 13.2 illustrates a GTRBAC policy. The periodicity constraint 1a specifies the enabling times of edu Grid role. For simplicity, we use *OffPeakTime* instead of their (I, P) forms. The periodicity constraint 1b allows the gov Grid role to be assigned to *users@nasa.gov* on *Mondays*, *Wednesdays*, and *Fridays*. Similarly, assignment in 1c allows *joe@pitt.edu* to assume the media Grid role everyday between 10 *am* and 3 *pm*. In 2a, the role gov Grid is *disabled* from July 1, 2006 to December 31, 2006, between 9 pm and 6 am. 2b specifies a duration constraint of 4 *hours* on the enabling time of the edu Grid role, but this constraint is valid for only 9 *hours* after the constraint c_1 has been enabled. Because of this, PSC Grid will be able to activate the edu Grid role at the most for 4 hours whenever the role is enabled. In row 3, we have a set of triggers. Trigger 3a indicates that constraint c_1 is enabled when the PSC Grid is enabled, which means, now, the edu Grid role can be enabled within the next 9 *hours*. Trigger 3b indicates that 10 *min* after the gov Grid role is enabled, the media Grid role is disabled.

An important extension included in the GTRBAC model is that of hybrid hierarchy. Within the GTRBAC framework, the following three hierarchy types have been identified: *permission-inheritance-only* hierarchy (*I*-hierarchy), *role-activation-only* hierarchy (*A*-hierarchy), and the combined *inheritance-activation* hierarchy (*IA*-hierarchy) [22]. Table 13.3 provides a brief definition of these hierarchies. Semantically, the predicate *can_be_acquired*(p, y, t) means that permission p can be acquired through role y at time t. The predicate

TABLE 13.3

Role Hierarchies in GTRBAC

Short Form	Notation	The Condition c Holds
I-hierarchy	$(x \geq_i y)$	$\forall p, (x \geq_i y) \land can\_be\_acquired(p, y, t) \rightarrow can\_be\_acquired(p, x, t)$
A-hierarchy	$(x \geq_a y)$	$\forall u, (x \geq_a y) \land can\_activate(u, x, t) \rightarrow can\_acquire(u, y, t)$
IA-hierarchy	$(x \geq y)$	$(x \geq y) \leftrightarrow (x \geq_i y) \land (x \geq_a y)$

can_activate(*u*, *x*, *t*) means that user *u* can activate role *x* at time *t*, and, the predicate *can_acquire*(*u*, *y*, *t*) means that user *u* can activate role *y* at time *t*. Separation of hierarchy types have been shown to be useful in supporting fine-grained separation of duty (SoD), cardinality, and temporal constraints on roles in a hierarchy.

13.2.2 The X-RBAC Language

X-RBAC, an XML-based specification language for RBAC model, provides a framework for expressing GTRBAC policies as well as interdomain role mappings in both loosely coupled and the federated multidomain environments [21]. In addition, it also allows specification of context-based constraints to capture context-based access requirements. Figure 13.2 shows the XML syntax for general policy specification. The key policy component definitions include XML Role Sheet (XRS), XML User Sheet (XUS), XML Permissions Sheet, (XPS), XML User-Role Assignment Sheet (XURAS), and XML Permission-Role Assignment Sheet (XPRAS), which are briefly described below. Each policy can include multiple constituent policies, thus facilitating specification of policies for multidomain environments. The relationship definition includes mapping specification between the global entities and the entities of the other domains.

An XUS is used to define credential types and users. Credential types defined are those that can be handled by the system when mapping unknown users to roles. Each attribute of a credential type may be defined as *mand*,

```
<!-- Policy Definition-->
<Policy [policy_id=(value)]>
    <PolicyName> (name) </PolicyName>
    <!-- XML User Sheet-->
    <!-- XML Role Sheet-->
    <!-- XML Permission Sheet-->
    <!-- XML User-Role Assignment-->
    <!-- XML Role-Permission Assignment-->
    [<!--Local Policy Definitions-->]
    [<!--Policy Relationship Definitions-->]
</Policy>
```

(a) **X-RBAC Policy Specification Format**

```
<!--Local Policy Definitions -->
<LocalPolicies>
    <!--(Local) Policy Definition -->
</LocalPolicies>

<!--Policy Relationship Definitions -->
<PolicyRelationships [prs_id=(id)][pt_id=(id)]>
    <!--Policy Relation Definitions-->
</PolicyRelationships>
```

(c) **Local Policies and Mapping Relations**

```
<!-- Periodicity Expression -->
<PeriodicTime pt_id="PT1"  pt_begin="2003-01-01" pt_end="2003-12-31">
    <TimeStart>
        <Day preset="Monday"/>
        <Hour hourSet="9"/>
    </TimeStart>
    <TimeDuration cal="Hours" len=12/>
</PeriodicTime>
```

(b) **Periodicity Expression**

FIGURE 13.2
X-RBAC syntax and components.

for mandatory, or as *opt*, for optional. User definition may simply define user name and user id, or additionally specify the assigned credentials that the user may carry. Credential type definition specifies the attribute list associated with a credential type. Credential expressions are of the form (*cred_type_id, cred_expr*), where *cred_type_id* is a unique credential type identifier and *cred_expr* is a set of attribute-value pairs. Permissions are defined in XPS in terms of objects and associated operations such as *read, write, delete, modify,* etc.

Role definitions for each role in XRS are a set of role attributes. Each role may have associated with it preconditions for its enabling, assignment, and activation. These preconditions are separately defined using the <EnablingCondition>, <AssignmentCondition>, and <Activation Condition> tags to capture associated GTRBAC event. Within a precondition tag, an *eType* attribute may be specified to indicate whether the precondition is for the enabling (activation) or the disabling (deactivation) of the associated role. For enabling/disabling preconditions, we use the periodic time expression as a condition. Semantically it means that the role is enabled if the current time instant is contained in the periodic time expression. An example of periodicity expression in the X-RBAC language is given in Figure 13.2(b), which specifies a periodic start time from Monday, 10 am for a duration of 12 hours within an interval beginning January 1, 2003 to December 31, 2003. We can define additional predicates to be used to express context-based conditions using the generic syntax for the logical condition. The XRS can also specify *Separation of Duty* (SoD) constraints. This is done by constructing a role set, and specifying a cardinality stating how many roles from the set may be assigned to (*Static SoD*), or activated by (*Dynamic SoD*) a user at the same time. The system administrator uses XURAS and XPRAS to specify the user-role and permission-role assignments. Keeping the user, role, and permission specifications separate from their mappings allows independent design and administration of the policy.

X-RBAC Mediation Policies. A policy definition can include local policy definitions using the XML syntax as depicted in Figure 13.2(a) and 13.2(c). Each policy may itself be a global policy over a set of local (or partner) domains. A relevant principle for mediation policies is the following scoping rule: *If a policy P becomes a local policy of a higher level policy, then P's local policy definitions and the policy relations are not known to the higher level policy.* This rule states that within a global policy definition, only the entities of its local policies and not those of constituent domains of these local policies are visible. This abstraction simplifies the metapolicy construction. However, if the higher level policy management must oversee the consistency of the overall federation, then this rule may need to be relaxed. With local policies included, we need to define the relationships among their policy entities with the global entities. Each global role may be mapped to a number of roles that may belong to the same or different local domains. For each mapping, a contextualized condition can be specified. For more details on X-RBAC, we refer the reader to [21].

13.3 A Framework for Trust-Based Secure Grid

Figure 13.3 shows the proposed trust-based secure interoperation framework for a grid environment. It is composed of two principle modules: the *Trust-Based Service Negotiation* (TBSN) module and the *Trust Sustenance and Evolution* (TSE) module, which are briefly overviewed below.

13.3.1 Trust-Based Service Negotiation (TBSN)

TBSN includes the following components: *service discovery, service broker, policy negotiation, trust negotiation,* and generation of an *export policy*. It specifies the services that will be exchanged between the interoperating domains and establishes a negotiated trust level for service access.

Service Discovery and Brokering. In a grid environment, the potential grid nodes advertise their services via service brokers. The service requesters, the VO layer or other nodes consult with the service broker to find the desired services. Service composition is needed when the functionalities of more than one service are required to satisfy a given request. The service broker can

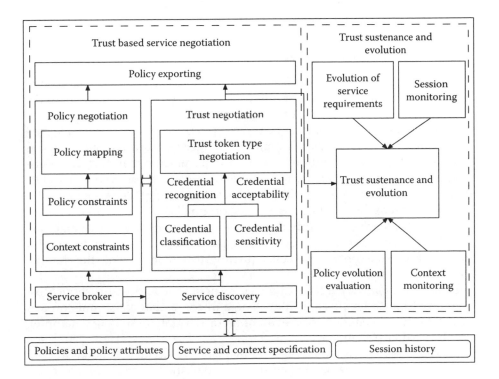

FIGURE 13.3
Trust-based secure interoperation framework.

be realized as a repository/directory where nodes publish services and requesters find services.

Trust Negotiation. Trust negotiation involves negotiation of the set of *trust tokens* that need to be disclosed based on the *trust token types* required for service access. A trust token type specifies a set of attributes and the range of their allowed values, while trust tokens represent any set of digital certificates that collectively can match a trust token type. For instance, a trust token type may indicate the requirement for proof of age to be above 18. Digital credentials that form valid trust tokens may include a passport, university ID, or driver's license. The negotiation phase may agree as to which of these credentials could be used as a trust token. Note that digital credentials used as a token may contain extra attributes and values that may have varying privacy protection requirements.

Policy Negotiation. A domain's fine-grained service requests can be represented as a set of permissions that a particular role within the requesting domain needs to access the services in the provider domain. Based on the service request, domains may attempt to establish a certain level of negotiated trust. At the same time, based on the achievable level of trust, a service request may be adjusted or negotiated. In our framework, the policy negotiation process is thus interleaved with the trust negotiation process. A key result of the TBSN phase is the establishment of policy mapping and the required set of trust tickets to be used by the interacting domains.

13.3.2   Trust Sustenance and Evolution

Trust sustenance refers to maintaining trust levels when domain characteristics change during the period of interoperation. Trust evolution refers to the change in trust levels because of changes in domain characteristics.

Evolution of Service Requirements. During a session, a new service requirement can arise or some services may no longer be required. Since trust is requirements-based, evolution of service requirements may trigger a decision on whether to sustain the trust value or reevaluate, or even renegotiate it. Changes in trust values could also be used to renegotiate services, for instance, to reduce the set of accesses given originally.

Context Monitoring. In highly dynamic environments, context changes are inevitable. Since trust levels are context-dependent, it is important to monitor the changes in the context and consequently sustain or calculate the changes to the trust level.

Policy Evolution Evaluation. Changes in policies could cause service usage/provision to be affected (like change in contextual constraints on services), leading to either trust reevaluation or renegotiation. Policy mapping will be particularly affected.

Session Monitoring. Anomalous and malicious behavior should be tracked and immediately recognized to adjust the established trust levels. Trust sustenance is usually associated with domain characteristics that vary in small scale and can be handled to gracefully end interoperation. Trust evolution is usually associated with more significant changes, like complete change of context, or access to highly sensitive information. In such cases, trust threshold levels are recomputed and if necessary, trust is renegotiated.

13.3.3 Trust Negotiation

In general, the interoperating grid domains need to negotiate the services they can provide each other. If any domain provides services worth less than it received, then it can pay some incentive to the domain that provided more services. Such *requirements-driven service negotiation* can be seen in practical applications and should be facilitated to support ad hoc partnerships between a pair of grid domains. Various cost factors may play a significant role as to how the negotiation may proceed. Table 13.4 lists the parameters for negotiation. Let d_x and d_y be service domains such that services requested by each are satisfied by the other after negotiation. Then, negotiation between domain d_x and d_y is said to converge when the condition $c \leq b + 1$ holds for both d_x and d_y.

Ideally, the cost incurred to a domain during interoperation should be less than the benefits and incentives it gets. Note that the condition for convergence may never occur as internal constraints on the services required or provided may restrict further negotiation. In such a situation, secure and desirable interoperation may not be possible. Trust negotiation is carried out simultaneously with service negotiation to enable establishment of interoperation. Typically, if two domains are involved in interoperation through exchange of services, each domain can request the other to disclose some information of a certain type as proof of trustworthiness. We introduce the notion of *trust token type* that indicates a set of attributes and the range of values within which they should be constrained.

Trust Token Type, Trust Token. We define these as follows. Let TT and T denote a *trust token type* and a *trust token*, respectively. Further, let $A = \{a_1, \ldots, a_n\}$ be a generic set of attributes, $Dom(a_i)$ be the evaluation domain of attribute a_i, and $A_1 \subseteq A$. Then, (i) $TT = (A_1, VS)$, where $VS = V_1, \ldots, V_{|A|}$ such that $V_i \subseteq Dom(a_i)$. (ii) $T = (A_1, V)$, where $v_i \in V$ is such that $v_i \in V_i \subseteq Dom(a_i)(i = 1..|A|)$. Further, a trust token T is said to satisfy a trust token type TT (denoted as

TABLE 13.4

Cost Parameters for Trust Negotiation

$m_{d_x}^{d_y}$	Cost incurred to d_y for policy mapping, to satisfy requirements of $d_x(d_x.SR)$
$r_{d_x}^{d_y}$	Cost incurred to d_y for resources used by d_x when using services provided by d_y (as per $d_x.SR$)
$i_{d_x}^{d_y}$	Incentives that d_y may receive (or lose) in the interoperation
$c_{d_x}^{d_y}$	Cost incurred by d_y for providing services to satisfy $d_x.SR$ $c_{d_x}^{d_y} = m_{d_x}^{d_y} + r_{d_x}^{d_y}$
$b_{d_x}^{d_y}$	Benefits for d_y when using service provided by d_x (as per $d_y.SR$)

$T \equiv TT$) if the following conditions hold: $\forall a_i \in TT.A$, $V_i \in TT.VS$, $[v_i \in T.V_i \wedge v_i \in V_i]$.

The provider domain demands the disclosure of credentials that verify a set of trust token types. Some typical examples of trust token types are (*age, greater than 18*) and (*nationality, residence, U.S. and U.S. Minor Islands, Pennsylvania*). Credentials are digitally signed endorsements of some attributes of an entity. They are basically attribute certificates, as specified in [27].

A trust token is constructed by selecting a set of candidate credentials that collectively satisfy the trust token type. It is possible that only a subset of the attributes endorsed by each credential is needed to satisfy the trust token type. A trust token can be defined as follows. Let TT be a trust token type, CA_i be certification authority, and $C = \{Cert_{CA_1}(A_1), \ldots, Cert_{CA_1}(A_n)\}$ be such that (*i*) each element of C has at least one unique $a \in TT.A$ (*ii*) the attribute set over all elements of $C \supseteq TT.A$.

Then $C_{CA}^{TT.A}$ represents a trust token generated by projecting over attribute set $TT.A$ of C and then certified by CA. If $C_{CA}^{TT.A} \equiv TT$, then $C_{CA}^{TT.A}$ is a valid trust token for TT. Note that $n = 1$ is possible in which case the certificate either exactly represents a trust token or a projection over its attributes is needed to generate a trust token.

This notion of trust tickets indicates that a trust token may need to be generated dynamically to satisfy the required trust token type. The requesting domain may decide to generate such an on-the-fly trust token using the credentials he has by creating a third-party certification. In such a case, trust will relate to who certifies the trust token. It is possible that the CA is the provider himself. In such a case, to satisfy the trust token type, the requester may simply submit a set of credential certificates. An issue here is the protection requirements of the extra attributes in the certificates, exposure of which is a risk that the requester may take based on the trust that it has on the provider and should be incorporated in the trust computation.

Trust Factors

Prior to negotiation, the interoperating grid domains also compute $tr_{S,C}^{d_y \rightarrow d_x}$, which denotes the trust d_x has with regards to d_y for services defined by S in context C. This is a value that is used to compute the payoff of a negotiation strategy. The computation of the overall trust values is the weighted sum of the recommended trust and direct trust values [14]. It is possible that a domain does not have both these values for another domain. The direct trust variables are historical satisfaction level (h) and risk (rk). Here, h indicates the cumulative level of satisfaction that a domain has had for another domain on their previous interactions and is computed based on session histories and older h values. Variable rk captures the risks associated with the desired interoperation. An example is the risk of too many claimed trust tokens being invalid. Another risk is that of services promised but not provided. The historical satisfaction level is also affected by the result of the verification of trust tokens in the earlier sessions. That is, if a domain presents valid trust tokens, then during actual cross-domain accesses, the historical satisfaction level will not be negatively affected. The sustenance of the direct trust is based on a family of functions, and can typically be a time-decaying value [14]. Recommended trust is determined by the recommendation value and the trust level for the recommender [16, 17, 20], denoted by $tr_{S,C}^{d_x \to d_y}$, where d_x is the recommending domain, and d_y is the recommender. Recommended trust can also be the result of a chain of recommendations, where each recommender assigns a trust value for the previous recommender [16]. The parameters that affect the trust relationships are context and the service specifications. The dependence of trust on contextual parameters like time and location have been mentioned in the literature [14, 19]. The trust levels may be different.

Trust-Level Computation. If S and C are the services provided by d_y and the corresponding contexts of interoperation, the trust level $tr_{S,C}^{d_y \to d_x}$ that d_y has on d_x, for services S in contexts C, is defined as follows.

$$tr_{S,C}^{d_y \to d_x} = (\alpha \times dtr_{S,C}^{d_y \to d_x}) + (\beta \times rtr_{S,C}^{d_y \to d_x}), \text{ where}$$

- $\alpha, \beta, \gamma, \delta, \psi, \lambda$ and ε are weights
- α is typically greater than β, as direct trust is usually more influential than recommended trust
- Very often α is the result of a time-decay function, which represents the degradation in the trust for a domain, due to the lack of interaction

$$dtr_{S,C}^{d_y \to d_x} = (\gamma \times h_{S,C}^{d_y \to d_x}) - (\delta \times rk_{S,C}^{d_y \to d_x}), \text{ where}$$

- $h_{S,C}^{d_y \to d_x}$ is the historical satisfaction level that d_y has for d_x
- $h_{S,C}^{d_y \to d_x}$ is bound by the previous risk levels as follows: $h_{S,C}^{d_y \to d_x} = \eta \times rk_{S,C}^{d_y \to d_x}$, where $0 \le \eta \le 1$

$$rtr_{S,C}^{d_y \to d_x} = (\psi \times tr_{S,C}^{d_y \to d_R}) - (\lambda \times r_{S,C}^{d_R \to d_x}), \text{ where}$$

- $rk_{S,C}^{d_y \to d_x}$ is the risk
- $r_{S,C}^{d_R \to d_x}$ is the recommendation given by d_R for domain d_x

$rk_{S,C}^{d_y \to d_x}$ is a complex parameter with a simple quantification done by computing a value from previous validations of trust tokens of the same type from the same domain. $tr_{S,C}^{d_y \to d_x}$ is computed for two purposes: (*i*) primarily to compute the payoff that is determined for each negotiation strategy; or (*ii*) to set a threshold (minimum) level on the trust that a domain must establish with the other. This facilitates trust token negotiation as well.

Ideally, the cost incurred to a domain during interoperation should be less than the benefits and incentives it gets. Note that the condition for convergence may never occur as internal constraints on the services required or provided may restrict further negotiation. In such a situation, secure and

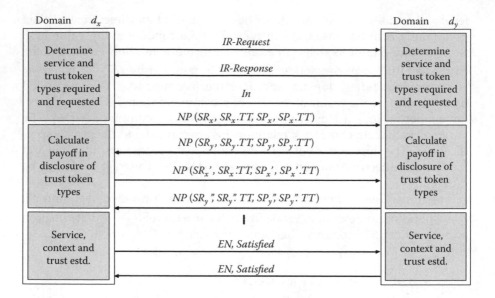

FIGURE 13.4
Protocol for service, context and trust negotiation.

desirable interoperation may not be possible, and the requestor domains will seek other provider domains.

Negotiation Protocol

Figure 13.4 describes a protocol for negotiation of services and trust tokens. Here, negotiation of services and associated trust tokens is done simultaneously. The messages exchanged by the domains are given in Table 13.5.

To determine the convergence point of the negotiation, we take the game-theoretic approach of defining payoffs for different strategies. Here trust

TABLE 13.5

Message Description for Trust Negotiation

Message	Syntax and Description
Interoperation Request/Response(**IR**)	⟨ **IR**, *Required (or Provided), Name, Service, Context*⟩ Such messages are sent by the initiator domain and the responder domains
Initiate Negotiation (**IN**)	⟨ **IN**, Accept ⟩ This is a message sent by the initiator to the domain(s) which it has selected from a set of domains that responded to its request, to start negotiation of services, context of service and trust token types required
Negotiation Proposal (**NP**)	⟨ **NP**, *Name, SR, SR.C, SR.TT, Sp, SP.C, SP.TT* ⟩ The negotiation messages exchanged between the domains
End Negotiation (**EN**)	⟨ **EN**, *Satisfied (or Not Satisfied)*⟩ This message is sent to end the negotiation either in satisfaction or disapproval

tokens are strategies, and each trust token has a different overall protection requirement. Based on the choice of trust tokens for disclosure, corresponding domains have gains or losses. The payoff for each domain is the linear sum of the payoffs from services and trust token negotiations, respectively. The trust token negotiation payoff is the difference between the trust level established and the *protection level* required of the trust tokens disclosed, as given below:

$$\phi'_{ij}(p_i^{d_x}, p_j^{d_y}) = \left((tr_{S,C}^{d_x \longrightarrow d_y} - ProtLev(d_x.T_i)), (tr_{S,C}^{d_y \longrightarrow d_x} - ProtLev(d_y.T_j)) \right)$$

The service negotiation payoff is the difference between the benefits from usage of services and the losses incurred through service exchange and service provision:

$$\phi''_{ij}(p_i^{d_x}, p_j^{d_y}) = \left(b_{d_y}^{d_x} - c_{d_y}^{d_x} - i_{d_y}^{d_x}, b_{d_x}^{d_y} - c_{d_x}^{d_y} - i_{d_x}^{d_y} \right)$$

Thus the overall negotiation payoff is given as

$$\phi_{ij}(p_i^{d_x}, p_j^{d_y}) = \phi'_{ij}(p_i^{d_x}, p_j^{d_y}) + \phi''_{ij}(p_i^{d_x}, p_j^{d_y}).$$

The negotiation is essentially modeled as a negotiation tree [8]. The different strategies used by the domains are the disclosure of different trust tokens that satisfy the other domain's requirements but have different protection requirements. It is reasonable to assume that protection requirement of a trust token is directly related to trust level desired. Traversal of the tree represents negotiation exchanges between the domains. Each domain computes the payoffs at the leaf nodes and selects a set of candidate payoffs. Using a goal-driven approach (goal being any of the candidate payoffs), the domains negotiate the payoffs. Ideally, both domains select the same candidate payoffs, because in game-theory-based negotiation, strategies are selected that optimize payoff for both parties. The candidate payoff values are selected through empirical studies. Consequently, backtracking is also facilitated in the negotiation if, say, d_y proposes a set of services and trust tokens that would lead to poor payoff for, say, d_x, then d_x will reject the proposal and d_y will have to go back and try another proposal.

13.4 Policy Mapping

Figure 13.5 illustrates the proposed policy framework. Assuming two domains interoperate, each domain first sends its service requirements to the other. Once the requirements have been received, the requests are fulfilled by identifying existing roles or creating roles with the requested permissions. Each domain generates a set of roles to be exported so that the requesting domain can activate these roles. The roles of requesting domain that are relevant are mapped to the exported roles as *A*-hierarchy relation. At this time, the provider domain can also establish activation conditions to capture any

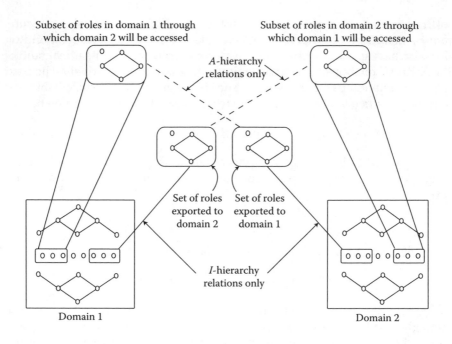

Subset of roles in domain 1 through which domain 2 will be accessed

Subset of roles in domain 2 through which domain 1 will be accessed

A-hierarchy relations only

Set of roles exported to domain 2

Set of roles exported to domain 1

I-hierarchy relations only

Domain 1

Domain 2

FIGURE 13.5
Policy mapping framework.

context-based mapping. The exported roles are themselves made *I*-seniors of other local roles that satisfy the requested accesses to ensure that the external entities do not activate other local roles. By using this *A* and *I* hierarchy structures, we prevent the transitivity of the activation semantics that is usually the underlying problem during secure interoperation [20].

Interdomain access specification When two domains interoperate, each domain first sends the service requirements to the other. The service requirements specified by the external domains must be converted into permissions. Once the requirements have been received and proper trust has been established, the requests are fulfilled by identifying the set of roles that can satisfy the requested permissions. The goal is to find a minimal set of roles that match the requested set of permissions. We represent a domain d_x's request set as $RQS = \{(d_x.r, P)\}$. A provider domain first finds a minimal role set for each $(d_x.r, P) \in RQS$ and then creates export policy. Mapping each $RQ \in RQS$ to a minimal set of roles, referred to as the Interdomain Role Mapping Problem (IDRM) has been shown to be *NP*-complete, which can be stated as follows.

For a given hybrid hierarchy $H = (R_y, F)$, where $R_y = \{r_1, r_2, \ldots r_n\}$ is a set of roles in the provider domain d_y, $F \subseteq \{\geq_i, \geq_a, \geq\}$ is a set of hierarchy relations, the IDRM problem is for each request $RQ = (d_x.r, P) \in RQS$, find the minimal set of roles R' in domain d_y, such that $P_{au}(R') = RQ.P, (R' \subseteq R)$, $d_x.r$ maps to R'.

In essence, request element $RQ = (d_x.r, P)$ indicates that role r in domain d_x needs to be mapped to the permission set P in domain d_y. The proposed policy framework finds the minimal set of roles R' that collectively have set P and creates an exported role r_e, which is made I-senior of the roles in R'. Role $d_x.r$ is now mapped to r_e by using A-hierarchy semantics (this is implicit for the X-RBAC relationship definition) indicating that anyone who can activate $d_x.r$ can also activate r_e in d_y by using the established trust tickets. The A-relation between $d_x.r$ and r_e can further be conditioned to allow context-based mapping. In particular, the extended request form $RQ = (d_x.r, P, C)$, where C defines the context (e.g., a periodicity expression (I, P)) used to capture the context-based interdomain access requirements. In such a case, C is transformed into a condition applied to the A-relation. Using A-relation from local role r to the exported role r_e also allows r_e to be used as a local role in d_x for policy analysis purpose. Similarly, r_e can be used for local policy analysis of domain d_y. We restrict the use of export role r_e of d_y only to the users external to d_y. For each request element $RQ \in RQS$, an export role is created for mapping. These export roles can also be arranged in a hierarchy based on a subset relation with regard to the permissions each provide. In generic case, where the requesting domain can include complex requests that also may include SoD types of constraints, a generic export policy may need to be created. This remains our future work.

The exported role r_e can also be used as it is or as a junior of another export role created to satisfy some requests from other domains. For a clean design, we adopt separate sets of export roles for separate interdomain interactions

---

Greedy-Search(R, RQ)
Input: R – a set of roles
Output: R^* – set of roles, such that $P_{au}(R^*) = RQ, (R^* \subseteq R)$
1   **for each** r in R
2   **if** $P_{au}(r) \subseteq R$
3          $R_1 \leftarrow r$
4            $R_1 \leftarrow \emptyset$
5   **while** $RQ \neq \emptyset$ **do**
6   **Find set** $V \in R_1 \setminus R^*$ **that maximizes** $P_{au}(V) \cap RQ$
7        $R^* \leftarrow R^* \cup V$
8        $RQ \leftarrow RQ \setminus V$
9   **return** R

---

so that when the interaction needs are no more required the exported roles can be deleted or if required maintained for supporting future similar requests.

In a monotype I-hierarchy, the role hierarchy can facilitate a top-down scan to solve the IDRM problem [13]. However, the presence of a hybrid hierarchy presents a more complicated and realistic model, in which a senior role may not have more permissions than a junior role (as illustrated in Example 13.2). In [13], the IDRM problem has been shown to be NP-complete by reducing the *Minimal Set Cover* (MSC) problem to the IDRM problem. There are well-known greedy search approximation algorithms with time complexity within $1 + ln|S|$ for MSC problem. **Greedy-Search**(R, RQ) is one such algorithm. The algorithm does not guarantee finding the optimal solution, R', however, it has been proved to give an H_n-approximation algorithm for the MSC problem. That is,

$$\frac{|R^*|}{|R'|} \leq H(max\{|H| : V \in R_1\}$$

where, $H(d)$ is the *dth harmonic number*, which is equal to $log(d) + O(1)$, and R^* is the minimal set of roles that satisfy the request [13]. Examples 13.2 and 13.3 illustrate the working of **Greedy-Search()** algorithm, and grid policy mapping.

Example 13.2

Interdomain Role-Mapping Example. Consider the hybrid hierarchy H shown in Figure 13.6. Consider an RQ that has the requested permissions = $\{p_1, p_2, p_3, p_4, p_5, p_6, p_7, p_8, p_{10}\}$. We apply the **Greedy-Search()** algorithm to find the minimal role set in H. The **Greedy-Search()** algorithm first constructs $R_1 = \{r_4, r_5, r_6, r_7, r_8, r_9, r_{10}, r_{12}, r_{13}, r_{14}, r_{15}, r_{16}, r_{17}\}$. The results for each step in the whole loop are as shown in Table 13.6. The solution $R^* = \{r_6, r_8, r_4, r_{10}\}$, with cardinality $|R^*| = 4$, returned by **Greedy-Search()** algorithm, is not optimal. The optimal solution, as can be seen is $R^* = \{r_4, r_7, r_{10}\}$ with cardinality $|R^*| = 3$. On the other hand, we have $H(max\{|V| : V \in X_1\}) = H(4) \approx 2.083$. So the upper-bound of cardinality of the solutions returned by the **Greedy-Search()** algorithm is $|R|*H(4) = 3*2.083 \approx 6.25$. Hence, the **Greedy-Search()** algorithm guarantees that at most a set of six roles can provide the required set of permissions.

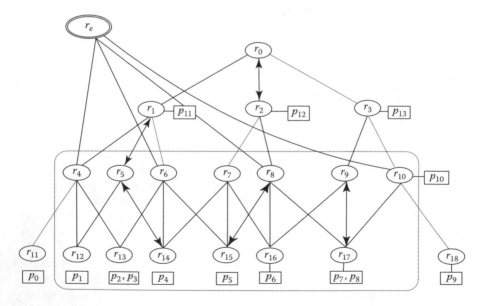

FIGURE 13.6
Role mapping using **Greedy-Search()** algorithm.

TABLE 13.6

Results for Each Step of Above Example

Step 1	Step 2	Step 3
$R^* = \emptyset$	$R^* = \{r_6\}$	$R^* = \{r_6, r_8\}$
$V = r_6$	$RQ = \{p_1, p_6, p_7, p_8, p_{10}\}$	$RQ = \{p_1, p_{10}\}$
	$V = r_8$	$V = r_4$

	Step 4	Step 5
	$R^* = \{r_6, r_8, r_4\}$	$R^* = \{r_6, r_8, r_4, r_{10}\}$
	$RQ = \{p_{10}\}$	$RQ = \{\emptyset\}$
	$V = r_{10}$	

Example 13.3

Multidomain Policy Mapping example. Consider the multidomain grid scenario shown in Figure 13.7. The University of Pittsburgh (UPitt) and the Pittsburgh Supercomputing Center (PSC) interoperate on a regular basis utilizing resources in terms of computing power, storage, and processes. A VO layer is created to allow shared resources to be utilized, and consists of two researcher roles r_1 and r_2.

Consider a scenario in which the role r_1 requires services from UPitt that are mapped to the requested permission set given in Example 13.2. That is,

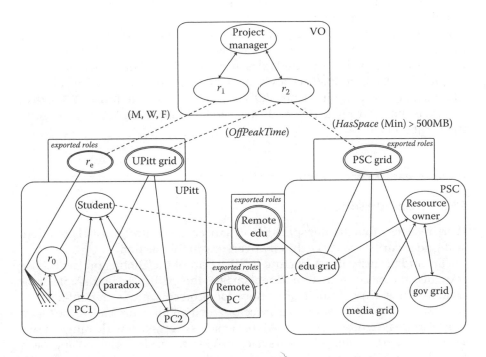

FIGURE 13.7
Policy mapping example for grid and SOA environment.

the hierarchy in Figure 13.6 is a subhierarchy of role Student. Role r_2 also requires a certain permissions set P_2 from the UPitt domain. The request set from VO layer to the UPitt domain can be specified as $RQS = \{(\text{VO}.r_1, P_1, C_1),(\text{VO}.r_2, P_2, C_2)\}$. Assume that P_1 is the set of requested permissions given in Example 13.2, P_2 is a set of permissions that are needed for role r_2 (it is assumed that the roles PC1 and PC2 can fulfill the requirements of P_2 permission set), $C_1 = [M, W, F]$ and $C_2 = [Off\,PeakTime]$ are temporal requirements for the two requests, respectively. In response to this service request from the VO layer, the **Greedy-Search**() is used in UPitt to process the request. As shown in Example 13.2, the export role r_e is created as *I*-senior to the roles that contain the required permissions to satisfy (VO.r_1, P_1, C_1). Similarly, the role UPitt Grid is created as *I*-senior to the roles that contain the required permissions that satisfy the request (VO.r_2, P_2, C_2). In Figure 13.7, exported roles are shown in double circles within the box attached to the domain.

Similarly, the VO layer may also request services from PSC to be made available to its users through role r_2. The PSC domain creates an export role PSC Grid that is *I*-senior of the roles edu grid, media grid and gov grid. Researcher r_2 can then access required grid services from the PSC domain through the export role PSC Grid.

Role mappings may also occur between the UPitt domain and the PSC domain for independent interactions between them. In Figure 13.7, Remote edu is an exported role in the PSC domain to which role student of the UPitt domain is mapped. Similarly, Remote PC role is an exported role to provide requested services to domain PSC.

Figure 13.8 depicts sample policy components that capture the mappings in VO's and PSC's policies. VO's policy will include roles r_1 and r_2 as global role that are mapped to exported roles in UPitt and PSC domains, as shown in Figure 13.8(a). Figure 13.8(b) shows the policy component of the PSC domain that includes the mapping from VO's role r_2 to its exported role PSC Grid, and its mapping from its role edu Grid to the exported role Remote PC in the UPitt domain.

13.5   Related Work

Work related to this chapter spans the areas of access control, trust negotiation, and grid security, which we briefly overview in this section.

The Grid Security Infrastructure (GSI), a part of the Globus project [2] has provided the basic security mechanisms for the grid including single sign-on algorithms, cross-domain authentication protocols, proxy credentials [23]. The Globus Toolkit provides a service-oriented architecture called the Open Grid Services Architecture (OGSA) that enables access to a wide range of services provided by heterogeneous systems. A key to service-oriented approach to grid security is the use of Web services technologies. *WS-Security* defines a standard set of Simple Object Access Protocol (SOAP) extensions, or message

```
<Policy policy_id="VO">
 <PolicyRelation pr_id="VO_Pitt_PSC">
  <GlobalToLocalMapping gMap_id="VO_Pitt">
   <RoleMapping>
    <MappedRole> r₁ </MappedRole>
    <MappedTo>
     <Role policy_id="Pitt"> rₑ
     </Role>
     <MappingCondition>
      <PeriodicTime pt_id="MWF"/>
     </MappingCondition>
    </MappedTo>
   </RoleMapping>
   <RoleMapping>
    <MappedRole> r₂ </MappedRole>
    <MappedTo>
     <Role policy_id="Pitt"> Upitt Grid
     </Role>
     <MappingCondition>
      <PeriodicTime pt_id="OffPeak"/>
     </MappingCondition>
    </MappedTo>
   </RoleMapping>
  </GlobalToLocalRoleMapping>
  <GlobalToLocalMapping gMap_id="VO_PSC">
   <RoleMapping>
    <MappedRole> r₂ </MappedRole>
    <MappedTo>
     <Role policy_id="PSC"> PSC Grid
     </Role>
    </MappedTo>
   </RoleMapping>
  </GlobalToLocalRoleMapping>
 </PolicyRelation>
</Policy>
```

(a) **VO's Policy Component**

```
<Policy policy_id="PSC">
 <PolicyRelation pr_id="PSC_VO_UPitt">
  <GlobalToLocalMapping gMap_id="PSC_VO" >
   <RoleMapping>
    <MappedRole> PSC Grid </MappedRole>
    <MappedFrom>
     <Role policy_id="VO"> r₂
     </Role>
     <MappingCondition>
      <Conditions op="AND">
       <LogicalExpression op="NOT">
        <Predicate>
         <Operation> InUse
         </Operation>
         <Parameter name="src" value="PC2">
        </Predicate>
       </LogicalExpression>
       <Predicate>
        <Operation> HasSpace </Operation>
        <Parameter name="min" value=">500MB">
       </Predicate>
      </Conditions>
     </MappingCondition>
    </MappedTo>
   </RoleMapping>
  </GlobalToLocalRoleMapping>
  <GlobalToLocalMapping gMap_id = "PSC_Pitt">
   <RoleMapping>
    <MappedRole> edu Grid </MappedRole>
    <MappedTo>
     <Role policy_id="Pitt"> Remote PC
     </Role>
    </MappedTo>
   </RoleMapping>
  </GlobalToLocalRoleMapping>
 </PolicyRelation>
</Policy>
```

(b) **PSC's Policy Component**

FIGURE 13.8
X-RBAC specification for Example 3.

headers that can be used to implement integrity and confidentiality in Web services applications [1]. *WS-Trust* describes a framework for trust models that enables Web services to securely interoperate [23]. *WS-Policy* provides a general-purpose model and syntax to describe and communicate the policies of a Web service [23]. *WS-Federation* describes how to manage the trust relationships in a heterogeneous federated environment including support for federated identities [23]. eXtensible Access Control Markup Language (XACML) allows the specification of access control policies, and supports the basic RBAC model. Several access control approaches to address the security requirements of a grid have been discussed in the literature, which include *Permis* [7], *Community Authorization Service* (CAS) [26], *Global Grid Forum* (GGF) Authorization Framework [13, 33, 38], *Privilege Management and Authorization Services* (PRIMA) [7] *Virtual Organization Membership Service* (VOMS) from the European *DataGrid* project [15, 33], the *JoVO* [33], *Shibboleth* [9], *Akenti* [34], and others [13, 33, 38]. The *Akenti system* enables multiple owners and administrators to define usage policies in a widely distributed system [34]. In CAS [26], resource providers grant access to community accounts as a whole. Lorch et al. propose an authorization service to support ad hoc collaborations using attribute certificates [24]. Similarly, Ramakrishnan et al. present an authorization infrastructure for component-based grid applications by providing authorization at the component interface [27]. *Sygn* is another grid access control mechanism that uses certificates and supports RBAC [29].

Several research efforts have been devoted to the topic of policy composition and secure interoperation in multidomain environment. In [30], an integer programming approach has been proposed to allow policy integration between multiple RBAC policies. The roles are mapped using the permission set associated with the corresponding roles. Our earlier work related to loosely coupled, secure interoperation has been discussed in [13], which has been adopted in this chapter. Other research efforts have been devoted to the topic of policy composition in multidomain environment [4, 11, 16].

Trust relationships among interoperating domains have been loosely divided into two parts: *negotiation of trust based on credentials* and *establishing trust based on peer-measured values* such as reputation and ranking [17]. Most of the existing literature on trust negotiation focuses on the negotiation of credentials, with little focus on the generic requirements of secure interoperation [8].

Several trust negotiation mechanisms have been proposed in the literature, such as *TrustServ* [32], *TrustBuilder* [37], *H-Trust* [6], *Trust-X* [1] and others [10, 17, 36]. *Trust-Serv* is a model-driven framework that uses state machines to represent and determine credential exchanges for access to resources [32]. Both *TrustBuilder* and *Trust-X* use credential disclosure trees and negotiation strategies to facilitate protection of credential information during negotiation. *TrustBuilder* defines families of disclosure trees to facilitate negotiation between entities [37]. The work has been augmented with the Generic Authorization and Access Control API (GAA-API) to obtain a composite framework for adaptive trust negotiation and access control [28]. The *Trust-X* system introduces the notion of *trust ticket* for efficient negotiation; if an entity already has a trust ticket for the current interoperating environment, then trust negotiation is not required [1]. *H-Trust* defines functions to establish, sustain, and evolve trust based on entity behavior history [6]. In [10], a trust establishment and sustenance framework for P2P systems has been presented where reputation is used as a basis for trust establishment. The sustenance is based on the concept of complaints, where peers can make complaints regarding other peers to reduce their rank. Reputation-based approach for grid systems has been presented where a trust index is calculated using fuzzy logic based on the success rate of a job and the defense capability of the domain. In other systems, reputation and negotiation have been combined for the negotiation of trust tokens between the interoperating domains. There are reputation-based models that calculate the reputation for every session. Similar to the reputation approach is the recommender approach, where trust is based on propagated recommendations of trusted parties [25, 31].

While these methods are quite specific in their approaches these approaches are: (1) primarily based on the client-server interaction model. *Trust-X* also attempts to propose a framework for P2P but the credential exchange is similar to that of the client server mode; (2) mainly based on credential exchange and do not handle credential types. Further, except for *TrustServ* and *Trust-X*, none of the existing approaches look at credentials of both the service provider and the service requestor; (3) they do not tackle the issue of access requirements-based negotiation between interoperating domains for policy mapping.

13.6 Conclusion

In this chapter, we have presented a trust-based framework for secure inter-operation for SOA-based grid environment. The key aspects of the framework include interleaved trust negotiation and policy mapping based on service requirements. The trust framework also places significant emphasis on the trust sustenance and evolution issues. The key result of the trust-based service-negotiation process is the negotiated trust level and the policy mappings to facilitate requirements-driven secure interoperation among grid peers. Policy mapping essentially included identifying roles that satisfy a given interdomain request and creating export policy that acts as policy interface for the partner domain to make cross-domain accesses. The framework presented is particularly aimed toward the P2P interactions over a service-based grid environment. The GTRBAC model and its extensions along with the X-RBAC language have been adopted to capture dynamic, context-based policy requirements. In this chapter, the challenging issues related to the integration and mapping of complex policies with SoD and cardinality constraints are not addressed.

Acknowledgment

This material is based upon work supported by the National Science Foundation Grant No. IIS-0545912.

Bibliography

1. E. Bertino, E. Ferrari, and A.C. Squicciarini. *Trust-x: a peer-to-peer framework for trust establishment*, Knowledge and Data Engineering, IEEE Transactions on 16 (2004), no. 7, 827–842.

2. R. Bhatti, E. Bertino, A. Ghafoor, and J.B.D. Joshi. *Xml-based specification for web services document security*, Computer 37 (2004), no. 4, 41–49.
3. Rafae Bhatti, Basit Shafiq, James B. D. Joshi, Elisa Bertino, and Arif Ghafoor. *X-gtrbac admin: A decentralized administration model for enterprise-wide access control*, ACM Transactions on Information and System Security (2005).
4. P. Bonatti, S.D.C. Vimercati, and P. Samarati. *An algebra for composing access-control policies*, ACM Transactions on Information and System Security 5 (2002), no. 1.
5. R. Buyya and S. Venugopal. *The gridbus toolkit for service-oriented grid and utility computing: An overview and status report*, First IEEE International Workshop on Grid Economics and Business Models (GECON 2004) (Seoul), IEEE Press, 2004.
6. Licia Capra. *Engineering human trust in mobile system collaborations*, Proceedings of the 12th ACM SIGSOFT International Symposium on Foundations of Software Engineering, ACM Press, Newport Beach, CA, 2004, pp. 107–116.
7. D Chadwick and A. Otenko. *The permis x.509 role based privilege management infrastructure*, SACMAT 2002 Conference, ACM Press, 2002, pp. 135–140.
8. Suroop M. Chandran, Korporn Panyim, and James B. D. Joshi. *A requirements-driven trust framework for secure interoperation in open environments*, The Fourth International Conference on Trust Management (iTrust-06) (Italy), 2006.
9. Linda A. Cornwall and et al. *Authentication and authorization mechanisms for multi-domain grid environments*, Journal of Grid Computing 22 (2004), no. 4, 301–311.
10. E. Damiani, S. de C. di Vimercati, S. Paraboschi, P. Samarati, and F. Violante. *A reputation-based approach for choosing reliable resources in peer-to-peer networks*, CCS'02 (Washington DC), 2002.
11. S. Dawson, S. Qian, and P. Samarati. *Providing security and interoperation of heterogeneous systems*, International Journal of Distributed and Parallel Databases (2000).
12. A. Detsch, L. P. Gaspary, M. P. Barcellos, and G. G. H. Cavalheiro. *Towards a flexible security framework for peer-to-peer-based grid computing*, Proceedings of the 2nd workshop on Middleware for Grid Computing, ACM Press, Toronto, Ontario, Canada, 2004, pp. 52–56.
13. Siqing Du and James B. D. Joshi. *Supporting authorization query and inter-domain role mapping in presence of hybrid role hierarchy*, The 11th ACM Symposium on Access Control Models and Technologies (U.S.), 2006.
14. O. W. Erwin and D. F. Snelling. *Unicore: A grid computing environment*, Lecture Notes in Computer Science, 2001.
15. Ian Foster, Carl Kesselman, and Steven Tuecke. *Anatomy of the grid: Enabling scalable virtual organizations*, International Journal of Supercomputer Applications 15 (2001), no. 3.
16. L. Gong and X. Qian. *Computational issues in secure interoperation*, IEEE Transaction on Software and Engineering 22 (1996), no. 1, 43–52.
17. M. Gupta, P. Judge, and M. Ammar. *A reputation system for peer-to-peer networks*, NOSSDAV'03 (Monterey, CA), 2003.
18. Hai Jin, Weizhong Qiang, Xuanhua Shi, and Deqing Zou. *Rb-gaca: A rbac-based grid access control architecture*, International Journal of Grid and Utility Computing 1 (2005), no. 1, 61–70.
19. J. Joseph, M. Ernest, and C. Fellenstein. *Evolution of grid computing architecture and grid adoption models*, IBM Systems Journal 43 (2004), no. 4, 624–645.

20. James Joshi, Arif Ghafoor, Walid G. Aref, and Eugene H. Spafford. *Digital government security infrastructure design challenges*, IEEE Computer 34 (2001), no. 2, 66–72.

21. James B. D. Joshi, Rafae Bhatti, Elisa Bertino, and Arif Ghafoor. *X-rbac an access-control language for multidomain environments*, IEEE Internet Computing 8 (2004), no. 6, 40–50.

22. J.B.D. Joshi, E. Bertino, U. Latif, and A. Ghafoor. *A generalized temporal role-based access control model*, Knowledge and Data Engineering, IEEE Transactions on 17 (2005), no. 1, 4–23.

23. Nitin V. Kanaskar, Umit Topaloglu, and Coskun Bayrak. *Globus security model for grid environment*, SIGSOFT Software Engineering Notes 30 (2005), no. 6, 1–9.

24. M. Lorch and D. Kafura. *Supporting secure ad-hoc user collaboration in grid environments*, 3rd International Workshop on Grid Computing (Baltimore), 2002, pp. 181–193.

25. John O'Donovan and Barry Smyth. *Trust in recommender systems*, Proceedings of the 10th International Conference on Intelligent User Interfaces, ACM Press, San Diego, CA, 2005, pp. 167–174.

26. L. Pearlman, V. Welch, I. Foster, C. Kesselman, and S. Tuecke. *The community authorization service: Status and future*, CHEP'03 (La Jolla, CA), 2003.

27. L. Ramakrishnan, H. Rehn, J. Alameda, R. Ananthakrishnan, M. Govindaraju, A. Slominski, K. Connelly, V. Welch, D. Gannon, R. Bramley, and S. Hampton. *An authorization framework for a grid-based component architecture*, 3rd International Workshop on Grid Computing (Baltimore, MD), Springer Press, 2002, pp. 169–180.

28. Tatyana Ryutov, Li Zhou, Clifford Neuman, Travis Leithead, and Kent E. Seamons. *Adaptive trust negotiation and access control*, Proceedings of the Tenth ACM Symposium on Access Control Models and Technologies, ACM Press, Stockholm, Sweden, 2005, pp. 139–146.

29. Ludwig Seitz, Jean-Marc Pierson, and Lionel Brunie. *Sygn: A certificate based access control in grid environments*, Tech. Report RR-LIRIS-2005-011, Laboratoire d'InfoRmatique en Images et Systmes d'information (LIRIS), 2005.

30. Basit Shafiq, James B. D. Joshi, Elisa Bertino, and Arif Ghafoor. *Secure interoperation in a multi-domain environment employing rbac policies*, IEEE Transactions on Knowledge and Data Engineering (2005).

31. Brian Shand, Nathan Dimmock, and Jean Bacon. *Trust for ubiquitous, transparent collaboration*, Wireless Networks 10 (2004), no. 6, 711–721.

32. Halvard Skogsrud, Boualem Benatallah, and Fabio Casati. *Model-driven trust negotiation for web services*, IEEE Internet Computing 7 (2003), no. 6, 45–52.

33. T.J. Smith and Lavanya Ramakrishnan. *Joint policy management and auditing in virtual organizations*, Grid Computing, 2003. Proceedings. Fourth International Workshop on Grid Computing, 2003, pp. 117–124.

34. M. Thompson, W. Johnston, S. Mudumbai, G. Hoo, K. Jackson, and A. Essiari. *Certificate-based access control for widely distributed resources*, Eight Usenix Security Symposium, 1999.

35. Liu Wenjie and Gu Guochang. *Security issues in grid environment*, Services Computing, 2004 (SCC 2004). Proceedings 2004 IEEE International Conference, 2004, pp. 510–513.

36. S. Ye, F. Makedon, and J. Ford. *Collaborative automated trust negotiation in peer-to-peer systems*, Fourth International Conference on Peer-to-peer Computing (2004), 108–115.

37. Ting Yu, Marianne Winslett, and Kent E. Seamons. *Supporting structured credentials and sensitive policies through interoperable strategies for automated trust negotiation*, ACM Transactions on Information and System Security 6 (2003), no. 1, 1–42.

38. G. Zhang and Manish Parashar. *Dynamic context-aware access control for grid applications*, Grid Computing, 2003. Proceedings Fourth International Workshop, 2003, pp. 101–108.

14

Distributed Computing Grids—Safety and Security

Mark Stephens, V. S. Sukumaran Nair and Jacob A. Abraham

CONTENTS

Abstract   Harnessing idle CPU cycles from PCs across a network (Internet or intranet) has proven to be an economically attractive solution for solving many problems. Research has shown that the average idle time for most PCs is over 90%, representing a virtually limitless source of untapped computing power. As such, distributed computing grids[1] have become an increasingly popular form of grid computing in research communities as well as in industry.

Likewise, industry leaders such as Sun Microsystems offer CPU cycles as a commodity with their Sun Grid Compute Utility services. Known as utility computing, this type of environment attempts to commoditize computational power by allowing consumers to purchase CPU cycles on demand.

A natural evolution is to combine these two paradigms, i.e., providing harnessed idle cycles as a commodity or managed service, governed by Service Level Agreement (SLA) contracts. However, such an enterprise presents difficult challenges with regard to safety and security. The complexity of these challenges is greatest when the resource owner, consumer, and broker are three distinct parties. This may introduce a motivational paradox such that all parties want to maximize utilization via cooperative use of resources, and yet have conflicting interests as to how and when the resources are utilized. This chapter will focus on the presentation of safety and security challenges in this environment, as well as solutions to address these challenges.

14.1   Introduction

In this section we provide a general introduction to distributed computing grid environments. This includes the power source (i.e., cycle stealing), different distributed computing grid environments, as well as a brief discussion of issues surrounding volatility.

14.1.1   Cycle-Stealing Paradigm

A distinguishing characteristic of cycle harvesting grids [1, 2] is that their computational power is generated via cycle stealing across a loosely connected, heterogeneous collection of resources, i.e., workstations, desktops, PCs, etc. Cycle stealing [3] refers to the type of computational paradigm where idle

---

[1] Also known as cycle harvesting grids, desktop grids, PC grids, etc.

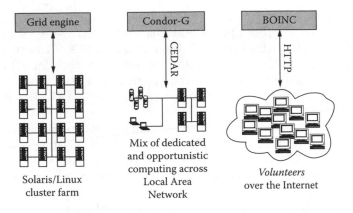

FIGURE 14.1
Three popular distributed resource management solutions.

CPU cycles of distributed computers are harnessed, i.e., perform work when desktop computers (Figure 14.1) are not being used by employees/students (intranet) or the general public (Internet). There is also an implied strict draconian contract [4] between the resource owner and the resource consumer: the owner is only willing to let the consumer harness idle CPU cycles on the owner's resource when it is idle.

"Loosely connected" implies that nodes are either connected internally to an organization via a LAN or externally via the global Internet, and that participation in the grid is not the primary purpose of each computational resource. Rather, the resources are standard desktop computers whose primary purpose is to be available when the owner requires their use for standard day-to-day desktop activities (e.g., word processing, email, etc.).

Undoubtedly this particular type of parallel computing architecture is growing in acceptance and popularity, albeit mostly in scientific and engineering circles. In the last decade we have seen many successful implementations, some with thousands to even millions of nodes; thus it is not surprising that several distributed computing grids are members of the top 500 supercomputer list[2]. However, it does not take this order of magnitude to be substantial, as recently as 2002 the top 500 supercomputer list included several such grids with 500 to 3000 nodes.

14.1.2 Volunteer Computing and Desktop Grids

Several distributed computing grids are hosted by nonprofit scientific organizations who harness idle cycles from supporters (i.e., volunteer computing). For example, SETI created SETI@home as an economical method for analyzing massive amounts of radio telemetry data collected from space. SETI@home

---
[2] www.top500.org

allows the general Internet public to volunteer their idle CPU cycles for scientific research [5]. Today SETI@home has harnessed over 1.6 million CPU years of donated idle CPU cycles and has realized an ROI of 1500:1. Similarly, the Folding@home PC grid is used to simulate protein folding. This critical research is used to study diseases such as Alzheimer's and Huntington's disease. The Folding@home project currently has over 164,442 active Internet connected nodes.

Some universities and companies have even created campus [6] or desktop grids [7] across an intranet to realize greater ROI on existing resources. In Chapter 5 of [7] David Johnson provides a comprehensive introduction to this type of grid with special emphasis on enterprise applications in industry. United Technologies created a desktop grid to leverage existing computers for solving complex modeling problems during off hours.

United Technologies reported an overall increase in computer utilization from 5% to 85%. Distributed Computing, Inc. developed a solution called CapCal (Capacity Calibration) to provide network and Web site performance and capacity testing.

CapCal used a cycle harvesting model where resource owners were paid $.30/hour for use of their idle CPU cycles. When idle, an owner's PC would launch agent software to put network traffic on the Web sites of customers in order to test performance and capacity.

Although there have been some examples of cycle harvesting grid deployments in industry, the widespread acceptance of this computing platform has not come to fruition. Even with the vast amounts of untapped computing power that distributed computing grids offer, there are significant challenges regarding safety and reliability, which appear to greatly inhibit its mainstream acceptance outside of academic or scientific organizations. In the next section, we will explore issues related to cycle stealing that serve as an impediment to widespread adoption of this technology.

14.1.3　Volatility of Cycle Harvesting Grid

Research estimates that the average wasted idle time for Internet/intranet connected PCs is over 90%. This represents an enormous amount of computing power that is currently not utilized, which could be used to advance scientific research or increase shareholder value by obtaining a greater return on investment (ROI). The reliance on cycle stealing for computational power is a cycle harvesting grid's greatest strength—and its greatest weakness. On the one hand, cycle stealing is very appealing since it is "free" untapped computing power that is being harnessed and used to execute tasks. On the other hand, the very nature of cycle stealing implies a great deal of volatility. Most regard this type of grid computing as best suited for large-scale, embarrassingly parallel problems [8], which require no guarantees with regard to result integrity for the consumer, resource safety for owners, or security provided by brokers. Specifically, skepticism with regard to this type of grid computing includes the following.

- *Safety:* Safety is the "avoidance of catastrophic consequences on the environment" [9], i.e., the avoidance of catastrophic consequences to the grid. Executing tasks on remote PCs opens the potential for many types of intentional and unintentional catastrophic consequences to the resource owners. For example, a single task infected with a virus that is on the server and ready for submission could very quickly proliferate and infect owner resources in the grid. Similarly, a single programming bug could unintentionally disrupt a resource owner's ability to perform his day-to-day work if it does not stop once the resource is no longer idle, or even inadvertently delete or corrupt important files or folders. Such "ill-behaved" tasks result in an annoyance to the resource owner or even the potential destruction of his documents, data, and information, which compromises the credibility of the entire grid.
- *Security:* Security is the "prevention of unauthorized access and/or handling of information" [9, 10]. There are two primary data security concerns in this type of grid environment: (1) unauthorized access by a task to sensitive or private data located on the owner's resource, and (2) unauthorized access by a resource user to input and output. Thus, mechanisms must be in place to discourage tasks from accessing private or sensitive information on nodes, as well as mechanisms to discourage resource owners from looking at the input and output data while a task is being processed on their PC.

These concerns must be addressed to make this type of grid computing a more dependable [11] parallel computing architecture, and thus gain more acceptance as a parallel computing platform. Likewise, once these challenges are addressed, the door is opened for a new form of e-Business where idle CPU cycles could be traded across the Internet as a commodity [11].

14.2   Safety and Security Terminology

We begin with a discussion of some basic techniques used to implement appropriate countermeasures to thwart attackers who attempt to compromise the overall safety and security of this type of environment.

- *One-Way Hash Functions.* Verifies that a message has not been altered during transmission.
- *Encryption.* Provides message secrecy.
- *Digital Signatures.* Authenticates the sender of a message.
- *Binary Sandboxing.* Enforces a security policy on an executable at runtime.

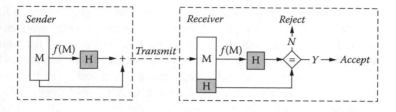

FIGURE 14.2
Message M is passed to the one-way hash function f(M) to produce the hash value H.

The following provides general descriptions of each of these mechanisms as well as an overview of their purpose.

14.2.1 One-Way Hash Functions

One-way hash functions [12] are algorithms that take a variable-length message, M, and create a fixed-length hash value, H (Figure 14.2). The one-way nature of the function is due to the fact that the domain (i.e., variable length) is infinite, whereas the range (i.e., fixed length) is finite; hence it is a many-to-one function and thus does not allow for the existence of an inverse function. Although different messages can map to the same hash value, collisions of (useful) plain text to the same hash value are improbable [12].

Hash values are typically attached to the associated message prior to being sent to a receiving party. The receiving party executes the same hash function and compares the results with the hash value that was sent by the sender. If the value is the same, the receiver has authenticated the message, meaning they can be assured the message was not altered. Note that this does not provide authentication of the sender, nor does it provide secrecy. Hash functions typically execute very quickly in software, due to the fact that encryption is not used. For distributed computing grids, hash codes should be sent with their associated binaries. Doing so verifies that the executable was not altered during transmission.

14.2.2 Encryption

There are two types of encryption, symmetric and asymmetric. Symmetric encryption is based on substituting and permuting [12] plaintext via an encryption algorithm and secret key over a sequence of rounds resulting ultimately with the ciphertext. A similar process using a decryption algorithm will generate the plaintext from the ciphertext. There are several popular symmetric encryption algorithms including DES, Triple DES, AES, Blowfish, etc., all of which rely on a single private key for encrypting and decrypting.

In contrast, asymmetric encryption (i.e., public key encryption) uses mathematical functions [12] to create the ciphertext, and requires a key pair (i.e., public and private keys). This approach resolves two important challenges, i.) secure private key exchange, and ii.) nonrepudiation. Asymmetric

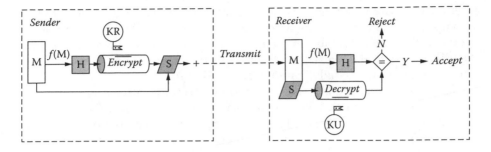

FIGURE 14.3
Value from one-way hash function f(M) is signed to authenticate sender.

encryption is computationally expensive, thus the primary applications for asymmetric encryption are key management and digital signatures [13]. Stallings provides an excellent introduction to cryptography [12], while Schneier provides a great applied view of the same [14].

14.2.3 Digital Signatures

Asymmetric encryption (i.e., public key encryption) can be used to digitally "sign" a message, M, thus authenticating the sender (Figure 14.3). This means the receiver can be assured the message is actually from the sender, and not a third party masquerading as the sender. The method requires a key pair: one key that is kept private by the sender, K_R, and the other that is distributed to the public, K_U. The entire message could be signed, or alternatively, a derived value is created from the message (e.g., hash code) and then signed by encrypting with the sender's private key, thus creating the signature, S. The message and signature are then sent together to the receiving party, who uses the public key to decrypt the signature. The receiver then creates the derived value from the message (via the same method as the sender) and the value is compared with the decrypted signature to verify that it was in fact sent from the sender. Note this method does not provide secrecy; it only authenticates the message was sent by the correct party.

14.2.4 Binary Sandboxing

Sandboxing was popularized by Java applets, which once downloaded are then sandboxed by the JVM in the browser. Microsoft has recently introduced a similar approach with its .NET platform via its virtual machine, the CLR (Common Language Runtime). The built-in sandboxing functionality provided by Java and .NET virtual machines greatly simplifies the ability to enforce security policies in distributed computing environments [15]. However, it is overly restrictive, requiring consumers to submit only tasks written in Java or .NET languages. A better method is to provide mechanisms that sandbox any binary executable, without specific requirements on languages, linking with special libraries, implementing specific interfaces, etc.

Binary sandboxing refers to any attempt to enforce a security policy on a binary executable by limiting access to resources on the PC. For example, if you can limit an executable to read and write to a specific directory tree, or restrict access to the registry, without requiring recompiling, relinking, etc., then you have successfully "sandboxed" the binary.

The task is launched by a monitor process that aids in security policy enforcement (Figure 14.4). Likewise, all task monitor processes (and their associated child process, i.e., the task) are started and monitored by a service/daemon (i.e., the owner resource manager). The owner resource manager is ultimately responsible for monitoring the execution of all task monitors and associated tasks. If a cycle harvesting grid is considered to be a distributed operating system [16], then the owner resource manager is analogous in nature to the Unix scheduler or swapper process (i.e., /sbin/init), which is the parent of all other Unix programs in execution [17].

Techniques for binary sandboxing in a Unix environment are well documented [18]. However, these techniques are not applicable for Windows, which has some intrinsic limitations at the operating system level. Alternative choices for binary sandboxing in a Windows environment include binary instrumenting. With binary instrumenting, the security policy is enforced by inserting instructions into the untrusted binary. A security engine accepts a policy-specification language [19, 20] and then uses a tool [21] to instrument the untrusted binary in such a manner as to enforce the security policy.

Another method that does not require rewriting the target binary is called API hooking [22]. This technique is commonly used by debugging tools to expose values of variables at runtime, as well as by rootkits [23] that use API hooking to subvert the normal control flow of an application. In doing so, the rootkit is able to hide directories/files, trap keystrokes, etc.

API hooking is a two-step process. First access is gained to the address space of the victim process. A favorite technique to accomplish this in the Windows environment is to use remote threads [23]. The purpose of the remote thread

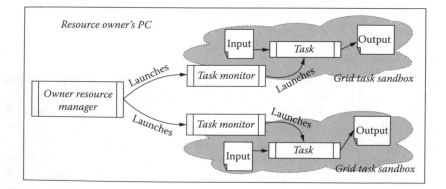

FIGURE 14.4
The owner resource manager launches task monitor processes to enforce security policies.

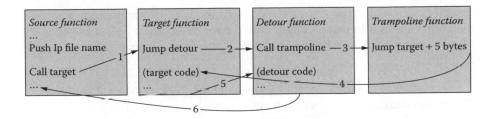

FIGURE 14.5
Source function calls target (1); however, first 5 bytes have been replaced with jump to a detour function (2). Detour function calls trampoline (3), which jumps to the first real instructions of target (4). Control goes back to detour function (5) prior to returning to source function (6).

is to initiate the second step, i.e., hook the victim process. A favorite technique to accomplish this is via inline function hooking. The remote thread stores and then writes over the first few bytes of target functions with a jump call to an alternate "detour" function, which has the same signature. The original first few bytes are saved so that the actual system call can be performed (i.e., the trampoline) [23]. Thus when the binary makes a system call (e.g., read/write to a file, etc.), the hook function is actually called instead (Figure 14.5).

When used as a method for binary sandboxing, the detour function can examine parameters prior to making the system call in order to see if a security policy breach is about to occur. Once the detour function determines that the system call may proceed, the trampoline is used to patch together the original system call. Likewise, in a similar fashion, any return data from the system call may also be examined.

The intent of this section was to introduce techniques that are used throughout the rest of this chapter. We will now review the general cycle harvesting grid environment by examining the different layers of distributed resource management as well as the different participants.

14.3   Distributed Resource Management

The distributed nature of cycle harvesting environments provides the power and flexibility that has become the hallmark of these environments. However, this quality is also the primary source of complexity. In this section, we examine the different layers required to manage these distributed resources, as well as the different participants, their concerns, and mutual interactions.

14.3.1   Resource Management Systems

The Resource Management System (RMS) acts as the central nervous system for a distributed computing grid. The RMS is responsible for resource discovery and reservation, executing consumer's tasks, and enforcing the owner's security policies for the resource. As such, the RMS is comprised of multiple components and services distributed over the grid. In [24] (Chapter 13), Miron Livney and Rajesh Raman describe the resource management layers as follows.

- *Local RM Layer.* Layer that represents the basic services a resource can provide (e.g., operating system, etc.).

- *Owner Layer.* Layer that represents the owner's interests, i.e., providing the owner's resource only when it is not being used by the owner, enforcement of same, accepting consumer tasks for execution, and advertising to resource brokers resource information (e.g., task accounting, usage behavior, etc.).

- *System Layer.* Layer that represents the broker's interests, i.e., accepting resource requests from consumers, resource discovery and reservation, ensuring that consumer tasks are ready for execution by the owner layer, notifying the owner layer of reservation.

- *Consumer Layer.* Layer that represents the consumer's interests, i.e., requesting resources from the system layer, providing expected QoS parameters (if applicable), and prioritizing the task queue.

- *Application RM Layer.* This layer represents the application-specific problem being addressed. This layer is responsible for distributing tasks and work units to the application layer, as well as accepting results from the tasks.

- *Application Layer.* Layer that represents the consumer's tasks, which are executing on an owner's resources. Tasks accept work units and provide results to the consumer via interaction with the application RM layer.

The physical location of each layer plays a critical role in determining the level of security mechanisms that should be put in place. If all layers are physically located behind a secure network, as could very well be the case in a desktop or campus grid, the threat of certain attacks may be mitigated. For example, the need to protect the system, customer, and application RM layers from denial-of-service attacks may be lessened, if these services are not exposed to the global Internet. Likewise, the need to secure data transmission between layers may also be reduced for the same reason.

14.3.2   Distributed Computing Participants

In the most general case, we categorize grid participants into three groups: resource owner, consumer, and broker. Each group could represent a single person, a group of individuals, or an organization, and is characterized as follows.

- *Resource Owner (RO).* The resource owner is the owner of the computational resource that is allowing cycles to be harvested when its resource is idle. The resource owner could be a volunteer across the public Internet (i.e., volunteer computing [5]), an organization that wishes to leverage employee workstations when the system is idle (i.e., desktop grid [7]), or an academic unit (e.g., department, school, etc.) that wants to leverage idle lab machines when students are not using them (i.e., campus grid [6]).

- *Resource Consumer (RC).* The resource consumer has one or more problems for which it requires computational resources to solve. For cycle stealing to be a viable solution, the consumer's problem should be parallel independent in nature [4, 25]. Such problems can be broken up into smaller subproblems, which can then be bundled into work units[3] and distributed to independent computational resources. The results are then provided at some later time back to the resource consumer.

- *Resource Broker (RB).* The resource broker has access to many resources, via some understood contract between the broker and owner. The resource broker is responsible for resource discovery, reservation, and allocation to consumers. When the consumer and broker are separate entities, the broker may expect a form of compensation whereas the consumer may expect QoS guarantees on utilization in the form of a Service Level Agreement contract [26, 27, 28].

The identity and trust relationships between participants also play critical roles in determining the level of security mechanisms to be put in place. For example, in the case of a desktop or campus grids all three participants are the same entity, i.e., company or college, respectively. In this type of environment, the need for protecting resources from malicious task execution may be limited to automated virus scanning and peer code reviews. Likewise, the threat of result sabotage [29] may also be limited in scope since the normal user of these resources are either employees or students; thus the risk level of being exposed engaging in subversive activity is very high, potentially leading to termination or expulsion. Finally, if a resource is compromised, the broker is not at risk of losing future resource participation, since all three are the same entity.

In contrast, when these participants are not the same entity, the level of trust is almost guaranteed to be much less, and thus the risk to each participant of malicious activity on behalf of another participant greatly increases [30]. As such, the level of safety and security mechanisms in place should increase so that guilty parties can be identified and disallowed from future participation.

14.3.3 Participant Concerns

When each participant is a separate entity and the RMS layers are geographically distributed across the Internet several concerns arise for each participant.

---

[3] See Berkeley Open Infrastructure for Network Computing at http://boinc.berkeley.edu

- *RO Concerns.* The owner of the resource has several concerns that must be addressed prior to allowing foreign tasks to be executed on his resource. First, the task is only allowed to execute if the owner is not using the system. Second, the task should be guaranteed not to be infected with any malicious parasitic code (e.g., virus, Trojan horse, rootkit, etc.) Lastly, the task itself must be guaranteed to adhere to a strictly enforced security policy [31]. Thus the task should be sandboxed so as to disallow starting additional executables and limiting access to resources (e.g., certain branches of the file system, registry, network, etc.).

- *RC Concerns.* The resource consumer needs to accept connections from the owners' resources across the Internet in order to distribute tasks and collect results. As such, the communication between the application RM layer and the application layer must include authentication provisions. Additionally, the consumer may require input data privacy as well as output data integrity (i.e., to prevent sabotage). If detection of either occurs, then no additional work units should be handed to the resource (i.e., owner blacklisting).

- *RB Concerns.* The resource broker needs to accept connections from available resources to learn of consumers, as well as accepting connections from consumers who want to make requests for resources, again both across the Internet. As such, the communication between the application RM layer and the application layer must include authentication provisions. Additionally, the consumer may require that input data be kept private and output data not be tampered with (i.e., sabotaged). If detection of either occurs, then no additional work units should be handed to the resource (i.e., owner blacklisting).

   The resource broker is responsible for finding resources from owners and providing those resources to the consumer. It is the responsibility of the resource broker to ensure the previously mentioned requirements regarding safety, security, QoS, etc., are enforced. Failure to do so will mean that distrustful resource owners will no longer provide resources and disgruntled consumers will no longer seek the broker's services. Likewise, resource brokers must also provide the accounting services required to compensate owners as well as bill consumers (i.e., grid economics). Finally, intruder and denial-of-service attack detection mechanisms must be in place to guarantee safety to resource providers and performance guarantees to resource consumers.

For the rest of the chapter we consider only the situation when the resource owner, consumer, and broker are three separate entities. This type of environment requires that the RMS layers are distributed across the Internet, thus maximizing the challenges to safety and security.

14.4 Participant Responsibilities and Interactions

Safety and security requirements must be balanced with performance considerations. The complexity of protocols as well as the type of encryption will affect the overall throughput of the grid. In this section we provide an example architecture along with lightweight security protocols in order to illustrate the responsibilities and interactions between participants.

14.4.1 Installation of Local Resource Manager

The first set of interactions occurs when the software comprising the owner layer is installed on the resource owner's PC (Figure 14.6).

Step 1. Installation of the initial owner layer software on the owner's resource is a critical first step. This installation package will contain several owner layer management applications including an RO resource management Windows service. The RO resource manager service is responsible for starting the owner layer applications after boot. Transfer of the owner layer installation package should be done over a secure channel such as SSL, which may require that the broker has been issued an X.509 certificate from a trusted certificate authority (CA).

Step 2. Once installed on the owner's resource, the owner layer establishes a secure connection with the system layer's authentication services, and a secret symmetric key is shared between the broker and owner via key exchange (e.g., Diffie-Hellman Key Exchange, etc.) [12]. Likewise, the owner layer also provides an asymmetric public key to the system layer, which can be used for authenticating the owner layer. The system layer creates a unique ID for the owner's resources, encrypts the ID with the shared key, and sends back this ID to the owner. The system layer then stores the shared key and unique ID in a participant database.

Step 3. The owner layer periodically submits usage profiling updates to the system layer. Profiling resource behavior is used by the system layer to measure the appropriate trust level of a resource, as well as modeling idle time frequency for utilization forecasting purposes.

14.4.2 Consumer Requests Resources from Broker

The second set of interactions occurs when a potential consumer requests resources from a broker. The request includes specific QoS parameters that must be met by the broker in order for the consumer to accept the contract (Figure 14.7).

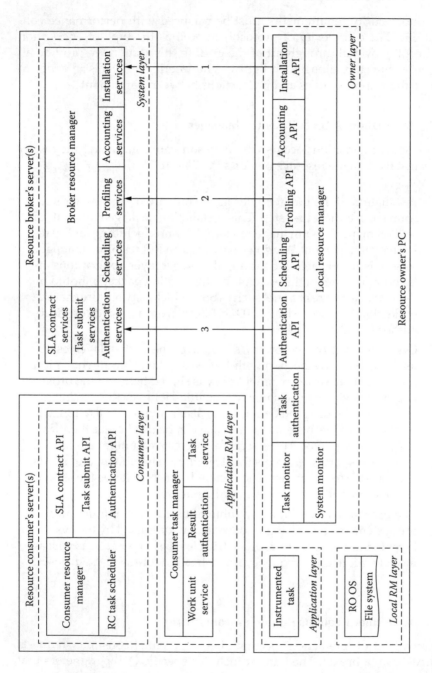

FIGURE 14.6

Installation of the local resource manager software on the owner's PC.

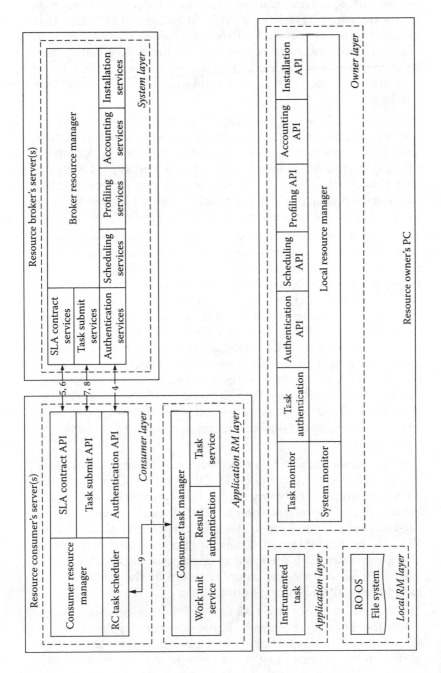

FIGURE 14.7
Consumer requests resources from broker with specific QoS parameters.

Step 4. A new consumer requires the use of resources managed by the broker. The relationship may be commercial in nature, whereby the broker will expect financial reimbursement for resource usage, or the relationship may be more of a collaborative nature whereby two organizations have agreed to share resources. A secret symmetric key is shared between the consumer and the broker. The system layer provides the consumer layer with a unique consumer ID. The system layer stores the shared key and unique ID in a participant database.

Step 5. The consumer layer requests resources from the system layer by sending the consumer's QoS requirements. The system layer makes a temporary reservation of resources and responds back with an SLA contract and a unique SLA contract ID encrypted with the shared symmetric key. The consumer may then accept or reject the proposed contract. If the contract is rejected or expires, then the temporary resource reservations are released.

Step 6. The consumer accepts the contract and sends the system layer a task service URL where owner resources can retrieve task binaries from the application RM layer. The broker makes the resource reservations permanent and creates two types of reservation certificates, a certificate pair for each resource and the consumer. The consumer's reservation certificate contains a contract ID and ephemeral owner ID (uniquely identifies the owner to the consumer for the duration of the contract only). This certificate is then encrypted with the consumer/broker shared key. The owner's reservation certificate contains the owner's ephemeral owner ID and consumer's task service URL. This certificate is then encrypted with the owner/broker shared key. These certificates are stored in a reservation database. Once the contract start time begins, these certificates are passed to the owner layer.

Step 7. The consumer submits to the broker a task binary, binary hash value, and associated contract ID. The contract ID is encrypted with a shared key. The binary hash value is created by hashing the binary appended with the contract ID. Although no encryption is used on the hash value per se, it acts as a cryptographic check sum since the contract ID is only known by the consumer and broker, and the contract ID is always sent encrypted. This hash value is used to guarantee that the binary was sent by the consumer and has not been altered during transmission.

Step 8. The system layer scans the task binary for virus infection, instruments the binary, and creates a digital signature for the newly instrumented binary by first creating a hash code of the instrumented binary and then encrypting the hash code with the broker's private key. The system layer sends back the newly instrumented task binary and associated digital signature to the consumer layer.

The consumer verifies that the instrumented task binary was sent by the broker.

Step 9. The consumer layer provides the instrumented task binary and digital signature to the application RM layer so that tasks can be distributed to reserved owner resources.

14.4.3 Broker Schedules Resources

The third set of interactions occurs when the broker schedules reserved resources for consumer tasks by notifying the reserved resources of the consumer's task URL (Figure 14.8).

Step 10. The system layer sends a scheduling request to all resources reserved for this contract. The request contains the reservation certificates (from step 6 in Section 14.5.2). The owner layer decrypts the owner certificate using the shared key. The owner now has the consumer's task service URL. In this manner, each resource can verify that the scheduling request is from the broker, which includes the consumer's task service URL for retrieving the instrumented binary.

Step 11. Once a reserved resource goes idle, the owner layer requests a task from the consumer's task service (provided via URL from step 10 in Section 14.5.3). The consumer responds back to the owner with the instrumented binary and digital signature of same.

Step 12. The owner layer verifies that the instrumented binary was signed by the broker and has not been altered. After the instrumented binary is authenticated the owner can safely launch the instrumented task.

Step 13. The task requests work units from the application RM layer. Depending on the sensitivity of the data, additional encryption can be applied to secure the transmission.

Step 14. The task reads and writes working files to a restricted location on the owner's resource (i.e., the sandbox). Again, depending on the data sensitivity, files can be encrypted as they are being written and decrypted while they are being read.

Step 15. The owner layer monitors the task usage of the owner's resource.

Step 16. The task periodically submits results to the application RM layer. Results should be authenticated as coming from a valid resource.

14.4.4 Resource Reports Results to Broker

The final set of interactions occurs between the resource and the broker when the resource sends the amount of cycles used by the consumer's task(s), as well as receiving any updates to the owner layer software (Figure 14.9).

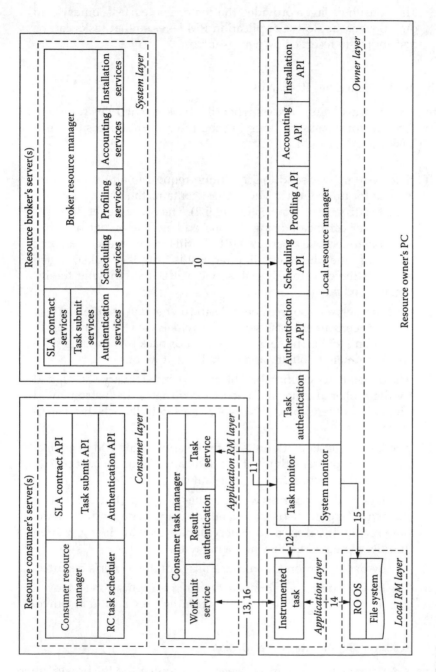

FIGURE 14.8
Broker schedules reserved resources to launch consumer tasks.

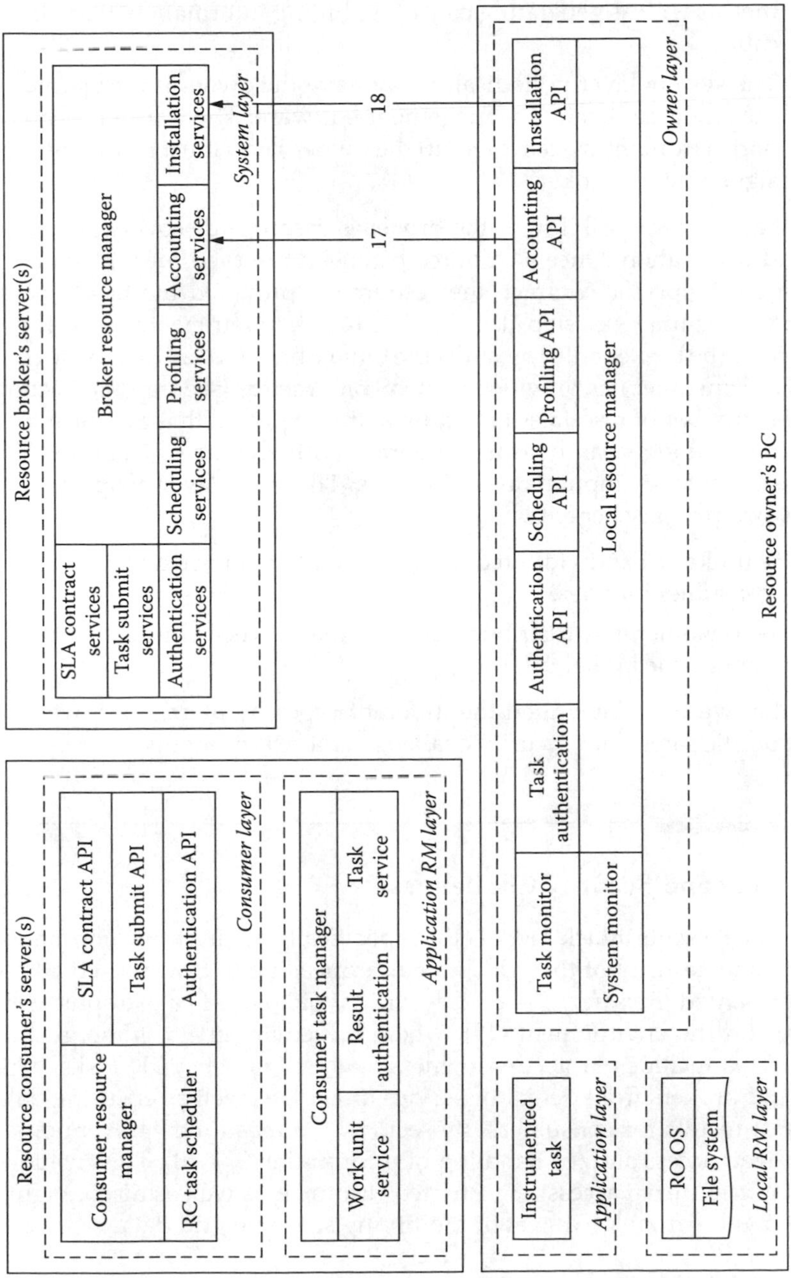

FIGURE 14.9
Resources report cycles used by consumer as well as install updates to owner layer software.

Step 17. The owner layer provides accounting information regarding consumer task utilization to the system layer. The owner layer uses this information to see if SLA utilization requirements are being met, as well as generating consumer billing information, if applicable.

Step 18. The system layer periodically may have the owner layer update the resource owner's management software. The transfer of binaries from the system layer to the owner layer must be digitally signed by the broker.

Note that the responsibility of the broker is resource discovery, management, and reservation. Once a resource has been reserved for a given consumer under a specific contract, the resource is provided the task service URL of the consumer (see step 10 in Section 14.5.3) to retrieve the task. Ergo, for each task, the broker only transmits the binary once that is back to the consumer. The consumer has provided via QoS parameters (see step 5 in Section 14.5.2) the number of resources he requires. It is expected that the consumer has allocated enough bandwidth to interact with resources, including the transmission of tasks, input data, and results. This extra layer of abstraction provides two primary benefits.

1. The broker's bandwidth requirements are not impacted by the size of consumer binaries.

2. The consumer may distribute any task that has been inspected and signed by the broker.

Now that we have illustrated the different types of interactions, we will discuss specific safety and security challenges more thoroughly.

14.5 Safety and Security Challenges

This section presents attack models that specifically attempt to compromise the safety and security of the grid. A countermeasure to thwart each attack will be presented. By safety we refer to the "avoidance of catastrophic consequences on the environment" [9], where the environment is the grid as a whole. This includes protecting owners against "ill-behaved" tasks, consumers and brokers from denial-of-service attacks, as well as ensuring data integrity of results for consumers. By security we mean the "prevention of unauthorized access and/or handling of information" [9, 10]. This includes preventing tasks from accessing restricted resources as well as discouraging resource owners from looking at or modifying sensitive grid data.

14.5.1 Establishing Trust over Time

One of the critical aspects of this type of environment is the profiling subsystem at the owner's layer. The profiling subsystem is responsible for providing usage behavior (e.g., monitoring frequency of resource idle time) as well as detecting potentially malicious activity on behalf of a resource owner. A "burn-in" period is expected wherein the owner layer is not provided actual consumer work and instead monitored by the system layer to evaluate behavior patterns. Only when enough data has been collected to allow reliable forecasting will a resource be reserved for use by a consumer. Likewise, if the owner layer detects any malicious attempts on behalf of the resource owner to compromise resulting data, then the trust relationship between broker and owner will diminish, potentially to the point of being blacklisted by the broker.

14.5.2 Opponent Attacks Owner Masquerading as Broker

In this attack model the opponent attacks the owner by attempting to masquerade as a valid broker. For example, the opponent might be attempting to trick resources into transferring and launching malicious executables instead of valid owner layer binaries. This would be an excellent way to trick owners into installing a rootkit [23].

The owner layer installation package should be handled via a secure channel such as SSL and require that the broker has a valid X.509 certificate issued by a public CA. All transferred binaries should also include a hash code of the binary encrypted with the broker's private key. This way the owner can be confident that the binaries were sent by the broker and have not been altered during transmission. Likewise, scheduling requests from the system layer to the owner layer contain an owner reservation certificate (see step 10 in Section 14.5.3), which includes the task service URL encrypted with a secret key shared between the broker and owner. This verifies that the scheduling request came from the broker and was not altered during transmission.

14.5.3 Opponent Attacks Owner Masquerading as Consumer

In this attack model the opponent attacks the owner by attempting to masquerade as a valid consumer. The opponent could be attempting to trick resources into transferring and launching malicious executables instead of instrumented tasks. These binaries may be infected with a virus, worm, Trojan horse, spyware, etc. To ensure that the resource is being provided a binary that the broker has instrumented and scanned for viruses, a digital signature is created by the broker (see step 8 in Section 14.5.2). Thus, when a consumer submits a task binary to the broker, the following steps are performed (Figure 14.10).

1. Previously the broker has created an asymmetric encryption key pair, B_R and B_U. B_U is known by all consumers and owners.
2. The broker performs automated virus scanning of the binary task, T.
3. The broker transforms the task to an instrumented task, i.e., f_i: $T \rightarrow T_i$.

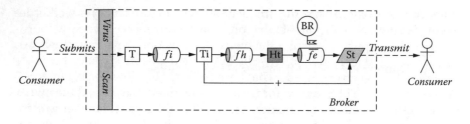

FIGURE 14.10
Consumer submits task to broker for virus scanning, instrumenting, and signing.

4. The broker creates a hash code H_t of the task T_i, i.e., $f_h{:}T_i \rightarrow H_t$.
5. The broker creates a digital signature S_t by encrypting H_t with B_R, i.e., $f_e{:}B_R, H_t \rightarrow S_t$.

This process occurs once per task when the consumer initially submits a task that will be executed on reserved resources. Likewise, additional tasks may be submitted by the consumer for the same contract. Once the broker completes this process, T_i and S_t are sent back to the consumer for distribution to reserved resources since it is the responsibility of the consumer to distribute tasks to resources, not the broker. The broker is only responsible for resource discovery, management, and reservation.

Whenever a resource requests a task from the consumer, the consumer transfers the task, T_i, and the signature, S_t. The owner then performs the following steps (Figure 14.11):

1. Decrypt signature S_t with public key B_U, recovering the hash code H_t, i.e., $f_e^{-1} : B_U , S_t \rightarrow H_t$
2. The owner creates a new hash code $\acute{H}_t$ of task T_i, i.e., $f_h{:}T_i \rightarrow \acute{H}_t$

If $H_t = \acute{H}_t$ then the node can accept the task with confidence that it was inspected by the broker and thus has been instrumented and scanned for virus infection. Otherwise, the task and associated hash value are discarded.

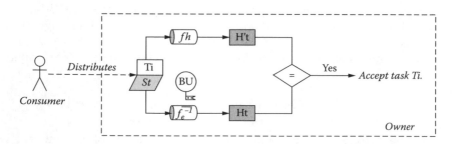

FIGURE 14.11
Owner receives task and digital signature from consumer and compares hash codes to ensure task was not tampered with after instrumentation by broker.

If the task is accepted, then the hash value H_t is saved with the task in order to counter the outside virus attack (see Section 14.6.8).

14.5.4 Opponent Attacks Consumer Masquerading as Owner

In this attack model the opponent attacks the consumer by attempting to masquerade as a valid owner. The opponent may be a saboteur attempting to trick the consumer to upload bogus results (see Section 14.6.11).

To thwart this attack the application layer must send the application RM layer the consumer reservation certificate, hash code H, and the data. The application layer creates the hash code, H, from the following:

1. Ephemeral owner ID,
2. Data (e.g., profiling, accounting, etc.)

The owner's ephemeral ID is part of the consumer's certificate, which was encrypted using the consumer/broker secret key. Upon receipt the application RM layer decrypts the certificate, retrieves the owner's ephemeral ID, and creates a hash code, H', of the ephemeral owner ID from the certificate and the data. If the hash codes match, then the consumer can be confident that the data came from a valid owner and has not been altered in transmission. Note that no encryption of the hash code is needed.

This does not protect against valid owners that turn saboteur. This attack is covered in Section 14.6.11.

14.5.5 Opponent Attacks Consumer Masquerading as Broker

In this attack model the opponent attacks the consumer by attempting to masquerade as a valid broker. The opponent may again be a saboteur attempting to trick the consumer to upload bogus results by first acting as a broker and then creating and signing bogus reservation certificates.

This attack is thwarted by authenticating the broker (see step 4 in Section 14.5.2). Once the broker's identity is authenticated, then it is not possible for an opponent to masquerade as a valid broker by creating and signing bogus reservation certificates. The problem then reduces to authenticating owners, which is covered in Section 14.6.4.

14.5.6 Opponent Attacks Broker Masquerading as Owner

In this attack model the opponent attacks the broker by attempting to masquerade as a valid owner. The opponent may again be a saboteur attempting to trick the broker into accepting bogus profiling data, account data, etc.; or attempting to trick the broker into disclosing consumer information (e.g., task service URL, etc.).

This attack is thwarted by the exchange of a secret key (see step 2 in Section 14.5.1). Profile and accounting data should include a cryptographic check sum that demonstrates that the data comes from a valid owner and has not been altered in transmission. An efficient way to do this is for the owner to create and send a hash code, H, for every transmission, as follows:

1. Secret key,
2. Data (e.g., profiling, accounting, etc.)

The owner ID is encrypted using the broker's public key and is sent along with the hash code and data. Upon receipt the system layer decrypts the owner's ID, pulls the owner's secret key from the participant database (see step 2 in Section 14.5.1) and creates a hash code, $\acute{H}$, of the secret key and data. If the hash codes match, then the broker can be confident that the data came from a valid owner and has not been altered in transmission. Note that no encryption of the hash code is needed.

Likewise, when distributing owner reservation certificates, the consumer's task service URL has been encrypted with the owner/broker secret key, thus preventing an opponent from extracting this information.

Finally, direct attacks against a consumer by an opponent masquerading as an owner is covered in Section 14.6.4. This does not cover the situation in which an opponent downloads and installs the owner layer installation package, only to eventually become a saboteur. This attack model will be discussed in Section 14.6.11.

14.5.7   Opponent Attacks Broker Masquerading as Consumer

In this attack model the opponent attacks the broker by attempting to masquerade as a valid consumer. The opponent may either be attempting to get free resources, or trick the broker into signing malicious tasks so that the opponent can distribute the tasks to resources.

Step 4 in Section 14.5.2 demonstrates the method for authenticating consumers. The consumer may be able to get free resources the first time around; however, the broker will blacklist the consumer from any further transactions. With regard to tricking the broker into signing malicious tasks, Section 14.6.3 demonstrates the infeasibility of this attack via automated virus scanning and binary instrumenting. Ergo, malicious tasks will never be signed by the broker, thus owners will never accept such tasks from a consumer's task service.

14.5.8   Thwarting Outside Virus Attack

In this attack model a virus has infected a task, after being distributed to a resource. Sections 14.6.2 and 14.6.3 demonstrate how to avoid initial virus propagation. This is a case where the task is infected by some other non-grid-related binary, which was inadvertently launched by the resource owner. Before starting any task, the Task Monitor should perform a subset of the steps outlined in Section 14.6.3 (Figure 14.12), i.e.:

1. Create a new hash code, $\acute{H}_t$, from the task T.
2. Compare $\acute{H}_t$ with the original H_t.

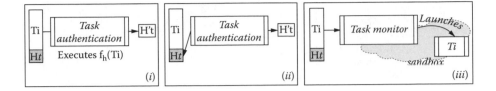

FIGURE 14.12
Task authentication (i) executes one-way hash function $f_h(T_i)$, then (ii) compares result $\acute{H}_t$ with H_t, and (iii) launches task T_i only after verifying task has not been altered (e.g., modified by a virus, etc.).

If $\acute{H}_t = H_t$ then the Task Monitor can launch the task with confidence that it has not been modified since distribution by the consumer's task service. If the hash codes do not match, then the task has been modified in some manner (e.g., virus infection) and the Task Monitor should delete the binary and request a new one from the consumer's task service.

14.5.9   Tasks That Refuse the Idle State

In this attack model a task refuses to go idle once the resource owner requires his resource back. Note that this does not imply some particular malicious behavior on behalf of the consumer. The majority of PCs on the Internet are Windows based, which does not have consistent support for suspending tasks temporarily. Windows does support an "idle" task priority; however, in practice it can be difficult to get all the threads of a Windows process to behave in a "nice" manner; and, in a cycle-stealing environment, we must guarantee that any task that is launched will stop consuming resources once the owner begins using his resources.

Since the task monitor launches each task, the tasks are each a child of a task monitor. And as any good parent, one of the parental responsibilities is to ensure their children are not an annoyance to others, especially to the resource owner. Once the task monitor identifies that the node is no longer in an idle state (i.e., by monitoring keyboard and mouse activity) then it will set the priority of its child to "idle." If this does not work, then the task monitor will send a request to the task's threads to stop execution. If this does not work, the parent terminates the ill-behaved child process (Figure 14.13).

14.5.10   Denial-of-Service Attacks Against Participants

Distributed resource management environments open the consumer and broker to DoS (Denial-of-Service) attacks. In this attack model, an attacker might masquerade as a valid owner that either continuously requests grid work packages (which they never plan to execute) and/or submits an endless stream of arbitrary bytes instead of actual results, accounting data, or profiling

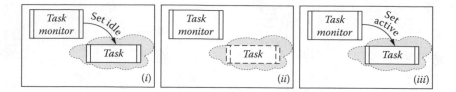

FIGURE 14.13
When the task monitor detects activity by the owner, the task monitor will (i) set the priority level of the grid task to "idle." If this succeeds then (ii) the task monitor allows the task to stay alive. After the task monitor determines the resource is idle then the monitor will (iii) reset the priority level of the task.

data. Likewise, the attacker might masquerade as a consumer and attempt to flood the broker, or vice versa.

The method for thwarting these types of attacks is to authenticate a session prior to accepting transmissions. This can be achieved by encrypting and signing session IDs, which are then exchanged prior to and after data transmissions. The length of the encrypted and signed session IDs are well known by all parties, hence a DoS attack during the session ID exchange is thwarted. Once the session ID exchange is complete, all subsequent transmissions will expect the session ID prior to accepting large streams of data.

14.5.11 Consumer Spoiler Attacks

Distributed computing grids are very susceptible to Byzantine faults as well as sabotage. A Byzantine fault implies that the owner's resource, or the task being executed, arbitrarily produces incorrect results. Sabotage means that a malicious resource owner (i.e., a saboteur) modifies or tampers with results prior to being sent back to the consumer. In [11] we propose that it is impossible to distinguish individual instances of sabotage from Byzantine faults encountered during task execution. From an error detection and recovery perspective these problems are equivalent; hence we collectively refer to both as a spoiler attack. Undetected spoiler attacks can have devastating effects on a project. If the tainted results are aggregated with other results, or if the tainted results are used as input for subsequent tasks, then the possibility exists for spoiling large batches of work. Unfortunately, cryptographic check sums alone do not provide 100% protection against spoiler attacks [29] since a binary code can be decompiled, algorithms analyzed, and memory can be examined during process execution.

Redundancy must be used to mask faults in any system [9]. One of the most common forms of redundant computing for fault tolerance is Triple Modular Redundancy (TMR), which was originally suggested by Von Neumann [9]. With TMR the same task is distributed to three different participants. Results from the three participants are then compared. Two out of three results that match are considered the canonical result that is then presented to the consumer.

Unfortunately, the direct application of TMR will decrease parallelism by a third. Alternately, the system might begin with distributing the task to just two nodes. If the results come back the same, that result is declared the canonical result and a third check is no longer necessary. If the results differ, then the application RM layer distributes the task again to a different resource, not in the first group. If the result returned matches one of the two original results, that result is declared canonical. If all three (or more) results differ, the same approach is repeated.

With this approach, there must be an upper limit on the number of times a task is submitted for checking. Once the threshold is reached, the task itself should be pulled from further scheduling until analysis can be conducted. If the threshold is reached, then this approach has probably uncovered one of three faults.

1. The task creating the result is exhibiting a Byzantine fault, which makes it produce erroneous output each time it is executed.
2. The task creating the results is fine; however, the output is nondeterministic, hence more sophisticated checking techniques may be required.
3. The logic used to compare results is flawed.

Again, the use of redundant computing carries a price, i.e., decreased levels of parallelism. To reduce the penalty of redundancy, a novel idea called "credibility-based fault tolerance" [29] uses spot checking in conjunction with redundancy to minimize sabotage while maximizing performance by limiting the use of redundancy. The general idea is to calculate credibility values for resources. Only when a certain credibility threshold is reached will work be accepted as canonical from a node. Continued accurate participation raises the credibility value for a node, whereas inaccurate result postings have a negative impact, which was discussed in Section 14.6.1. Thus the broker can quantitatively measure the credibility of the entire system (i.e., the credibility threshold). An alternative probabilistic verification process is offered by Germain-Renaud and Nathalie Playez [32].

Additionally, blacklisting can be used to greatly increase the effectiveness of the countermeasures discussed for both denial-of-service attacks as well as spoiler attacks. Blacklisting means to take action based on the knowledge that a specific communication session is from an attacker (either denial-of-service or spoiler attack). The only difference between attacks (i.e., denial-of-service or spoiler attack) is what action to take. If the attacker is attempting a denial-of-service attack, then the best approach is to simply terminate the session, thus ending the attack. However, if the attacker is a saboteur then it is best not to "tip your hand." Rather, one can accept the results and then discard them. Otherwise, the saboteur is likely to simply rejoin the grid, at which point one has to catch him again. Likewise, if the spoiler attack is unintentional (i.e., a Byzantine fault) then the culprit code should be fixed and redistributed, at which point the nodes can be taken off the blacklist, with their credibility rating rolled back to a point prior to distributing the faulty code.

The greatest vulnerabilities of these countermeasures are authentication of resources and the ability to quickly identify a spoiler attack. Vulnerabilities with owner authentication were discussed in Sections 14.6.4 and 14.6.6. Probabilistically, the timeliness of identifying a spoiler attack is directly correlated to the amount of spot checking and redundancy used. Again a cost/risk analysis should be performed to determine what acceptable threshold of redundancy vs. safety is adequate. Of course this threshold is not static, and can be adjusted to increase or decrease the level of spot checking, redundancy, etc., until the appropriate balance is achieved.

14.5.12   Restricting Access to Resources

In this attack model a task is again acting in an "ill-behaved" manner. This time the task is attempting to access a resource outside of its sandbox. This includes access to registry entries, specific directories, opening socket connections, or any other activity that violates the security policy associated with the grid user who submitted the task.

The binary sandboxing presented in Section 14.3.3 is specifically designed to restrict access of tasks. The process of instrumenting and/or API hooking will thwart the attempts of a task to violate the security policy.

The vulnerability of this technique is signing a task for which an opponent has purposely coded around the instrumentation and/or API hooking. This vulnerability is addressed by authenticating the consumer and establishing trusted relationships with same.

14.5.13   Securing Input and Output Results

In this attack model the grid user wants to discourage the owner from examining sensitive input and output data from tasks, or attempting to modify the same [33]. This assumes the grid work data contains sensitive information that the grid user would prefer resource owners to view, or the grid user suspects that saboteurs might tamper with results.

The task itself can be designed to apply cryptographic check sums before writing data to files. Doing so assures the consumer that output data was not manually modified by the owner prior to uploading to the application RM layer.

Likewise, the API hooking approach presented in Section 14.3.3 can also be used to address runtime decryption of input streams to the task as well as runtime encryption of output streams from the task, or to apply a cryptographic check sum. The novelty of this approach is that tasks are completely unaware that the input and output data was encrypted. Once the data is presented to the task, it has already passed through a stream-cipher filter that is applied by the hook. Likewise, when the task calls a write system call (presenting the output in plaintext) the hook intercepts this call and encrypts the results and/or provides a cryptographic check sum prior to making the system call to write to the file. Again, this approach is very unobtrusive, since it does not put any special implementation requirements on the consumer (e.g., special coding, recompiling, relinking, etc.).

Each task has an associated security policy file, which is read by the task monitor prior to launching the task being launched. The security policy for that task provides the task monitor with a security policy ID. At runtime, the task monitor contacts the consumer and presents the security policy ID. In return the broker presents a symmetric stream-cipher key. Now the task monitor can decrypt the input file stream before presenting the data to the task. Likewise, it can apply the same technique (albeit in reverse) for encrypting output data from the task. Or if required, the task monitor can apply a cryptographic check sum to alert the consumer if a saboteur has tampered with the results.

The vulnerability of this approach is the data in memory is plaintext. There are many tools that can be used to view data in memory while a process is in execution, including most debuggers. However, most of these tools also use API hooking. Thus one approach would be to use the same techniques used by rootkit detection tools to discover if a task has already been hooked [23]. In essence, the task monitor can set "tripwires" to monitor owner behavior. Any detected attempt to expose or alter task data will diminish the trust relationship of that owner, which ultimately would lead to the broker blacklisting the owner.

14.6 Conclusion

The growing interest in distributed computing grid technology from scientific, academic, and business sectors demonstrates wide applicability for solving many types of problems. Increasing dependability, or the trustworthiness of a system [9], requires increasing four primary attributes: safety, security, reliability, and availability [9, 10]. In this chapter we provide an overview of distributed computing grid environments as well as addressing challenges with regard to safety and security. We demonstrate that the level of complexity is greatest when the participants are three discrete groups, i.e., owner, broker, and consumer. We also address the need to balance appropriate safety and security protocols with the overhead associated with same. Our future efforts will focus on increasing the last two attributes of dependability, i.e., reliability and availability.

Once issues with dependability [9] are addressed, brokers can guarantee specific levels of quality of service (e.g., as with utility computing). This allows a new e-commerce market to emerge whereby idle CPU cycles are traded across the Internet as a commodity, thus allowing organizations to tap into large amounts of computing power on demand, similar to the electric power grid.

Bibliography

1. J. Basney, M. Livny, and T. Tannenbaum. High Throughput Computing with Condor, *HPCU News* Vol. 1 No. 2 (1997).
2. R. Raman, M. Livny, and M. Solomon. Matchmaking: Distributed Resource Management for High Throughput Computing, *Proceedings of the Seventh IEEE International Symposium on High Performance Distributed Computing*, 1998.
3. K. Dong Ryu and J. K. Hollingsworth. Unobtrusiveness and Efficiency in Idle Cycle Stealing for PC Grids, *Proc. of the 18th International Parallel and Distributed Processing Symposium*, 2004.
4. A. L. Rosenberg. Guidelines for Data-Parallel Cycle Stealing in Networks of Workstations, *Journal of Parallel and Distributed Computing*, Vol. 59, No. 1 (1999) pp. 31–53.
5. D.P. Anderson. Public Computing: Reconnecting People to Science, *Conference on Shared Knowledge and the Web* (2003).
6. A. Ashok et al. Building a Campus Grid: Concepts and Technologies, *NSF Middleware Initiative (NMI) Integration Testbed Case Study Series: Supplemental Documentation* (2005).
7. A. Abbas, eds. *Grid Computing: A Practical Guide to Technology and Applications*, Charles River Media, Inc., 2004.
8. B. Wilkinson and M. Allen. *Parallel Programming Techniques and Applications Using Networked Workstations and Parallel Computers*, Prentice Hall, 1999.
9. P. Jalote. *Fault Tolerance in Distributed Systems*, PTR Prentice Hall, 2004.
10. J. C. Laprie. Dependable Computing and Fault Tolerance: Concepts and Terminology, *15th International Symposium on Fault Tolerant Computing Systems* (1985) pp. 2–11.
11. M. Stephens and S. Nair. Internet Distributed Computing for e-Business, *Proc. 7th International Conference on Electronic Commerce Research* (2004) pp. 221–233.
12. W. Stallings. *Cryptography and Network Security, Principles and Practices (Third Edition)*, Prentice Hall, 2003.
13. W. Diffie. The First Ten Years of Public-Key Cryptography, *Proceedings of the IEEE*, Vol. 76, No. 5 (1998), pp. 560–577.
14. B. Schneier. *Applied Cryptography (Second Edition)*, John Wiley & Sons, Inc., 1996.
15. L. F. G. Sarmenta, S. J. V. Chua, P. Echevarria, J. M. Mendoza, R. Santos, S. Tan, and R. P. Lozada. Bayanihan Computing .NET: Grid Computing with XML Web Services, *Cluster Computing and the Grid 2nd IEEE/ACM International Symposium CCGRID2002* (2002).
16. A. Tanenbaum. *Distributed Operating Systems*, Prentice Hall, 1995.
17. W. Richard Stevens. *Advanced Programming in the UNIX Environment*, Addison Wesley Longman, Inc., 1993.
18. I. Goldberg, D. Wagner, R. Thomas, and E. Brewer. A Secure Environment for Untrusted Helper Applications (Confining the Wily Hacker), *Proc. of 1996 USENIX Security Symposium* (1996).
19. L. Bauer, J. Ligatti, and D. Walker, Types and Effects for Non-interfering Program Monitors, *Software Security—Theories and Systems Mext-NSF-JSPS International Symposium ISSS* (2002).
20. J. Ligatti, L. Bauer, and D. Walker, Edit Automata: Enforcement Mechanisms for Run-time Security Policies, *International Journal of Information Security*, Vol. 4, No. 1–2 (2005), pp. 2–16.

21. A. Srivastava, A. Edwards, and H. Vo. Vulcan: Binary transformation in a distributed environment, *Technical Report MSR-TR-2001-50* Microsoft Research.
22. J. Richter. *Programming Applications for Microsoft Windows, Fourth Edition,* Microsoft Press, 1999.
23. G. Hoglund and J. Butler. *Rootkits—Subverting the Windows Kernel,* Addison-Wesley, 2005.
24. I. Foster and C. Kesselman, eds. *The Grid: Blueprint for a New Computing Infrastructure,* Morgan Kaufmann, 1999.
25. A. L. Rosenberg. Guidelines for Data-Parallel Cycle-Stealing in Networks of Workstations II: On Maximizing Guaranteed Output, *International Journal of Foundations of Computer Science,* Vol. 11, No. 1 (1999), pp. 183–204.
26. C. Kenyon and G. Cheliotis. Creating Services with Hard Guarantees from Cycle-Harvesting Systems, *Proc. of the 3rd International Symposium on Cluster Computing and the Grid* (2003).
27. D. A. Menasce and E. Casalicchio. QoS in Grid Computing. *Internet Computing, IEEE,* Vol. 8, No. 4 (2004), pp. 85–87.
28. M. Stephens and S. Nair. Providing Service Level Agreements for Idle CPU Cycles from an Internet PC Grid, *Proc. 16th IASTED International Conference on Parallel and Distributed Computing and Systems* (2004).
29. L. F. G. Sarmenta. Sabotage-Tolerance Mechanisms for Volunteer Computing Systems, *Proc. of ACM/IEEE International Symposium on Cluster Computing and the Grid (CCGrid'01)* (2001).
30. I. Foster, C. Kesselman, G. Tsudik, and S. Tuecke. A Security Architecture for Computational Grids, *Proc. of the 5th ACM Conference on Computer and Communications Security* (1998).
31. A. R. Butt, S. Adabala, N. H. Kapadia, R. Figueiredo and J. A. B. Fortes. Fine-grain access control for securing shared resources in computational grids, *Proc. of International Parallel and Distributed Processing Symposium, IPDPS 2002,* pp. 206–213.
32. C. Germain-Renaud and N. Playes. Result Checking in Global Computing Systems, *Proceedings of the 17th Annual International Conference on Supercomputing* (2003).
33. A. Chien, B. Calder, S. Elbert, and K. Bhatia. Entropia: Architecture and Performance of an Enterprise Desktop Grid System, *Journal of Parallel and Distributed Computing,* Vol. 63 (2003), pp. 597–610.

Part IV

Security in Pervasive Computing

15

Security Solutions for Pervasive Healthcare

Krishna Venkatasubramanian and Sandeep K.S. Gupta

CONTENTS

Abstract Pervasive healthcare systems use pervasive computing technologies, e.g., wearable medical sensors with wireless interconnects, to increase the modalities and spatiotemporal dimensions in which healthcare services can be provided for improving patient outcomes. Security is very important in pervasive healthcare systems to protect sensitive health information that it collects and manages; therefore, they have to maintain data confidentiality, integrity of data, and provide strong authentication features, thereby controlling unauthorized access of personal health information. This chapter presents an overview of security solutions for pervasive healthcare systems, focusing primarily on three aspects: 1) securing data collected by medical sensors, 2) controlling access to health information managed by the pervasive healthcare system, and 3) legislative framework available for securing healthcare systems.

15.1 Introduction

The goal of pervasive healthcare (PH) is to use pervasive computing technologies to provide round-the-clock healthcare outside the confines of traditional medical establishments, such as hospitals and medical clinics, but rather in their homes and outdoors. Traditional model for health management consists of *observing symptoms, visiting a doctor, getting treatment*. Pervasive healthcare aims to change this model into one that provides healthcare facilities to individuals anywhere and at any time. It uses large-scale deployment of sensing and communication (wired and wireless) technologies to monitor patients continuously. This allows it to deliver accurate health information to the medical professionals, thereby stimulating timely diagnosis and treatment for health problems.

Pervasive healthcare, therefore, by facilitating improved patient-caregiver interaction, has the potential to provide accurate, timely, and error-free care to all. This is particularly useful nowadays since the population is aging rapidly; medical institutions are facing shortages of medical staff; cost of healthcare is skyrocketing; and incidences of medical errors are at an all-time high [11].

Significant advances in communication and sensing technologies has led to the development of intelligent handheld and wearable devices (such as PDAs, cell phones, smart watches, clothes, and bands) that have made it possible to implement a wide range of solutions for PH systems. The health management capability of pervasive healthcare systems makes them ideal for many diverse applications including [1] the following.

- **Mobile telemedicine**: Provides the ability to monitor, diagnose, and treat patients from a distance. This reduces the chances of medical errors and enables timely treatment of patients by providing accurate, real-time, and complete health information to the medical professional. Example usage scenarios include monitoring patients in remote rural locations and reacting immediately in response to a medical emergency (dispatching an ambulance), and providing patient monitoring and treatment for postoperative care.

- **Disaster response**: Provides the ability to respond effectively to disasters, where the numbers of patients far exceeds the number that can be handled by the available medical staffs. Using an appropriate pervasive healthcare system, patients can be automatically monitored and doctors' attention can be brought to only those patients who are critical, thereby improving the effectiveness of the response.

- **Pervasive access to patient health data**: Pervasive healthcare systems are designed to collect data from patients over long periods of time. These data are stored in an organized manner so that they can be studied by the patients' caregivers to provide better care. Such large data sets can be useful for studying issues such as response to medicine, demographics of people with specific ailments, possible improvements in the care, improvement in medicine, alternative treatments and diagnosis.

- **Lifestyle management**: Pervasive healthcare systems have the ability to provide personalized care. For example, it can be used by people to improve their health by developing specialized meal and exercise plans.

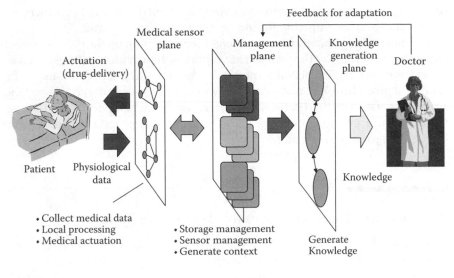

FIGURE 15.1
A generic pervasive healthcare system model.

Figure 15.1 presents a generic model for a pervasive healthcare system. Conceptually, the model consists of three main planes: the **medical sensor plane**, the **management plane**, and the **knowledge generation plane**. The sensor plane provides the capability to incorporate a large number of various types of medical sensors in a pervasive healthcare system. These sensors may have the capability to continuously or intermittently monitor various physiological parameters, such as EKG (electrocardiogram), blood pressure, body temperature, galvanic skin resistance, and motion detection of various body limbs. The sensors may be placed on the patient[1] (wearable) or inside the patient's body. In some cases, these sensors may also have actuation capabilities and can perform tasks, such as drug delivery, under the control of the management plane.

The management plane provides an infrastructure for managing the health data collected by the sensors. It takes raw health data from the sensors, and organizes them into a structured format by generating an Electronic Patient Record (EPR). An EPR collects health data concerning a single patient in a manner that is easy to store and access. It also stores useful information about the patient to assist in better understanding of the data. Further, the management plane provides intelligent indexing and mining capabilities for fast retrieval of pertinent health data and information from EPRs. In addition, the management plane provides functionalities to direct the sensor plane to collect specific stimuli based on the current requirements or actuate specific

---

[1] The term patient is used interchangeably with the term individuals, to denote any individual who is wearing medical sensors on his body for health monitoring.

treatment. Computational devices such as PDAs, cell phones, PCs, and servers are employed in the implementation of the management plane.

The knowledge generation plane is used for reasoning on the data collected and stored (in EPRs) by the previous two planes. It provides features, such as detection of the occurrence of a medical emergency, the failure of a specific treatment procedure, inconsistencies between the proposed diagnosis and the symptoms. This capability gives the caregivers feedback pertaining to their diagnoses and treatment, allowing them to make appropriate adjustments through the management plane.

15.2   Security Threats in Pervasive Healthcare Systems

A pervasive healthcare system collects and manages health data in an electronic format—EPRs—as compared to the largely paper-based records of today. The usage of EPRs, however, imposes many security risks to the health data that did not exist before with paper-based records. This may lead to unauthorized access and tampering of sensitive health data of patients. The reasons for this newfound vulnerability are:

1. Paper-based health data storage is highly centralized and any copying of this information is tedious and a time-consuming process. With EPRs kept on networked systems for availability reasons, it is accessible from anywhere and is very easy to copy [2].

2. More and more sensitive information is being included in a patient's EPR for faster and easier retrieval. Examples include HIV status, psychiatric records, and genetic information [2].

3. The networked nature of pervasive healthcare systems allows the EPRs to be moved across administrative or even national boundaries with ease, thereby circumventing any local legal issues [2].

Therefore the ability of pervasive healthcare systems to continuously collect, exchange, store, and reason, based on electronic health data poses many avenues of abuse of privacy and security. Some of the more probable threats to the pervasive healthcare systems include:

1. Unauthorized access to health data.

2. Deliberate alteration of health data of specific patients, leading to incorrect diagnosis and treatment.

3. Deliberate generation of false alarms or suppression of real alarms raised by the system in case of emergencies.

4. Economic and social discrimination of patients (insurance companies offering health insurance with high premiums to people who have certain chronic problems).

Recently there has been a significant increase in concern, in the popular press and masses, over privacy issues, relating to the electronic health data. Therefore, the viability and long-term success of the technology depends upon addressing the aforementioned threats [2]. Security and privacy preservation in pervasive healthcare has not been investigated in much depth before, and thus provides ample avenues for research. Section 15.3 presents the security solutions for a PH system focusing on preserving the security of health data collected and maintained by the system.

15.3   Security Solutions for Pervasive Healthcare

Security is essential for any system. In the context of pervasive healthcare systems, it is even more important because these systems deal with health information maintained within the EPRs. *The principal idea behind securing pervasive healthcare systems is to preserve patient privacy.* To ensure this, care needs to be taken to prevent all unauthorized access to EPRs in the system.

The notion of providing security in the domain of pervasive healthcare is not different from traditional systems and relies on the maintenance of three basic properties. **Data Integrity**: All information provided is accurate, complete, and has not been altered (during transit and storage) in any way. **Data Confidentiality**: Information is only disclosed to those who are authorized to see it. **Authentication**: To ensure correctness of claimed identity of communicating entities. Here, we present security solutions of pervasive healthcare systems that focus on protecting health data from three different aspects:

- **Securing Medical-Sensor Communication**: Individual medical sensors, used in a pervasive healthcare system, have very small form factors and therefore have limited capabilities. Hence, in general, a complex, computation-intensive security mechanism (such as Public Key Infrastructure (PKI)) is not suitable for securing medical-sensor communication in the context of pervasive healthcare.

- **Controlling Access to EPRs**: An important property of a medical system is that patients have a high level of control over deciding who accesses their health information. Pervasive healthcare systems use EPRs to store pertinent health information about patients. As many organizations, such as pharmacies, insurance agencies, drug companies, and caregivers, need to gain access to patient EPRs for their own economic needs and to provide better service (e.g., improved drugs and competitive insurance rates), patients should be able to easily control access to their EPRs so that personally identifiable sensitive health information is not released.

- **Legislative Solutions**: Realizing the importance of a legal framework for protecting sensitive medical information stored as EPRs, the U.S. Congress proposed a Health Information Portability and Accountability Act (HIPAA) in 1996. All technical solutions are required to address the recommendation proposed by HIPAA and a basic understanding of its provisions is required.

Further, there are two additional issues associated with pervasive healthcare systems; security of wireless communication and physical security of handheld devices. Pervasive healthcare systems make extensive use of wireless communication technologies such as WLAN and cellular phones to communicate health data collected by medical-sensor networks [7]. However, both these communication technologies have many security vulnerabilities. The security problems primarily relate to poor encryption algorithms (Wired Equivalent Protocol (WEP))[8, 9, 10] and session management (GSM) [24]. However, the next generation of both technologies have addressed the issues (with 802.11i and 3G systems, respectively)[11, 24].

To provide pervasive health monitoring, the PH systems make use of portable handheld devices that are used by both patients and caregivers. Such devices may store sensitive health information about the patient and cause a serious privacy breach, if stolen or misplaced. Therefore, physical security of the devices involved also has to be considered. Some solutions for this problem include user-device authentication (using biometric [4, 12, 13], RFID [14], and e-tokens [27]), and use of smart cards [15, 17, 16]. However, these issues are outside the scope of our presentation, and are mentioned here solely for completeness reasons.

15.3.1   Sensor Networks Security in Healthcare

In this section we present issues relating to securing communication between medical sensors used in a pervasive healthcare system. In recent years several promising clinical prototypes for implantable and wearable health-monitoring sensors have started to emerge [7]. These devices are being used for continuous monitoring of patients over long periods of time. Much of the work so far has gone into their design to make them stable, biocompatible, power efficient, and reliable. However, as these sensors are used for collecting health data from patients, ensuring that they do so in a secure manner is equally important.

Security for generic sensor networks has been a prime topic of research over the last couple of years and large numbers of interesting results have been obtained. However, security issues for medical sensors are largely an unexplored area. We need a slightly different outlook while addressing security issues for medical sensors primarily because of the environment (i.e., the human body) in which they are placed. One of the most important requirements of medical sensors is that they should not hinder the day-to-day activities of the person who is wearing them. This requires the sensors to be extremely small in size and weight. The computation and communication capabilities of the medical sensors are therefore more constrained than generic sensors. As security adds overhead to the system, care needs to be taken to ensure that this overhead is minimized in case of medical sensors.

One of the first works to address the issue of security for implantable and wearable medical sensors is [3]. It advocates the use of the human body itself as a means of generating cryptographic keys for securing intersensor communication. As the human body is an extremely dynamic environment, it can produce many specific physiological values that are time variant and not easy to guess (are random and from a large range of values). Using these for cryptographic purposes provides strong security and eliminates key distribution. Both the sender and receiver can now measure the physiological values from their environment and use them for security purposes, when they want to communicate [3].

The principal idea behind this scheme is for the senders and receivers to measure previously agreed-upon physiological values (PV) simultaneously. The synchrony in measurement is required because the values of the PVs are time variant. Once the values are measured, say, the values are K_s and K_r for the sender and receiver, respectively,[2] to send a confidential message, the sender first generates a random session key $K_{session}$, encrypts the payload with it [$C = E_{K_{session}}(Data)$], and then hides the $K_{session}$ using K_s by computing a one-time pad on it ($\gamma = K_{session} \oplus K_s$). It also computes a Message Authentication Code (MAC) on the encrypted message C using the $K_{session}$ ($mac = MAC(K_{session}|C)$) to allow verification and to maintain message integrity. The sender then transmits the message [C, γ, mac] to the receiver, which then uses K_r to obtain $K'_{session}$ from γ ($K'_{session} = \gamma \oplus K_r$). Due to the dynamic nature of the human body, the values of K_s and K_r may not be the same, resulting in the derived $K'_{session} \neq K_{session}$.

In [3, 6], the authors contend that values of PVs measured from the same individual are very close, and any discrepancies in their values are treated as analogous to communication errors. Error correction code (such as majority encoding) is then used to correct the difference. Therefore the receiver performs error correction on $K'_{session}$, yielding $K''_{session}$ ($K''_{session} = f(K'_{session})$, where f is the error correction code). The receiver now computes its own version of the MAC using $K''_{session}$, $mac' = MAC(K''_{session}|C)$. If the values of mac and mac' are identical, then the receiver decrypts C to obtain $Data$, otherwise,

---

[2] The values being measured may not be same at both ends because the values are analog in nature and some discrepancy may arise.

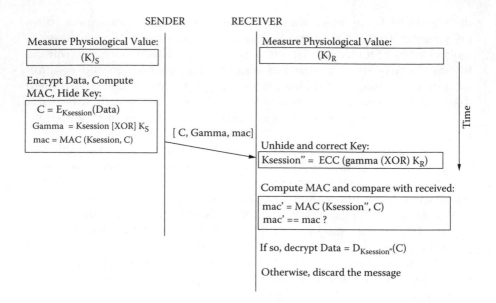

FIGURE 15.2
Secure communication, in body sensor networks, using physiological values.

it discards the message received (Figure 15.2). The pseudocode for this process is given below:

DATA_PROCESS()
1. measure chosen PVs at both sender (K_s) and receiver (K_r) simultaneously
2. if (DataToSend)
3. $C \leftarrow E_{K_{session}}(Data)$
4. $\gamma \leftarrow K_{session} \oplus K_s$
5. $mac = MAC(K_{session}|C)$
6.   send($C \parallel \gamma \parallel mac$)
7. end if
8. if (DataToReceive)
9. $K''_{session} \leftarrow f(K_r \oplus \gamma)$
10. $mac' = MAC(K''_{session}, C)$
11.   if ($mac == mac'$)
12.   Data $= D_{K''_{session}}(C)$
13.   else
14.   reject data received
15.   end if
16. end if

The choice of PVs is an important issue here. Not all PVs possess the time variance and randomness that is required to effectively hide $K_{session}$. For example, if we choose, blood glucose whose value in humans normally ranges between 64–140 mg/dl [28], as PV, irrespective of its time variance, the range of values is so small that it is vulnerable to brute-force attacks. For similar reasons, the use of PVs like blood pressure and heart rate directly is also not advised. In [4] and [5], the use of more complex PVs such as Inter-Pulse-Interval

(IPI) and Heart-Rate Variation (HRV) have been proposed as suitable PVs for securing implanted biomedical sensor communication, respectively. In both cases the PVs, i.e., HRV and IPI signals, were encoded to 128-bit values. Values of any two measurements of these PVs were found to vary considerably, when measured from two different individuals and were very similar when measured from the same individual. Further, the values varied with time and were not predictable. Similarly EKG (electrocardiogram), which has been shown to uniquely identify individuals [29], can also be used here. More work, however, needs to be done to identify other PVs for this purpose. The use of PVs as security primitives for intersensor communication in pervasive healthcare ensures the confidentiality of the data (through encryption), integrity (through MAC), and effectively authenticates the communicating sensors because of the uniqueness of the PVs[3] with respect to the individuals in whom they are measured.

15.3.2   Controlling Access to EPRs

Preserving the privacy of the information collected and maintained as EPRs by a networked and distributed architecture, like that of PH, is very important. This is especially true, when such information may be accessed by entities other than the patient's caregiver and family, such as pharmacies, insurance companies, and drug manufacturers, for their economic and service needs. Since sharing of patients' health information requires their informed consent, pervasive healthcare systems need access control schemes to capture and enforce the specific needs of each patient. In this section we address the issue of authorization in accessing EPR of patients within a pervasive healthcare system using access control mechanisms.

Preliminaries

One of the frequently used techniques for access control in healthcare systems is Role-Based Access Control (RBAC) [18, 19, 20]. RBAC, first proposed in [22] and [23], is a mechanism for access control that organizes users [in the system] into specific groups called *roles*. Roles are groups of users formed based on the functions they perform within the system. For example, all users who are doctors within a hospital will be assigned the role of *doctor* and all the nurses in the hospital system will have the role *nurses* assigned to them. RBAC further assigns access privileges to these roles, instead of to each individual user. This decoupling of users' identity from the privileges associated with them provides a greater level of scalability, as opposed to Access Control List (ACL)-based access control schemes that maintain lists of privileges for different users with respect to the resources within the system. The primary advantage, therefore, of an RBAC-based system is its ability to reduce complexity and

---

[3] In [5], it has been shown that the values of HRV measured from two different individuals vary by as much as 80 bits of Hamming distance. However, two measurements of HRV from the same individual vary slightly (3–8 bits of Hamming distance).

the effort for managing access to large-scale systems. RBAC further defines role hierarchies in order to allow management of relationships between roles within the organization. For example the role of a doctor in a hospital could be a parent role for the role of cardiologist, as a cardiologist is a form of doctor. All privileges associated with the role doctor are inherited by the role cardiologist as well.

Extension of RBAC for Controlling Access to Medical Information

As mentioned before, for healthcare information, the patients themselves must be able to define who can and cannot access their EPRs. In [25] it is argued that access control schemes used in healthcare environments should support two types of policy expression: *general consent qualified by explicit denial (GC-ED)*, and *general denial qualified by explicit consent (GD-EC)*. Example of the former could be a rule, such as all physicians except Dr. X, and of the latter could be no physician except Dr. X. GD-EC is required in scenarios where access needs to be tightly restricted, where it is more convenient to block all users except a few who are explicitly provided access. GC-ED on the other hand is useful for efficiency purposes, for example, using the GC-ED mechanism, a hospital could specify a default set of policies specifying who are prevented by accessing a patient's health information, and the patients can then modify this according to their needs.

In RBAC, as roles can execute only those privileges that are assigned to them and no other, by nature it can easily express GD-EC scenarios. Implementing GC-ED in RBAC is tedious because we would need to define a role explicitly listing all users who need to be given access. If access is to be prohibited for only a handful of users, this role will be very tedious to populate. Using constraints with RBAC has been defined as a means of denying exercising of privileges for a role that would otherwise be allowed. However, constraints do not provide an elegant solution especially with the role hierarchies of RBAC [18]. For example if a constraint is applied to the role of a clinician, then its child role (doctor) will also inherit the constraints of the role. Therefore we will not be able to easily execute a policy of the form *provide access to all clinicians except Dr. X*.

In [18], the authors have proposed a solution to this problem by presenting a simple extension of RBAC. In their healthcare access control model, a patient's access policy is recorded and enforced through a consumer-centric role called *care-team role* (CTR). A CTR consists of four main components: list of roles who are allowed access to the patient's health information, list of roles who are denied access to the patient's health information, the access privileges, and administrative information about the CTR such as its ID and description. Figure 15.3 shows the CTR structure, where all doctors and nurses (except Nurse Y) were allowed read access to the patient's EKG and radiology reports. Similarly, all radiologists are also prevented from reading the patient's EKG and radiology reports. It further needs to be noted that all roles for which access is denied override all roles that are equal or more general. Therefore, if all doctors are prohibited from access, then all clinicians

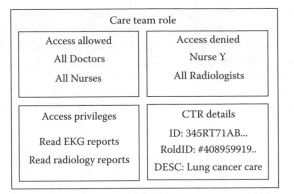

FIGURE 15.3
Care team role structure [18].

would be prohibited, too, unless they are explicitly mentioned in the list of roles who are allowed access. The presence of two lists ensures the implementation of both GD-EC (through the access-allowed list), and GC-ED (through the access-denied list). Figure 15.4 shows the relationship between roles and permissions when using CTR given in Figure 15.3 [18]. Here all nurses are assigned privileges to read both EKG and radiology reports, while a particular nurse Y is denied permission.

Context Awareness in Controlling Access to Medical Information

The aforementioned model modified the RBAC model to support scenarios requiring GC-ED, apart from GD-EC, to facilitate easy expression of patients' wishes regarding access to their EPRs. In [20] the RBAC model is extended in a different way, by introducing the element of context in access control

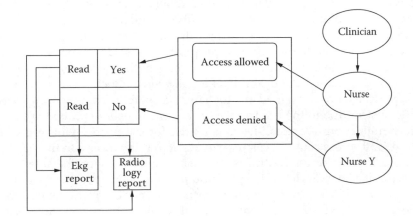

FIGURE 15.4
CTR usage [18].

decisions. The authors argue that access privileges to information in the EPRs for different roles are not static and may vary with the system context. For example if Doctor X had the privilege to access a patient's record now, he may not have one the next day because of his reassignment to another patient (change of context). The ability to describe contexts is an important feature of their work, which is done in the form of a regular expression. Its schema is generic enough to allow the expression of complex context information [20]:

$Context = Clause_1 \cup Clause_2 \cup \ldots \cup Clause_i \ldots$
$Clause = Condition_1 \cap Condition_2 \cap \ldots \cap Condition_i \ldots$
$Condition = Context\ Type < OP > Value$

Here Context Type defines a specific property of the system, such as the time and the location; OP is the operator, such as $\leq, \geq, \neq$. Given this means of expressing contextual information, the system provides access to users (e.g., caregivers, pharmacists) based on *authorization policies* (AP), which is defined as a triple $< R, P, C >$ where R is a role, P is the requested permission, and C is the given context (described using the schema given above). When users want to obtain access, they present a *data access request* (DAR) of the form $< U, P', CR >$ where U is the user ID, P is the permission required, and CR is the actual values of the context types. Access is provided to the user if and only if the actual values presented with the context types (CR) of the DAR evaluates as *true* in the context description (C), $P = P'$ and $U \in R$.

Controlling Access for Managing Medical Emergencies

The Context-Aware Role-Based Access Control (CA-RBAC) was designed for taking context information into consideration when providing access to patients' EPRs. The access provided to the health information is, however, reactive in nature, that is, access is provided only on explicit request from the users. Though adaptive in nature, it only takes into account the current context of the users, compares it with existing rules about privileges to be assigned in such contexts before providing appropriate access. However, what it does not consider is the occurrence of critical events in the system and providing access for this change. An example of a critical event includes a heart attack for a patient whose assigned doctor is not available. The resulting effects on the system due to the occurrence of the critical event is called *criticality*. Timely mitigation of criticalities is essential for the proper working of the system, and access control systems can assist in this process. In the previous example, a smart access control mechanism should therefore be able to find other qualified doctors in the hospital and provide them appropriate access to the patient's EPR. This mitigates the effects of the critical event (heart attack).

Here, if a CA-RBAC model were used, it would not have provided access to the patient's EPRs to any doctor other than the one assigned to the patient without explicit consent of any kind. In [19], the authors have proposed a novel access control model for handling access control in such emergencies called Criticality Aware Access Control (CAAC). CAAC is designed to provide *proactive* access to handle system emergencies. By proactive we mean facilitating continuous monitoring of the system for critical events and in event of observing one, automatically providing an alternate set of access privileges to selected users without any prompting or request.

As the access in CAAC is provided automatically, care has to be taken that it is not provided for longer than absolutely required, for minimizing misuse. Therefore any access provided to users in response to an emergency is temporary and is rescinded after a specific amount of time. The value of this time duration is limited by the window of opportunity (W_o) of the emergency.[4] Every emergency has a duration, called the window of opportunity, associated with it. This is the maximum time before which mitigative measures have to be initiated and completed for controlling the emergency. If the emergency is not handled within W_o then irreparable damage could ensue, for example, in the event of a heart attack, the window of opportunity for controlling it is 1 hour (in most cases); if not controlled by this time, it may not be possible to save the patient's life. In normal circumstances the CAAC model degenerates to Context Aware RBAC similar to [20]; however, in case of emergencies, the system suspends the Context Aware RBAC model and implements the CAAC model. Figure 15.5 shows the execution model of CAAC [19]. When the system observes a critical event, it moves into a CAAP (Criticality Aware Access Policies) mode where the system implements an alternate set of access policies to facilitate effective mitigation.

15.3.3 Legislative Solutions

Apart from technical solutions proposed, an equally important means of ensuring security of information collected in a pervasive healthcare system is legislative. With a growing rise in the digitization and electronic exchange of medical records, the Health Insurance Portability and Accountability Act (HIPAA) was passed in the U.S. Congress August 21, 1996, to address information portability and security issues that emerge from this trend. The scope of HIPAA is threefold: 1) To simplify the administrative overhead in collecting, managing, and accessing EPRs; 2) To prevent healthcare fraud and abuse; 3) Tax-related group plans and revenue offset provisions [21]. Though the scope of HIPAA is considerably larger, we will focus on the HIPAA privacy and security rules that are designed for the prevention of fraud and abuse of medical data. Before proceeding further we need to define the notion of *covered*

---

[4] The actual duration is defined as the earliest time when either one of the following is true: the criticality has been successfully mitigated, the window of opportunity has expired, or all the mitigative actions that could possibly be taken have been executed and nothing more can be done irrespective of the presence or absence of criticality.

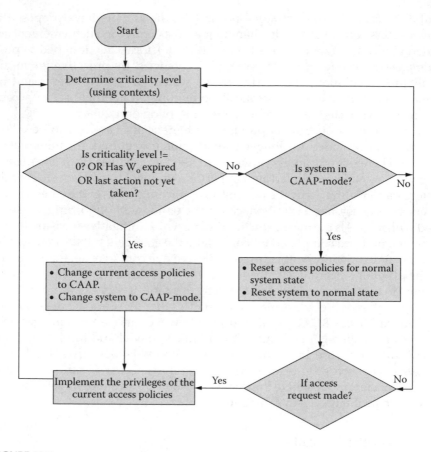

FIGURE 15.5
Execution model for CAAC [19].

entities (CE). A CE includes all entities that deal with collection, storage, and management of health information, health plan providers, healthcare clearinghouses, and healthcare providers. All HIPAA regulations apply to CEs only.

The HIPAA privacy rule concerns itself with defining policies for information flow, rights of patients to access, review, and change their medical data. It defines the notion of personally identifiable health information (PHI) (which can be contained in either electronic, paper, or oral form) and requires that it be protected. It further proposes methods for releasing such information by: 1) removing all identification information from it such as name, geographic locations, telephone numbers, medical record numbers, and health plan IDs; 2) releasing a limited dataset for research purposes, public health and healthcare operations. It further provides certain rights to the individuals concerning their health information, such as 1) the right to receive notice for any information that a CE released; 2) the right to request restriction on the use and disclosure of health information; 3) the right to access and amend one's medical records; 4) the right to audit disclosure of medical data [21]. However, covered entities have the power to apply discretion on honoring any of these rights of individuals. Further, the rule also limits the types of restrictions that an individual can impose on the medical data. The privacy rule states that the CEs must protect PHI irrespective of how they are generated by implementing safeguards to prevent their improper disclosure. It further obligates all the covered entities to provide training to all their workforce members to ensure compliance with HIPAA privacy rules [21].

The HIPAA security rule complements the privacy rule and provides recommendations for the implementation of *administrative, physical,* and *technical* safeguards by covered entities to ensure the availability, confidentiality, and integrity of all *electronic health records.* The administrative safeguards recommend 1) implementation of policies and procedures to prevent, detect, and contain security violations; 2) designation of an individual responsible for managing security; 3) implementation of policies and procedures for ensuring only authorized workforce staff have access to Electronic PHI (EPHI); 4) development of a security awareness and training program for the CE's entire workforce; 5) implementation of policies and procedures for reporting, responding, and managing security incidents; 6) implementation of policies and procedures for disasters and emergencies that damage information systems containing EPHI; 7) ensuring all business associates who create, receive, maintain, or transmit EPHI on behalf of the CE will safeguard EPHI [26]. The physical safeguards further recommend 1) implementing policies, procedures, and processes that limit physical access to electronic information systems ensuring authorized access only; 2) implementing procedures that specify appropriate use of data access devices (e.g., PCs, PDAs) and characterize physical environment of workstations that can access EPHI; 3) implementing physical safeguards for all data access devices that can access EPHI in order to limit access to authorized users; 4) implementing policies, procedures, and processes for receipt and removal of hardware and electronic media that contain EPHI in and out of CE and movement of those items within the CE [26]. Finally the technical safeguards include provisions for developing and implementing policies, procedures, and process for electronic information systems, which 1) ensures access control; 2) maintains audit trails; 3) maintains data integrity; 4) enforces authentication; and 5) ensures transmission security [26].

With the development of pervasive healthcare systems and pervasive availability of information, prevention of fraud and abuse will become an ever greater issue. The introduction of HIPAA law provides framework for ensuring the security of medical data, especially the electronic versions, and maintaining patients' privacy. Care needs to be taken to ensure that every pervasive healthcare system maintains the security by implementing comprehensive solutions that are both technically sound and legislatively compliant.

15.4 Conclusions

In this chapter, we presented an overview of security solutions for pervasive healthcare systems. We began by motivating the need for security in pervasive healthcare systems and what are associated challenges. We then presented security solutions for a pervasive healthcare system, starting with the security issues associated with medical sensors, which collect health data from individuals, followed by access control issues that help control the entities that can access EPRs that store the health data in the system. Finally we presented the complementary legislative aspect to providing security in pervasive healthcare systems by discussing the HIPAA security and privacy rules and their provisions for ensuring privacy and security of electronic health data.

15.5 Acknowledgments

This work is supported in part by a gift from Mediserve Inc. and grants #ANI-0196156 and #CNS-0617671 from the National Science Foundation.

Bibliography

1. U. Varshney. Pervasive Healthcare, *IEEE Computer*, Vol. 36, No. 12 (2003), pp. 138–140.
2. For the Record: Protecting Electronic Health Information, *National Academy Press*, 1997.
3. S. Cherukuri, K. Venkatasubramanian, and S.K.S. Gupta, BioSec: A Biometric-Based Approach for Securing Communication in Wireless Networks of Biosensors Implanted in the Human Body, *Workshop on Wireless Security and Privacy (WiSPr)*, International Conference on Parallel Processing Workshops, 2003, Taiwan.
4. S. Bao and Y.T. Zhang. A New Symmetric Cryptosystem of Body Area Sensor Networks for Telemedicine, *6th Asian-Pacific Conference on Medical and Biological Engineering*, Japan, 2005.
5. S. Bao, Y.T. Zhang, and Lian-Feng Shen. Physiological Signal-Based Entity Authentication for Body Area Sensor Networks and Mobile Healthcare Systems, *27th IEEE Conference on Engineering in Medicine and Biology*, Shanghai, China, 2005, pp. 2455–2458.
6. A. Juels and M. Wattenberg. A Fuzzy Commitment Scheme, *ACM Conference on Computer and Communications Security*, 1999, pp. 28–36.
7. K. Van Laerhoven et al. Medical Healthcare Monitoring with Wearable and Implantable Sensors, *The 3rd International Workshop on Ubiquitous Computing for Pervasive Healthcare Applications*, Nottingham, U.K., 2004.

8. N. Cam-Winget, R. Housley, D. Wagner, and J. Walker. Security flaws in 802.11 data link protocols. *Commununications of ACM*, Vol. 46, No. 5, May 2003, pp. 35–39.

9. N. Borisov, I. Goldberg, and D. Wagner. Intercepting mobile communications: the insecurity of 802.11, *7th Annual Iinternational Conference on Mobile Computing and Networking*, Rome, Italy, 2001, pp. 180–189.

10. S.R. Fluhrer, I. Mantin, and A. Shamir. Weaknesses in the Key Scheduling Algorithm of RC4, *8th Annual International Workshop on Selected Areas in Cryptography*, pp. 1–24, 2001.

11. V. Stanford. Pervasive Health Care Applications Face Tough Security Challenges, *IEEE Pervasive Computing*, Vol. 8, No. 12, April-June 2002.

12. A.C. Weaver. Biometric Authentication, *IEEE Computer*, Vol. 39, No. 2, pp. 96–97, Feb 2006.

13. U. Uludag, S. Pankanti, S. Prabhakar, and A.K. Jain. Biometric cryptosystems: issues and challenges, *IEEE Special Issue on Enabling Security Technologies for Digital Rights Management*, Vol. 92, No. 6, June 2004, pp. 948–960.

14. K. Fishkin and J. Lundell. RFID in Healthcare, *RFID: Applications, Security, and Privacy, S. Garfinkel and B. Rosenberg*, Ed. Addison-Wesley, 2005, pp. 211–228.

15. Y. Yang, X. Han, F. Bao, R.H. Deng. A smart-card-enabled privacy preserving E-prescription system, *IEEE Transactions on Information Technology in Biomedicine*, Vol. 8, No. 1, pp. 47–58, March 2004.

16. E.S. Hall, D.K. Vawdrey, C.D. Knutson, and K. Archibald. Enabling remote access to personal electronic medical records, *IEEE Engineering in Medicine and Biology Magazine*, Vol. 22, No. 3 pp. 133–139, May-June 2003.

17. A.T.S. Chan, J. Cao, H. Chan, and G. Young. A Web-Enabled Framework for Smart Card Application in Health Services, *Communications of the ACM*, Vol. 44, No. 9, September 2001, pp. 77–82.

18. J. Reid, I. Cheong, M. Henricksen, and J. Smith, A Novel Use of RBAC to Protect Privacy in Distributed Health Care Information Systems, *8th Australasian Conference on Information Security and Privacy (ACISP 2003)*, pp. 403–415, 2003, Wollongong, Australia.

19. S.K.S. Gupta, T. Mukherjee, and K. Venkatasubramanian. Criticality Aware Access Control Model for Pervasive Applications, *4th IEEE Conference on Pervasive Computing (PERCOM)*, Pisa, Italy, 2006.

20. J. Hu and A.C. Weaver. A Dynamic, Context-Aware Security Infrastructure for Distributed Healthcare Applications, *Pervasive Security, Privacy and Trust (PSPT2004)*, Boston, MA, August 2004.

21. S.J. Dwyer III, A.C. Weaver, and K. Knight Hughes. Health Insurance Portability and Accountability Act, *Security Issues in the Digital Medical Enterprise, Society for Computer Applications in Radiology*, 2nd edition, April 2004.

22. R. Sandhu, E.J. Coyne, H.L. Feinstein and C.E. Youman. Role Based Access Control Models, *IEEE Computer*, Feb 1996, pp. 38–47.

23. D.F. Ferraiolo and D.R. Kuhn. Role Based Access Control, *15th National Computer Security Conference*, 1992.

24. F. Adelstein, S.K.S. Gupta, G.G. Richard III, L. Schwiebert. Fundamentals of Mobile and Pervasive Computing, *McGraw-Hill*, December 2004.

25. R. Clark. e-Consent: A critical element of trust in e-business, *15th Bled Electronic Commerce Conference. e-Reality: Constructing the e-Economy—Research Volume*, 2002.

26. Health Insurance Portability and Accountability Act: Security Rule, www.securityfocus.com/infocus/1764.
27. Unified Authentication Tokens. http://www.verisign.com/products-services/security-services/unified-authentication/index.html/.
28. Medline Plus Medical Encyclopedia, *U.S National Library of Medicine*, http://www.nlm.nih.gov/medlineplus/encyclopedia.html.
29. L. Biel, O. Pettersson, L. Philipson, and P. Wide. ECG Analysis: A New Approach in Human Identification, *IEEE Transaction on Instrumentation and Measurement*, Vol. 50, No. 3, June 2001, pp. 808–812.

16

Wireless Sensor Network Security: A Survey

John Paul Walters, Zhengqiang Liang, Weisong Shi and Vipin Chaudhary

CONTENTS

Abstract   As wireless sensor networks continue to grow, so does the need for effective security mechanisms. Because sensor networks may interact with sensitive data and/or operate in hostile unattended environments, it is imperative that these security concerns be addressed from the beginning of the system design. However, due to inherent resource and computing constraints, security in sensor networks poses different challenges from traditional network/computer security. There is currently enormous research potential in the field of wireless sensor network security. Thus, familiarity with the current research in this field will greatly benefit researchers. With this in mind, we survey the major topics in wireless sensor network security, and present the obstacles and the requirements in the sensor security, classify many of the current attacks, and finally list their corresponding defensive measures.

16.1   Introduction

Wireless sensor networks are quickly gaining popularity due to the fact that they are potentially low-cost solutions to a variety of real-world challenges [1]. Their low cost provides a means to deploy large sensor arrays in a variety of conditions capable of performing both military and civilian tasks. But sensor networks also introduce severe resource constraints due to their lack of data storage and power. Both of these represent major obstacles to the implementation of traditional computer security techniques in a wireless sensor network. The unreliable communication channel and unattended operation make security defenses even harder. Indeed, as pointed out in [65], wireless sensors often have the processing characteristics of machines that are decades old (or longer), and the industrial trend is to reduce the cost of wireless sensors while maintaining similar computing power. With that in mind, many researchers have begun to address the challenges of maximizing the processing capabilities and energy reserves of wireless sensor nodes while also securing them against attackers. All aspects of the wireless sensor network are being examined, including secure and efficient routing [15, 41, 62, 79], data aggregation [22, 33, 54, 68, 75, 91], group formation [6, 42, 69], and so on.

In addition to those traditional security issues, we observe that many general-purpose sensor network techniques (particularly the early research) assumed that all nodes are cooperative and trustworthy. This is not the case for most, or much of, real-world wireless sensor networking applications, which require a certain amount of trust in the application in order to maintain proper network functionality. Researchers therefore began focusing on building a sensor trust model to solve the problems beyond the capability of cryptographic security [23, 49, 48, 50, 70, 80, 90, 92]. In addition, there are many attacks designed to exploit the unreliable communication channels and unattended operation of wireless sensor networks. Furthermore, due to the inherent unattended feature of wireless sensor networks, we argue that physical attacks on sensors play an important role in the operation of wireless sensor networks. Thus, we include a detailed discussion of the physical attacks and their corresponding defenses [3, 4, 30, 34, 43, 71, 74, 84, 85, 88], topics typically ignored in most of the current research on sensor security.

We classify the main aspects of wireless sensor network security into four major categories: *the obstacles to sensor network security, the requirements of a secure wireless sensor network, attacks,* and *defensive measures.* The organization then follows this classification. For the completeness of the chapter, we also give a brief introduction of related security techniques, while providing appropriate citations for those interested in a more detailed discussion of a particular topic.

The remainder of this chapter is organized as follows. In Section 16.2, we summarize the obstacles for the sensor network security. The security requirements of a wireless sensor network are listed in Section 16.3. The major attacks in a sensor network are categorized in Section 16.4, and we outline the corresponding defensive measures in Section 16.5. Finally, we conclude the chapter in Section 16.6.

16.2 Obstacles of Sensor Security

A wireless sensor network is a special network that has many constraints compared to a traditional computer network. Due to these constraints it is difficult to directly employ the existing security approaches to the area of wireless sensor networks. Therefore, to develop useful security mechanisms while borrowing the ideas from the current security techniques, it is necessary to know and understand these constraints first [10].

16.2.1 Very Limited Resources

All security approaches require a certain amount of resources for the implementation, including data memory, code space, and energy to power the sensor. However, currently these resources are very limited in a tiny wireless sensor.

- **Limited Memory and Storage Space** A sensor is a tiny device with only a small amount of memory and storage space for the code. In order to build an effective security mechanism, it is necessary to limit the code size of the security algorithm. For example, one common sensor type (TelosB) has a 16-bit, 8-MHz RISC CPU with only 10 K RAM, a 48-K program memory, and a 1024-K flash storage [14]. With such a limitation, the software built for the sensor must also be quite small. The total code space of TinyOS, the de facto standard operating system for wireless sensors, is approximately 4 K [32], and the core scheduler occupies only 178 bytes. Therefore, the code size for all security-related codes must also be small.

- **Power Limitation** Energy is the biggest constraint to wireless sensor capabilities. We assume that once sensor nodes are deployed in a sensor network, they cannot be easily replaced (high operating cost) or recharged (high cost of sensors). Therefore, the battery charge taken with them to the field must be conserved to extend the life of the individual sensor node and the entire sensor network. When implementing a cryptographic function or protocol within a sensor node, the energy impact of the added security code must be considered. When adding security to a sensor node, we are interested in the impact that security has on the life span of a sensor (i.e., its battery life). The extra power consumed by sensor nodes due to security is related to the processing required for security functions (e.g., encryption, decryption, signing data, verifying signatures), the energy required to transmit the security-related data or overhead (e.g., initialization vectors needed for encryption/decryption), and the energy required to store security parameters in a secure manner (e.g., cryptographic key storage).

16.2.2   Unreliable Communication

Certainly, unreliable communication is another threat to sensor security. The security of the network relies heavily on a defined protocol, which in turn, depends on communication.

- **Unreliable Transfer** Normally the packet-based routing of the sensor network is connectionless and thus inherently unreliable. Packets may get damaged due to channel errors or dropped at highly congested nodes. The result is lost or missing packets. Furthermore, the unreliable wireless communication channel also results in damaged packets. Higher channel error rate also forces the software developer to devote resources to error handling. More importantly, if the protocol lacks the appropriate error handling it is possible to lose critical security packets. This may include, for example, a cryptographic key.
- **Conflicts** Even if the channel is reliable, the communication may still be unreliable. This is due to the broadcast nature of the wireless sensor network. If packets meet in the middle of a transfer, conflicts

will occur and the transfer itself will fail. In a crowded (high-density) sensor network, this can be a major problem. More details about the effect of wireless communication can be found in [1].

- **Latency** The multihop routing, network congestion, and node processing can lead to greater latency in the network, thus making it difficult to achieve synchronization among sensor nodes. The synchronization issues can be critical to sensor security where the security mechanism relies on critical event reports and cryptographic key distribution. Interested readers may refer to [78] on real-time communications in wireless sensor networks.

16.2.3 Unattended Operation

Depending on the function of the particular sensor network, the sensor nodes may be left unattended for long periods of time. There are three main caveats to unattended sensor nodes:

- **Exposure to Physical Attacks** The sensor may be deployed in an environment open to adversaries, bad weather, and so on. The likelihood that a sensor suffers a physical attack in such an environment is therefore much higher than the typical PCs, which are located in a secure place and mainly face attacks from a network.

- **Managed Remotely** Remote management of a sensor network makes it virtually impossible to detect physical tampering (i.e., through tamper-proof seals) and physical maintenance issues (e.g., battery replacement). Perhaps the most extreme example of this is a sensor node used for remote reconnaissance missions behind enemy lines. In such a case, the node may not have any physical contact with friendly forces once deployed.

- **No Central Management Point** A sensor network should be a distributed network without a central management point. This will increase the vitality of the sensor network. However, if designed incorrectly, it will make the network organization difficult, inefficient, and fragile.

Perhaps most importantly, the longer that a sensor is left unattended the more likely that an adversary has compromised the node.

16.3 Security Requirements

A sensor network is a special type of network. It shares some commonalities with a typical computer network, but also poses unique requirements of its own as discussed in Section 16.2. Therefore, we can think of the requirements of a wireless sensor network as encompassing both the typical network requirements and the unique requirements suited solely to wireless sensor networks.

16.3.1   Data Confidentiality

Data confidentiality is the most important issue in network security. Every network with any security focus will typically address this problem first. In sensor networks, the confidentiality relates to the following [10, 65]:

- A sensor network should not leak sensor readings to its neighbors. Especially in a military application, the data stored in the sensor node may be highly sensitive.
- In many applications nodes communicate highly sensitive data, e.g., key distribution, therefore it is extremely important to build a secure channel in a wireless sensor network.
- Public sensor information, such as sensor identities and public keys, should also be encrypted to some extent to protect against traffic analysis attacks.

The standard approach for keeping sensitive data secret is to encrypt the data with a secret key that only intended receivers possess, thus achieving confidentiality.

16.3.2   Data Integrity

With the implementation of confidentiality, an adversary may be unable to steal information. However, this does not mean the data is safe. The adversary can change the data, so as to send the sensor network into disarray. For example, a malicious node may add some fragments or manipulate the data within a packet. This new packet can then be sent to the original receiver. Data loss or damage can even occur without the presence of a malicious node due to the harsh communication environment. Thus, data integrity ensures that any received data has not been altered in transit.

16.3.3   Data Freshness

Even if confidentiality and data integrity are assured, we also need to ensure the freshness of each message. Informally, data freshness suggests that the data is recent, and it ensures that no old messages have been replayed. This requirement is especially important when there are shared-key strategies employed in the design. Typically shared keys need to be changed over time. However, it takes time for new shared keys to be propagated to the entire network. In this case, it is easy for the adversary to use a replay attack. Also, it is easy to disrupt the normal work of the sensor, if the sensor is unaware of the new key change time. To solve this problem a nonce, or another time-related counter, can be added into the packet to ensure data freshness.

16.3.4 Availability

Adjusting the traditional encryption algorithms to fit within the wireless sensor network is not free, and will introduce some extra costs. Some approaches choose to modify the code to reuse as much code as possible. Some approaches try to make use of additional communication to achieve the same goal. What is more, some approaches force strict limitations on the data access, or propose an unsuitable scheme (such as a central point scheme) in order to simplify the algorithm. But all these approaches weaken the availability of a sensor and sensor network for the following reasons:

- Additional computation consumes additional energy. If no more energy exists, the data will no longer be available.
- Additional communication also consumes more energy. What is more, as communication increases so, too, does the chance of incurring a communication conflict.
- A single point failure will be introduced if using the central point scheme. This greatly threatens the availability of the network.

The requirement of security not only affects the operation of the network, but also is highly important in maintaining the availability of the whole network.

16.3.5 Self-Organization

A wireless sensor network is typically an ad hoc network, which requires every sensor node be independent and flexible enough to be self-organizing and self-healing according to different situations. There is no fixed infrastructure available for the purpose of network management in a sensor network. This inherent feature brings a great challenge to wireless sensor network security as well. For example, the dynamics of the whole network inhibits the idea of preinstallation of a shared key between the base station and all sensors [21]. Several random key predistribution schemes have been proposed in the context of symmetric encryption techniques [13, 21, 37, 53]. In the context of applying public key cryptography techniques in sensor networks, an efficient mechanism for public key distribution is necessary as well. In the same way that distributed sensor networks must self-organize to support multihop routing, they must also self-organize to conduct key management and build trust relation among sensors. If self-organization is lacking in a sensor network, the damage resulting from an attack or even the hazardous environment may be devastating.

16.3.6 Time Synchronization

Most sensor network applications rely on some form of time synchronization. In order to conserve power, an individual sensor's radio may be turned off for periods of time. Furthermore, sensors may wish to compute the end-to-end delay of a packet as it travels between two pairwise sensors. A more collaborative sensor network may require group synchronization for tracking applications, etc. In [24], the authors propose a set of secure synchronization protocols for sender-receiver (pairwise), multihop sender-receiver (for use when the pair of nodes are not within single-hop range), and group synchronization.

16.3.7   Secure Localization

Often, the utility of a sensor network will rely on its ability to accurately and automatically locate each sensor in the network. A sensor network designed to locate faults will need accurate location information in order to pinpoint the location of a fault. Unfortunately, an attacker can easily manipulate non secured location information by reporting false signal strengths, replaying signals, etc.

A technique called verifiable multilateration (VM) is described in [81]. In multilateration, a device's position is accurately computed from a series of known reference points. In [81], authenticated ranging and distance bounding are used to ensure accurate location of a node. Because of distance bounding, an attacking node can only increase its claimed distance from a reference point. However, to ensure location consistency, an attacking node would also have to prove that its distance from another reference point is shorter [81]. Since it cannot do this, a node manipulating the localization protocol can be found. For large sensor networks, the SPINE (Secure Positioning for sensor NEtworks) algorithm is used. It is a three-phase algorithm based upon verifiable multilateration [81].

In [47], SeRLoc (Secure Range-Independent Localization) is described. Its novelty is its decentralized, range-independent nature. SeRLoc uses locators that transmit beacon information. It is assumed that the locators are trusted and cannot be compromised. Furthermore, each locator is assumed to know its own location. A sensor computes its location by listening for the beacon information sent by each locator. The beacons include the locator's location. Using all of the beacons that a sensor node detects, a node computes an approximate location based on the coordinates of the locators. Using a majority-vote scheme, the sensor then computes an overlapping antenna region. The final computed location is the "center of gravity" of the overlapping antenna region [47]. All beacons transmitted by the locators are encrypted with a shared global symmetric key that is preloaded to the sensor prior to deployment. Each sensor also shares a unique symmetric key with each locator. This key is also preloaded on each sensor.

16.3.8   Authentication

An adversary is not just limited to modifying the data packet. It can change the whole packet stream by injecting additional packets. So the receiver needs to ensure that the data used in any decision-making process originates from the correct source. On the other hand, when constructing the sensor network, authentication is necessary for many administrative tasks (e.g., network re-programming or controlling sensor node duty cycle). From the above, we can see that message authentication is important for many applications in sensor networks. Informally, data authentication allows a receiver to verify that the data really is sent by the claimed sender. In the case of two-party communication, data authentication can be achieved through a purely symmetric mechanism: the sender and the receiver share a secret key to compute the message authentication code (MAC) of all communicated data.

Adrian Perrig et al. propose a key-chain distribution system for their μTESLA secure broadcast protocol [65]. The basic idea of the μTESLA system is to achieve asymmetric cryptography by delaying the disclosure of the symmetric keys. In this case a sender will broadcast a message generated with a secret key. After a certain period of time, the sender will disclose the secret key. The receiver is responsible for buffering the packet until the secret key has been disclosed. After disclosure the receiver can authenticate the packet, provided that the packet was received before the key was disclosed. One limitation of μTESLA is that some initial information must be unicast to each sensor node before authentication of broadcast messages can begin.

Liu and Ning [51, 52] propose an enhancement to the μTESLA system that uses broadcasting of the key-chain commitments rather than μTESLA's unicasting technique. They present a series of schemes starting with a simple predetermination of key chains and finally settling on a multilevel key-chain technique. The multilevel key-chain scheme uses predetermination and broadcasting to achieve a scalable key distribution technique that is designed to be resistant to denial-of-service attacks, including jamming.

16.4 Attacks

Sensor networks are particularly vulnerable to several key types of attacks. Attacks can be performed in a variety of ways, most notably as denial-of-service attacks, but also through traffic analysis, privacy violation, physical attacks, and so on. Denial-of-service attacks on wireless sensor networks can range from simply jamming the sensor's communication channel to more sophisticated attacks designed to violate the 802.11 MAC protocol [64] or any other layer of the wireless sensor network.

Due to the potential asymmetry in power and computational constraints, guarding against a well-orchestrated denial-of-service attack on a wireless sensor network can be nearly impossible. A more powerful node can easily jam a sensor node and effectively prevent the sensor network from performing its intended duty.

We note that attacks on wireless sensor networks are not limited to simply denial-of-service attacks, but rather encompass a variety of techniques including node takeovers, attacks on the routing protocols, and attacks on a node's physical security. In this section, we first address some common denial-of-service attacks and then describe additional attacks, including those on the routing protocols as well as an identity-based attack known as the Sybil attack.

16.4.1   Background

Wood and Stankovic define one kind of denial-of-service attack as "any event that diminishes or eliminates a network's capacity to perform its expected function" [88]. Certainly, denial-of-service attacks are not a new phenomenon. In fact, there are several standard techniques used in traditional computing to cope with some of the more common denial-of-service techniques, although this is still an open problem in the network security community. Unfortunately, wireless sensor networks cannot afford the computational overhead necessary to implement many of the typical defensive strategies.

What makes the prospect of denial-of-service attacks even more alarming is the projected use of sensor networks in highly critical and sensitive applications. For example, a sensor network designed to alert building occupants in the event of a fire could be highly susceptible to a denial-of-service attack. Even worse, such an attack could result in the deaths of building occupants due to the nonoperational fire detection network.

Other possible uses for wireless sensors include the monitoring of traffic flows, which may include the control of traffic lights, and so forth. A denial-of-service attack on such a sensor network could prove very costly, especially on major roads.

For this reason, researchers have spent a great deal of time both identifying the various types of denial-of-service attacks and devising strategies to subvert such attacks. We now describe some of the major types of denial-of-service attacks.

16.4.2   Types of Denial-of-Service Attacks

A standard attack on wireless sensor networks is simply to jam a node or set of nodes. Jamming, in this case, is simply the transmission of a radio signal that interferes with the radio frequencies being used by the sensor network [88]. The jamming of a network can come in two forms: constant jamming and intermittent jamming. Constant jamming involves the complete jamming of the entire network. No messages are able to be sent or received. If the jamming is only intermittent, then the nodes are able to exchange messages periodically, but not consistently. This, too, can have a detrimental impact on the sensor network as the messages being exchanged between nodes may be time sensitive [88].

Attacks can also be made on the link layer itself. One possibility is that an attacker may simply intentionally violate the communication protocol, e.g., ZigBee [94] or IEEE 801.11b (Wi-Fi) protocol, and continually transmit messages in an attempt to generate collisions. Such collisions would require the retransmission of any packet affected by the collision. Using this technique it would be possible for an attacker to simply deplete a sensor node's power supply by forcing too many retransmissions.

At the routing layer, a node may take advantage of a multihop network by simply refusing to route messages. This could be done intermittently or constantly with the net result being that any neighbor who routes through the malicious node will be unable to exchange messages with, at least, part of the network. Extensions to this technique include intentionally routing messages to incorrect nodes (misdirection) [88].

The transport layer is also susceptible to attack, as in the case of flooding. Flooding can be as simple as sending many connection requests to a suscep-tible node. In this case, resources must be allocated to handle the connection request. Eventually a node's resources will be exhausted, thus rendering the node useless.

16.4.3 The Sybil Attack

Newsome et al. describe the Sybil attack as it relates to wireless sensor net-works [59]. Simply put, the Sybil attack is defined as a "malicious device illegitimately taking on multiple identities" [59]. It was originally described as an attack able to defeat the redundancy mechanisms of distributed data storage systems in peer-to-peer networks [18]. In addition to defeating dis-tributed data storage systems, the Sybil attack is also effective against routing algorithms, data aggregation, voting, fair resource allocation, and foiling mis-behavior detection. Regardless of the target (voting, routing, aggregation), the Sybil algorithm functions similarly. All of the techniques involve utilizing multiple identities. For instance, in a sensor network voting scheme, the Sybil attack might utilize multiple identities to generate additional "votes." Simi-larly, to attack the routing protocol, the Sybil attack would rely on a malicious node taking on the identity of multiple nodes, and thus routing multiple paths through a single malicious node.

16.4.4 Traffic Analysis Attacks

Wireless sensor networks are typically composed of many low-power sensors communicating with a few relatively robust and powerful base stations. It is not unusual, therefore, for data to be gathered by the individual nodes where it is ultimately routed to the base station. Often, for an adversary to effectively render the network useless, the attacker can simply disable the base station. To make matters worse, Deng et al. demonstrate two attacks that can identify the base station in a network (with high probability) without even understanding the contents of the packets (if the packets are themselves encrypted) [16].

A rate monitoring attack simply makes use of the idea that the nodes closest to the base station tend to forward more packets than those farther away from the base station. An attacker need only monitor which nodes are sending packets and follow those nodes that are sending the most packets. In a time correlation attack, an adversary simply generates events and monitors to whom a node sends its packets. To generate an event, the adversary could simply generate a physical event that would be monitored by the sensor(s) in the area (turning on a light, for instance) [16].

16.4.5   Node Replication Attacks

Conceptually, a node replication attack is quite simple: an attacker seeks to add a node to an existing sensor network by copying (replicating) the node ID of an existing sensor node [63]. A node replicated in this fashion can severely disrupt a sensor network's performance: packets can be corrupted or even misrouted. This can result in a disconnected network, false sensor readings, etc. If an attacker can gain physical access to the entire network he can copy cryptographic keys to the replicated sensor and can also insert the replicated node into strategic points in the network [63]. By inserting the replicated nodes at specific network points, the attacker could easily manipulate a specific segment of the network, perhaps by disconnecting it altogether.

16.4.6   Attacks Against Privacy

Sensor network technology promises a vast increase in automatic data collection capabilities through efficient deployment of tiny sensor devices. While these technologies offer great benefits to users, they also exhibit significant potential for abuse. Particularly relevant concerns are privacy problems, since sensor networks provide increased data collection capabilities [28]. Adversaries can use even seemingly innocuous data to derive sensitive information if they know how to correlate multiple sensor inputs. For example, in the famous "panda-hunter problem" [61], the hunter can imply the position of pandas by monitoring the traffic.

The main privacy problem, however, is not that sensor networks enable the collection of information. In fact, much information from sensor networks could probably be collected through direct site surveillance. Rather, sensor networks aggravate the privacy problem because they make large volumes of information easily available through remote access. Hence, adversaries need not be physically present to maintain surveillance. They can gather information in a low-risk, anonymous manner. Remote access also allows a single adversary to monitor multiple sites simultaneously [11]. Some of the more common attacks [11, 28] against sensor privacy are:

- **Monitor and Eavesdropping** This is the most obvious attack on privacy. By listening to the data, the adversary could easily discover the communication contents. When the traffic conveys the control information about the sensor network configuration, which contains potentially more detailed information than accessible through the location server, the eavesdropping can act effectively against the privacy protection.
- **Traffic Analysis** Traffic analysis typically combines with monitoring and eavesdropping. An increase in the number of transmitted

packets between certain nodes could signal that a specific sensor has registered activity. Through the analysis on the traffic, some sensors with special roles or activities can be effectively identified.

- **Camouflage** Adversaries can insert their nodes or compromise the nodes to hide in the sensor network. After that these nodes can masquerade as a normal node to attract the packets, then misroute the packets, e.g., forward the packets to the nodes conducting the privacy analysis.

It is worth noting that, as pointed out in [64], the current understanding of privacy in wireless sensor networks is immature, and more research is needed.

16.4.7 Physical Attacks

Sensor networks typically operate in hostile outdoor environments. In such environments, the small form factor of the sensors, coupled with the unattended and distributed nature of their deployment make them highly susceptible to physical attacks, i.e., threats due to physical node destructions [86]. Unlike many other attacks mentioned above, physical attacks destroy sensors permanently, so the losses are irreversible. For instance, attackers can extract cryptographic secrets, tamper with the associated circuitry, modify programming in the sensors, or replace them with malicious sensors under the control of the attacker [85]. Recent work has shown that standard sensor nodes, such as the MICA2 motes, can be compromised in less than one minute [30]. While these results are not surprising given that the MICA2 lacks tamper-resistant hardware protection, they provide a cautionary note about the speed of a well-trained attacker. If an adversary compromises a sensor node, then the code inside the physical node may be modified.

16.5 Defensive Measures

Now we are in a position to describe the measures for satisfying security requirements, and protecting the sensor network from attacks. We start with *key establishment in wireless sensor networks*, which lays the foundation for the security in a wireless sensor network, followed by *defending against DoS attacks, secure broadcasting and multicasting, defending against attacks on routing protocols, combating traffic analysis attacks, defending against attacks on sensor privacy, intrusion detection, secure data aggregation, defending against physical attacks,* and *trust management.*

16.5.1 Key Establishment

One security aspect that receives a great deal of attention in wireless sensor networks is the area of key management. Wireless sensor networks are unique (among other embedded wireless networks) in this aspect due to their size, mobility, and computational/power constraints. Indeed, researchers envision wireless sensor networks to be orders of magnitude larger than their traditional embedded counterparts. This, coupled with the operational constraints described previously, makes secure key management an absolute necessity in most wireless sensor network designs. Because encryption and key management/establishment are so crucial to the defense of a wireless sensor network, with nearly all aspects of wireless sensor network defenses relying on solid encryption, we first begin with an overview of the unique key and encryption issues surrounding wireless sensor networks before discussing more specific sensor network defenses.

Background

Key management issues in wireless networks are not unique to wireless sensor networks. Indeed, key establishment and management issues have been studied in depth outside of the wireless networking arena. Traditionally, key establishment is done using one of many public key protocols. One of the more common is the Diffie-Hellman public key protocol, but there are many others.

Most of the traditional techniques, however, are unsuitable in low-power devices such as wireless sensor networks. This is due largely to the fact that typical key-exchange techniques use asymmetric cryptography, also called public key cryptography. In this case, it is necessary to maintain two mathematically related keys, one of which is made public while the other is kept private. This allows data to be encrypted with the public key and decrypted only with the private key. The problem with asymmetric cryptography, in a wireless sensor network, is that it is typically too computationally intensive for the individual nodes in a sensor network. This is true in the general case, however, [25, 29, 55, 87] show that it is feasible with the right selection of algorithms.

Symmetric cryptography is therefore the typical choice for applications that cannot afford the computational complexity of asymmetric cryptography. Symmetric schemes utilize a single shared key known only between the two communicating hosts. This shared key is used for both encrypting and decrypting data. The traditional example of symmetric cryptography is DES (Data Encryption Standard). The use of DES, however, is quite limited due to the fact that it can be broken relatively easily. In light of the shortcomings of DES, other symmetric cryptography systems have been proposed including 3DES (Triple DES), RC5, AES, and so on [73].

An analysis of the various ciphers is presented in [44] with a summary of their results shown in Table 16.1. The table ranks ciphers in several ways: key setup efficiency, encryption/decryption efficiency, data memory, and code memory (Table 16.1). Further, each cipher is both size optimized and speed optimized. Skipjack is the overall winner, where either the size optimized or speed optimized proves best in every category. While RC5 and RC6 rank fairly high in code size, their poor performance in both encryption/decryption

TABLE 16.1

A Summary of Cipher Performance from [44]. Subscript z Means Size-Optimized, s Means Speed-Optimized

Code Memory		Data Memory		En/decryption Efficiency		Key Setup Efficiency	
CBC	OFB	CBC	OFB	CBC	OFB	Encryption	Decryption
Skipjack$_z$	Skipjack$_z$	Skipjack$_z$	Skipjack$_z$	Skipjack$_s$	Skipjack$_s$	Skipjack$_z$	Skipjack$_z$
RC5$_z$	Skipjack$_s$	MISTY1$_z$	MISTY1$_z$	Rijndael$_s$	Rijndael$_s$	Skipjack$_s$	Skipjack$_s$
Skipjack$_s$	RC5$_z$	Skipjack$_s$	Skipjack$_s$	Twofish$_s$	Twofish$_s$	MISTY1$_s$	MISTY1$_s$
RC6$_z$	RC6$_z$	MISTY1$_s$	MISTY1$_s$	MISTY1$_s$	MISTY1$_s$	MISTY1$_z$	MISTY1$_z$
RC6$_s$	RC6$_s$	KASUMI$_z$	KASUMI$_s$	Camellia$_s$	Camellia$_s$	Rijndael$_s$	KASUMI$_s$
RC5$_s$	Rijndael$_z$	KASUMI$_s$	KASUMI$_z$	MISTY1$_z$	Rijndael$_z$	KASUMI$_s$	KASUMI$_z$
MISTY1$_z$	RC5$_s$	RC5	RC5	KASUMI$_s$	MISTY1$_z$	Rijndael$_z$	Rijndael$_s$
MISTY1$_s$	MISTY1$_z$	Twofish$_z$	Twofish$_s$	Rijndael$_z$	Skipjack$_z$	KASUMI$_z$	Twofish$_s$
Rijndael$_z$	Rijndael$_s$	Twofish$_s$	Twofish$_z$	Skipjack$_z$	KASUMI$_s$	Twofish$_s$	Rijndael$_z$
Twofish$_z$	MISTY1$_s$	RC6$_z$	Rijndael	KASUMI$_z$	KASUMI$_z$	Twofish$_z$	Twofish$_z$
KASUMI$_z$	KASUMI$_z$	Rijndael$_z$	RC6	Twofish$_z$	Twofish$_z$	Camellia$_s$	Camellia$_s$
Rijndael$_s$	KASUMI$_s$	RC6$_s$	Camellia$_z$	Camellia$_z$	Camellia$_z$	Camellia$_z$	Camellia$_z$
KASUMI$_s$	Twofish$_z$	Rijndael$_s$	Camellia$_s$	RC6$_s$	RC6$_s$	RC5$_s$	RC5$_s$
Twofish$_s$	Twofish$_s$	Camellia$_z$		RC5$_s$	RC5$_s$	RC6$_s$	RC6$_s$
Camellia$_z$	Camellia$_z$	Camellia$_s$		RC6$_z$	RC6$_z$	RC5$_z$	RC5$_z$
Camellia$_s$	Camellia$_s$			RC5$_z$	RC5$_z$	RC6$_z$	RC6$_z$

efficiency and key setup lower their rankings significantly. Both MISTY1 and KASUMI rate as average in all categories except for code memory [44].

One major shortcoming of symmetric cryptography is the key exchange problem. Simply put, the key exchange problem derives from the fact that two communicating hosts must somehow know the shared key before they can communicate securely. So the problem that arises is how to ensure that the shared key is indeed shared between the two hosts who wish to communicate and no other rogue hosts who may wish to eavesdrop. How to distribute a shared key securely to communicating hosts is a nontrivial problem since predistributing the keys is not always feasible.

Key Establishment and Associated Protocols

Random key predistribution schemes have several variants [13, 21, 37, 53]. Eschenauer and Gligor propose a key predistribution scheme [21] that relies on probabilistic key sharing among nodes within the sensor network. Their system works by distributing a key ring to each participating node in the sensor network before deployment. Each key ring should consist of a number of randomly chosen keys from a much larger pool of keys generated offline. An enhancement to this technique utilizing multiple keys is described in [13]. Further enhancements are proposed in [19, 53] with additional analysis and enhancements provided by [37].

Using this technique, it is not necessary that each pair of nodes share a key. However, any two nodes that do share a key may use the shared key to establish a direct link to one another. Eschenauer and Gligor show that, while not perfect, it is probabilistically likely that large sensor networks will enjoy shared-key connectivity. Further, they demonstrate that such a technique can be extended to key revocation, rekeying, and the addition/deletion of nodes.

The LEAP protocol described by Zhu et al. [93] takes an approach that utilizes multiple keying mechanisms. Their observation is that no single security requirement accurately suits all types of communication in a wireless sensor network. Therefore, four different keys are used depending on with whom the sensor node is communicating. Sensors are preloaded with an initial key from which further keys can be established. As a security precaution, the initial key can be deleted after its use in order to ensure that a compromised sensor cannot add additional compromised nodes to the network.

In PIKE [12], Chan and Perrig describe a mechanism for establishing a key between two sensor nodes that is based on the common trust of a third node somewhere within the sensor network. The nodes and their shared keys are spread over the network such that for any two nodes A and B, there is a node C that shares a key with both A and B. Therefore, the key establishment protocol between A and B can be securely routed through C.

Huang et al. [36] propose a hybrid key establishment scheme that makes use of the difference in computational and energy constraints between a sensor node and the base station. They posit that an individual sensor node possesses far less computational power and energy than a base station. In light of this, they propose placing the major cryptographic burden on the base station where the resources tend to be greater. On the sensor side, symmetric-key operations are used in place of their asymmetric alternatives. The sensor and the base station authenticate based on elliptic curve cryptography. Elliptic curve cryptography is often used in sensors due to the fact that relatively small key lengths are required to achieve a given level of security.

Huang et al. also use certificates to establish the legitimacy of a public key. The certificates are based on an elliptic curve implicit certificate scheme [36]. Such certificates are useful to ensure both that the key belongs to a device and that the device is a legitimate member of the sensor network. Each node obtains a certificate before joining the network using an out-of-band interface.

Public Key Cryptography

Two of the major techniques used to implement public key cryptosystems are RSA and elliptic curve cryptography (ECC) [73]. Traditionally, these have been thought to be far too heavyweight for use in wireless sensor networks. Recently, however, several groups have successfully implemented public key cryptography (to varying degrees) in wireless sensor networks.

In [29] Gura et al. report that both RSA and elliptic curve cryptography are possible using 8-bit CPUs with ECC, demonstrating a performance advantage over RSA. Another advantage is that ECC's 160-bit keys result in shorter messages during transmission compared to the 1024-bit RSA keys. In particular Gura et al. demonstrate that the point multiplication operations in ECC are an order of magnitude faster than private-key operations within RSA, and are comparable (though somewhat slower) to the RSA public key operation [29].

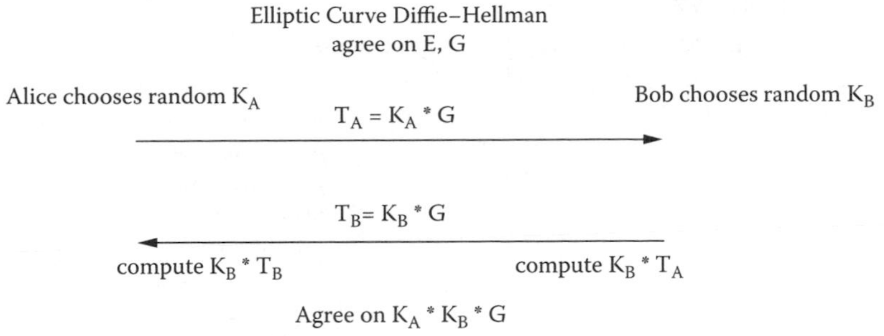

Elliptic Curve Diffie–Hellman
agree on E, G

Alice chooses random K_A Bob chooses random K_B

$$T_A = K_A * G$$

$$T_B = K_B * G$$

compute $K_B * T_B$ compute $K_B * T_A$

Agree on $K_A * K_B * G$

FIGURE 16.1
The Diffie-Hellman elliptic curve key exchange algorithm [55].

In [87], Watro et al. show that portions of the RSA cryptosystem can be successfully applied to actual wireless sensors, specifically the UC Berkeley MICA2 motes [32]. In particular, they implemented the public operations on the sensors themselves while offloading the private operations to devices better suited for the larger computational tasks. In this case, a laptop was used.

The TinyPK system described by [87] is designed specifically to allow authentication and key agreement between resource-constrained sensors. The agreed-upon keys may then be used in conjunction with the existing cryptosystem, *TinySec* [39]. To do this, they implement the Diffie-Hellman key exchange algorithm and perform the public key operations on the Berkeley motes.

The Diffie-Hellman key exchange algorithm used in [55] is depicted in Figure 16.1. In this case, a point G is selected from an elliptic curve E, both of which are public. A random integer K_A is selected, which will act as the private key. The public key (T_A in the case of Alice from Figure 16.1) is then $T_A = K_A * G$. Bob performs a similar set of operations to compute $T_B = K_B * G$. Alice and Bob can now easily compute the shared secret using their own private keys and the public keys that have been exchanged. In this case, Alice computes $K_A * T_B = K_A * K_B * G$ while Bob computes $K_B * T_A = K_B * K_A * G$. Because $K_A * T_B = K_B * T_A$, Alice and Bob now share a secret key.

As stated above, the elliptic curve cryptography shows promise over that of RSA due to its efficiency compared to the private-key operations of RSA. Further, using ECC, the key length required to securely transmit TinySec keys can be as small as 163 bits rather than the 1024 bits required in RSA. In [55], Malan et al. demonstrate a working implementation of Diffie-Hellman based on the Elliptic Curve Discrete Logarithm Problem (Figure 16.1). And while key generation is by no means fast or inexpensive (34.161 seconds to generate a public/private-key pair and 34.173 seconds to generate a shared secret with Diffie-Hellman [55]), it is sufficient for infrequent use in generating keys in the TinySec protocols.

TABLE 16.2

Sensor Network Layers and DoS Attacks/Defenses [88]

Network Layer	Attacks	Defenses
Physical	Jamming	Spread-spectrum, priority messages, lower duty cycle, region mapping, mode change
Link	Tampering	Tamper-proof, hiding
	Collision	Error correcting code
	Exhaustion	Rate limitation
	Unfairness	Small frames
Network and routing	Neglect and greed	Redundancy, probing
	Homing	Encryption
	Misdirection	Egress filtering, authorization monitoring
	Black holes	Authorization, monitoring, redundancy
Transport	Flooding	Client Puzzles
	Desynchronization	Authentication

16.5.2   Defending Against DoS Attacks

In Table 16.2 the most common layers of a typical wireless sensor network are summarized along with their attacks and defenses. Since denial-of-service attacks are so common (see Section 16.4), effective defenses must be available to combat them. One strategy in defending against the classic jamming attack is to identify the jammed part of the sensor network and effectively route around the unavailable portion. Wood and Stankovic [88] describe a two-phase approach where the nodes along the perimeter of the jammed region report their status to their neighbors who then collaboratively define the jammed region and simply route around it.

To handle jamming at the MAC layer, nodes might utilize a MAC admission control that is rate limiting. This would allow the network to ignore those requests designed to exhaust the power reserves of a node. This, however, is not foolproof as the network must be able to handle any legitimately large traffic volumes.

Overcoming rogue sensors that intentionally misroute messages can be done at the cost of redundancy. In this case, a sending node can send the message along multiple paths in an effort to increase the likelihood that the message will ultimately arrive at its destination. This has the advantage of effectively dealing with nodes that may not be malicious, but rather may have simply failed as it does not rely on a single node to route its messages.

To overcome the transport layer flooding denial-of-service attack, Aura, Nikander, and Leiwo suggest using the client puzzles posed by Juels and Brainard [5] in an effort to discern a node's commitment to making the connection by utilizing some of their own resources. Aura et al. advocate that a server should force a client to commit its own resources first. Further, they suggest that a server should always force a client to commit more resources up front than the server. This strategy would likely be effective as long as the client has computational resources comparable to those of the server.

16.5.3   Secure Broadcasting and Multicasting

The wireless sensor network research community has progressively reached a consensus that the major communication pattern of wireless sensor networks is broadcasting and multicasting, e.g., 1-to-N, N-to-1, and M-to-N, instead of the traditional point-to-point communication on the Internet. Next we examine the current state of research in secure broadcasting and multicasting. As we will see, in wireless sensor networks, a great deal of the security derives from ensuring that only members of the broadcast or multicast group possess the required keys in order to decrypt the broadcast or multicast messages. Because of this, most of the work presented in 16.5.1 is still applicable. Here, however, we will address those schemes that have been specifically designed to support broadcasting and multicasting in wireless sensor networks.

Traditional Broadcasting and Multicasting

Traditionally, multicasting and broadcasting techniques have been used to reduce the communication and management overhead of sending a single message to multiple receivers. In order to ensure that only certain users receive the multicast or broadcast, encryption techniques must be employed. In both a wired and wireless network this is done using cryptography. The problem then is one of key management. To handle this, several key management schemes have been devised: centralized group key management protocols, decentralized management protocols, and distributed management protocols [69].

In the case of the centralized group key management protocols, a central authority is used to maintain the group. Decentralized management protocols, however, divide the task of group management among multiple nodes. Each node that is responsible for part of the group management is responsible for a certain subset of the nodes in the network. In the last case, distributed key management protocols, there is no single key management authority. Therefore, the entire group of nodes are responsible for key management [69].

In order to efficiently distribute keys, one well-known technique is to use a logical key tree. Such a technique falls into the centralized group key management protocols. This technique has been extended to wireless sensor networks in [45, 46, 66]. While centralized solutions are often not ideal, in the case of wireless sensor networks a centralized solution offers some utility. Such a technique allows a more powerful base station to offload some of the computations from the less powerful sensor nodes.

Secure Multicasting

Di Pietro et al. describe a directed diffusion-based multicast technique for use in wireless sensor networks that also takes advantage of a logical key hierarchy [66]. In a standard logical key hierarchy a central key distribution center is responsible for disbursing the keys throughout the network. The key distribution center, therefore, is the root of the key hierarchy while individual nodes make up the leaves. The internal nodes of the key hierarchy contain keys that are used in the rekeying process [66].

Directed diffusion is a data-centric, energy-efficient dissemination technique that has been designed for use in wireless sensor networks [38]. In directed diffusion, a query is transformed into an interest (due to the data-centric nature of the network). The interest is then diffused throughout the network and the network begins collecting data based on that interest. The dissemination technique also sets up certain gradients designed to draw events toward the interest. Data collected as a result of the interest can then be sent back along the reverse path of the interest propagation [38].

Using the above-mentioned directed diffusion technique, Di Pietro et al. enhance the logical key hierarchy to create a directed diffusion-based logical key hierarchy. The logical key hierarchy technique provides mechanisms for nodes joining and leaving groups where the key hierarchy is used to effectively rekey all nodes within the leaving node's hierarchy [66]. The directed diffusion is also used in node joining and leaving. When a node declares an intent to join, for example, a join "interest" is generated that travels down the gradient of "interest about interest to join" [66]. When a node joins, a key set is generated for the new node based on keys within the key hierarchy.

Kaya et al. discuss the problem of multicast group management in [42]. In this case, nodes are grouped based on locality and attach to a security tree. However, they assume that nodes within the mobile network are somewhat more powerful than a traditional sensor in a wireless sensor network.

Secure Broadcasting

Lazos and Poovendran describe a tree-based key distribution scheme that is similar to [66]. They suggest a routing aware-based tree where the leaf nodes are assigned keys based on all relay nodes above them. They argue that their technique, which takes advantage of routing information, is more energy efficient than routing schemes that arbitrarily arrange nodes into the routing tree. They propose a greedy routing-aware key distribution algorithm [45].

In [46], Lazos and Poovendran use a similar technique to [45], but instead use geographic location information (e.g., GPS) rather than routing information. In this case, however, nodes (with the help of the geographic location system) are grouped into clusters with the observation that nodes within a cluster will be able to reach one another with a single broadcast. Using the cluster information, a key hierarchy is constructed as in [45].

16.5.4 Defending Against Attacks on Routing Protocols

Routing in wireless sensor networks has, to some extent, been reasonably well studied. However, most current research has focused primarily on providing the most energy-efficient routing. There is a great need for both secure and energy-efficient routing protocols in wireless sensor networks as attacks such as the sinkhole, wormhole, and Sybil attacks demonstrate [35, 40, 59]. As wireless sensor networks continue to grow in size and utility, routing security must not be an afterthought, but rather they must be included as part of the overall sensor network design. This section describes the current state of routing security as it applies to wireless sensor networks.

Background

Because wireless sensors are designed to be widely distributed power and computationally constrained networks, efficient routing protocols must be used in order to maximize the battery life of each node. There are a variety of routing protocols in use in wireless sensor networks, so it is not possible to provide a single security protocol that will be able to secure each type of routing protocol. Before introducing several techniques used to provide secure routing in wireless sensor networks, we will begin with a general overview of several routing protocols that are currently in use. An excellent discussion on many of the attacks on routing protocols is also discussed in [40].

In general, packet-routing algorithms are used to exchange messages with sensor nodes that are outside of a particular radio range. This is different than sensors that are within radio range where packets can be transmitted using a single hop. In such single-hop networks security is still a concern, but is more accurately addressed through secure broadcasting and multicasting.

The first packet-routing algorithm is based on node identifiers similar to traditional routing. In this case, each sensor is identified by an address and routing to/from the sensor is based on the address. This is generally considered inefficient in sensor networks, where nodes are expected to be addressed by their location, rather than their identifier.

As a consequence of the distaste of routing based on node identifiers, geographic routing protocols have been introduced [7, 41]. One common routing protocol, GPSR [41] allows nodes to send a packet to a region, rather than a particular node. Such a routing protocol lends itself nicely to the concept of data-centric networks. A data-centric network is one in which data are stored by name in the sensor network. Data with the same name are stored at the same node. In fact, data need not be stored anywhere near the sensor responsible for generating the data. When searching the network, searches are therefore based on the data's general name, rather than the identity responsible for holding the data. Security specific to this type of network is discussed in [79].

Techniques for Securing the Routing Protocol

Deng, Han, and Mishra describe an intrusion-tolerant routing protocol, IN-SENS, that is designed to limit the scope of an intruder's destruction and route despite network intrusion without having to identify the intruder [15].

They note that an intruder need not be an actual intrusion on the sensor network, but might simply be a node that is malfunctioning for no particularly malicious reason. Identifying an actual intruder versus a malfunctioning node can be extremely difficult, and for this reason Deng et al. make no distinction between the two. The first technique they describe to mitigate the damage done by a potential intruder is to simply employ the use of redundancy. In this case, as described previously under denial of service, multiple identical messages are routed between a source and destination. A message is sent once along several distinct paths with the hope that at least one will arrive at the destination. To discern which, if any, of the messages arriving at the destination are authentic, an authentication scheme can be employed to confirm the message's integrity [15].

Deng et al. also make use of an assumed asymmetry between base stations and wireless sensor nodes. They assume that the base stations are somewhat less resource constrained than the individual sensor node. For this reason, they suggest using the base station to compute routing tables on behalf of the individual sensor nodes. This is done in three phases. In the first phase, the base station broadcasts a request message to each neighbor, which is then propagated throughout the network. In the second phase, the base station collects local connectivity information from each node. Finally, the base station computes a series of forwarding tables for each node. The forwarding tables will include the redundancy information used for the redundant message transmission described above.

There are several possible attacks that can be made on the routing protocol during each of the three stages described above. In the first phase, a node might spoof the base station by sending a spurious request message [15]. A malicious node might also include a fake path(s) when forwarding the request message to its neighbors. It may not even forward the request message at all.

To counter this, Deng et al. use a scheme similar to μTESLA where one-way key chains are used to identify a message originating from the base station.

Tanachaiwiwat et al. present a novel technique named TRANS (Trust Routing for Location Aware Sensor Networks) [79]. The TRANS routing protocol is designed for use in data-centric networks. It also makes use of a loose-time synchronization asymmetric cryptographic scheme to ensure message confidentiality. In their implementation, μTESLA is used to ensure message authentication and confidentiality. Using μTESLA, TRANS is able to ensure that a message is sent along a path of trusted nodes while also using location-aware routing. The strategy is for the base station to broadcast an encrypted message to all of its neighbors. Only those neighbors who are trusted will possess the shared key necessary to decrypt the message. The trusted neighbor(s) then adds its location (for the return trip), encrypts the new message with its own shared key and forwards the message to its neighbor closest to the destination. Once the message reaches the destination, the recipient is able to authenticate the source (base station) using the MAC that will correspond to the base station. To acknowledge or reply to the message, the destination node can simply forward a return message along the same trusted path from which the first message was received [79].

One particular challenge to secure routing in wireless sensor networks is that it is very easy for a single node to disrupt the entire routing protocol by simply disrupting the route discovery process. Papadimitratos and Haas propose a secure route discovery protocol that guarantees, subject to several conditions, that correct topological information will be obtained [62]. This scenario is somewhat similar to the TRANS protocol mentioned above. The security relies on the MAC (message authentication code) and an accumulation of the node identities along the route traversed by a message. In so doing, a source can discover the sensor network topology as each node along the route from source to destination appends its identity to the message. In order to ensure that the message has not been tampered with, a MAC is constructed and can be verified both at the destination and the source (for the return message from the destination).

A related problem is the concept of wormholes in a sensor network. A wormhole attack is one in which a malicious node eavesdrops on a packet or series of packets, tunnels them through the sensor network to another malicious node, and then replays the packets. This can be done to misrepresent the distance between the two colluding nodes. It can also be used to more generally disrupt the routing protocol by misleading the neighbor discovery process [40].

Often additional hardware, such as a directional antenna [34], is used to defend against wormhole attacks. This, however, can be cost-prohibitive when it comes to large-scale network deployment. Instead, Wang and Bhargava use a visualization approach to identifying wormholes [83]. They first compute a distance estimation between all neighbor sensors, including possible existing wormholes. Using multidimensional scaling, they then compute a virtual layout of the sensor network. A surface smoothing strategy is then used to adjust for roundoff errors in the multidimensional scaling. Finally, the shape of the resulting virtual network is analyzed. If a wormhole exists within the network, the shape of the virtual network will bend and curve toward the offending nodes. Using this strategy the nodes that participate in the wormhole can be identified and removed from the network. If a network does not contain a wormhole, the virtual network will appear flat [83].

Defending Against the Sybil Attack

To defend against the Sybil attack described previously in Section 16.4.3, the network needs some mechanism to validate that a particular identity is the only identity being held by a given physical node [59]. Newsome et al. describe two methods to validate identities, direct validation and indirect validation. In direct validation a trusted node directly tests whether the joining identity is valid. In indirect validation, another trusted node is allowed to vouch for (or against) the validity of a joining node [59]. Newsome et al. primarily describe direct validation techniques, including a radio resource test. In the radio test, a node assigns each of its neighbors a different channel on which to communicate. The node then randomly chooses a channel and listens. If the node detects a transmission on the channel it is assumed that the node transmitting on the channel is a physical node. Similarly, if the node does not detect a transmission on the specified channel, the node assumes that the identity assigned to the channel is not a physical identity.

Another technique to defend against the Sybil attack is to use random key predistribution techniques. The idea behind this technique is that with a limited number of keys on a keyring, a node that randomly generates identities will not possess enough keys to take on multiple identities and thus will be unable to exchange messages on the network due to the fact that the invalid identity will be unable to encrypt or decrypt messages.

16.5.5   Detecting Node Replication Attacks

In [63], Parno et al. describe two algorithms: randomized multicast and line-selected multicast. Randomized multicast is an evolution of a node broadcasting strategy. In the simple node broadcasting strategy each sensor propagates an authenticated broadcast message throughout the entire sensor network. Any node that receives a conflicting or duplicated claim revokes the conflicting nodes [63]. This strategy will work, but the communication cost is far too expensive. In order to reduce the communication cost, a deterministic multicast could be employed where nodes would share their locations with a set of witness nodes. In this case, witnesses are computed based on a node's ID. In the event that a node has been replicated on the network, two conflicting locations will be forwarded to the same witness who can then revoke the offending nodes [63]. But since a witness is based on a node's ID, it can easily be computed by an attacker who can then compromise the witness nodes. Thus, securely utilizing a deterministic multicast strategy would require too many witnesses and the communication cost would be too high.

Randomized multicast improves upon the insecurity of deterministic multicast by randomly choosing the witnesses. In the event that a node is replicated two sets of witness nodes are chosen. Assuming a network of size n, if each node derives $\sqrt{n}$ witnesses then the birthday paradox suggests that there will likely be at least one collision [63]. In the event that a collision is detected, the offending nodes can easily be revoked by propagating a revocation throughout the network. Unfortunately, the communication cost of the randomized multicast algorithm is still $O(n^2)$, too high for large networks.

The line-selected multicast algorithm seeks to further reduce the communication costs of the randomized multicast algorithm. It is based upon rumor routing described in [8]. The idea is that a location claim traveling from source s to destination d will also travel through several intermediate nodes. If each of these nodes records the location claim, then the path of the location claim through the network can be thought of as a line segment [63]. In this case the destination of the location claims is one of the randomly chosen witnesses described in the multicast algorithm. As the location claim routes through the network toward a witness node, the intermediate sensors check the claim. If the claim results in an intersection of a line segment then the nodes originating the conflicting claims are revoked. The line selected multicast algorithm reduces the communication cost to $O(n\sqrt{n})$ as long as each line segment is of length $O(\sqrt{n})$ nodes. The storage cost of the line-selected multicast algorithm is $O(\sqrt{n})$ [63].

16.5.6 Combating Traffic Analysis Attacks

Strategies to combat the traffic analysis attacks previously described are possible. Deng et al. propose using a random walk forwarding technique that occasionally forwards a packet to a node other than the sensor's parent node [16]. This would make it difficult to discern a clear path from the sensor to the base station and would help to mitigate the rate monitoring attack, but would still be vulnerable to the time correlation attack. To defend against the time correlation attack, Deng et al. suggest a fractal propagation strategy [16]. In this technique a node will (with a certain probability) generate a fake packet when its neighbor is forwarding a packet to the base station. The fake packet is sent randomly to another neighbor who may also generate a fake packet. These packets essentially use a time-to-live (TTL) to decide when forwarding should stop. This effectively hides the base station from time correlation attacks. Since traffic analysis is closely related to privacy violation, we discuss traffic analysis in the next subsection.

16.5.7 Defending Against Attacks on Sensor Privacy

Regarding the attacks on privacy mentioned earlier, there exist effective techniques to counter many of the attacks levied against a sensor. Here we describe several common techniques [28].

Anonymity Mechanisms

Location information that is too precise can enable the identification of a user, or make the continued tracking of movements feasible. This is a threat to privacy. Anonymity mechanisms depersonalize the data before the data is released, which present an alternative to privacy policy-based access control. Researchers have discussed several approaches using anonymity mechanisms, for example, Gruteser and Grunwald [26] analyze the feasibility of anonymizing location information for location-based services in an automotive telematics environment; Beresford and Stajano [6] independently evaluate anonymity techniques for an indoor location system based on the *Active Bat*.

Total anonymity is a difficult problem given the lack of knowledge concerning a node's location. Therefore, a tradeoff is required between anonymity and the need for public information when solving the privacy problem. In [27, 28, 67, 76], three main approaches are proposed:

- **Decentralize Sensitive Data** The basic idea of this approach is to distribute the sensed location data through a spanning tree, so that no single node holds a complete view of the original data.

- **Secure Communication Channel** Using secure communication protocols, such as SPINS [65], the eavesdropping and active attacks can be prevented.

- **Change Data Traffic** Depatterning the data transmissions can protect against traffic analysis. For example, inserting some bogus data can intensively change the traffic pattern when needed.

- **Node Mobility** Making the sensor movable can be effective in defending privacy, especially the location. For example, the *Cricket* system [67] is a location-support system for in-building, mobile, location-dependent applications. It allows applications running on mobile and static nodes to learn their physical location by using listeners that hear and analyze information from beacons spread throughout the building. Thus the location sensors can be placed on the mobile device as opposed to the building infrastructure, and the location information is not disclosed during the position determination process and the data subject can choose the parties to which the information should be transmitted.

Policy-Based Approaches

Policy-based approaches are currently a hot approach to address the privacy problem. The access control decisions and authentication are made based on the specifications of the privacy policies. In [57], Molnar and Wagner present the concept of private authentication, and give a general scheme for building private authentication with work logarithmic in the number of tags in (but not limited by) RFID (radio frequency identification) applications. In the automotive telematics domain, Duri and colleagues [20] propose a policy-based framework for protecting sensor information, where an in-car computer can act as a trusted agent. Snekkenes [77] presents advanced concepts for specifying policies in the context of a mobile phone network. These concepts enable access control based on criteria such as time of the request, location, speed, and identity of the located object. Myles and colleagues [58] describe an architecture for a centralized location server that controls access from client applications through a set of validator modules that check XML-encoded application privacy policies. Hengartner and Steenkiste [31] point out that access control decisions can be governed by either room or user policies. The room policy specifies who is permitted to find out about the people currently in a room, while the user policy states who is allowed to get location information about another user.

Information Flooding

Ozturk et al. propose antitraffic analysis mechanisms to prevent an outside attacker from tracking the location of a data source, since that information will release the location of sensed objects [61]. The randomized data routing mechanism and phantom traffic generation mechanism are used to disguise the real data traffic, so that it is difficult for an adversary to track the source of data by analyzing network traffic. Based on flooding-based routing protocols, Ozturk et al. have developed comparable methods for single path routing to try to solve the privacy problems in sensor network.

- **Baseline Flooding** In the baseline implementation of flooding, every node in the network only forwards a message once, and no node retransmits a message that it has previously transmitted. When a message reaches an intermediate node, the node first checks whether it has received and forwarded that message before. If this is its first time, the node will broadcast the message to all its neighbors. Otherwise, it just discards the message.

- **Probabilistic Flooding** In probabilistic flooding, only a subset of nodes within the entire network will participate in data forwarding, while the others simply discard the messages they receive. One possible weakness of this approach is that some messages may get lost in the network and as a result affect the overall network connectivity. However, as [61] explain later in this section, this problem does not appear to be a significant factor.

- **Flooding with Fake Messages** The previous flooding strategies can only decrease the chances of a privacy violation. An adversary still has a chance to monitor the general traffic and even the individual packets. This observation suggests that one approach to alleviate the risk of source-location privacy breaching is to augment the flooding protocols to introduce more sources that inject fake messages into the network. By doing so, even if the attacker captures the packets, he will have no idea whether the packets are real.

- **Phantom Flooding** Phantom flooding shares the same insights as probabilistic flooding in that they both attempt to direct messages to different locations of the network so that the adversary cannot receive a steady stream of messages to track the source. Probabilistic flooding is not very effective in achieving this goal because shorter paths are more likely to deliver more messages. Therefore, Ozturk et al. [61] suggest enticing the attacker away from the real source and toward a fake source, called the phantom source. In phantom flooding, every message experiences two phases: (1) a walking phase, which may be a random walk or a directed walk, and (2) a subsequent flooding meant to deliver the message to the sink. When the source sends out a message, the message is unicast in a random fashion within the first h_{walk} hops (referred to as random walk phase). After the h_{walk} hops, the message is flooded using the baseline flooding technique (referred to as flooding phase).

Similar mechanisms are also used to disguise an adversary from finding the location of a base station by analyzing network traffic [29]. One key problem for these antitraffic analysis mechanisms is the energy cost incurred by anonymization.

Another strategy used to mask location information from eavesdroppers is presented in [89]. They propose a two-way greedy random-walk strategy GROW (Greedy Random Walk). In this case, the random walk is taken from both the source and the sink. The sink first initiates an N-hop random walk. The source then initiates an M-hop random walk. Once the source packet reaches an intersection of these two paths, it is forwarded through the path created by the sink. Local broadcasting is used to detect when the two paths intersect. In order to minimize the chance of backtracking along the random walk, the nodes are stored in a bloom filter as the walk progresses. At each stage, the intermediate nodes are checked against the bloom filter to ensure that backtracking is minimized [89].

16.5.8   Intrusion Detection

We now turn to the area of intrusion detection in wireless sensor networks. It is important to note that in this section we cover intrusion detection as it applies to detecting attacks on the sensor network itself, rather than the popular intrusion detection application being researched for such uses as perimeter monitoring, and so forth.

With that in mind, we note that intrusion detection is not necessarily a category unto itself, but rather has its place in nearly every aspect of sensor network security. Many secure routing schemes attempt to identify network intruders, and key establishment techniques are used in part to prevent intruders from overhearing network data.

Despite the necessity of effective intrusion detection schemes for wireless sensor networks, a good solution has not yet been devised. Of course, this is due largely to the resource constraints present in wireless sensor networks. However, resource constraints are not the only reason. Another problem is that researchers have not yet been able to develop methods of reliably detecting intruders in sensor networks. As such, it is difficult to define characteristics (or signatures) that are specific to a network intrusion as opposed to the normal network traffic that might occur as the result of normal network operations or malfunctions resulting from the environment change.

Background on Intrusion Detection

Traditionally, intrusion detection has focused on two major categories: anomaly-based intrusion detection (AID), and misuse intrusion detection (MID) [72]. Anomaly-based intrusion detection relies on the assumption that intruders will demonstrate abnormal behavior relative to the legitimate nodes.

Thus, the object of anomaly-based detection is to detect intrusion based on unusual system behavior. Typically this is done by first developing a profile of the system in normal use. Once the profile has been generated it can be used to evaluate the system in the face of intruders.

The advantage of using an anomaly-based system is that it is able to detect previously unknown attacks based only upon knowing that the system behavior is unusual. This is particularly advantageous in wireless sensor networks where it can be difficult to boil an attack down to a signature. However, such flexible intrusion detection comes at a cost. The first is that the anomaly-based approach is susceptible to false positives. This is due largely to the fact that it can be difficult to define normal system behaviors. To help combat this, new profiles can be taken of the network to ensure that the profile in use is up to date. However, this takes time. And further, even with the most up-to-date profile possible, it can still be difficult to discern unusual, but legitimate, behavior from an actual intrusion. Another fault in the anomaly-based intrusion detection techniques is that the computational cost of comparing the current system activity to the profile can be quite high [72]. In the case of a wireless sensor network, such added computation can severely impact the longevity of the network.

In systems based on misuse intrusion detection, the system maintains a database of intrusion signatures. Using these signatures, the system can easily detect intrusions on the network. Further, the system is less prone to false positives as the intrusion signatures are narrowly defined. Such narrowly defined signatures, while leading to fewer false positives, also imply that the intrusion detection system will be unable to detect unknown attacks. This problem can be somewhat mitigated by maintaining an up-to-date signature database. However, since it can be difficult to characterize attacks on wireless sensor networks, such databases may be inherently limited and difficult to generate. An advantage, however, is that the misuse intrusion detection system requires less computation in order to identify intruders as the comparison of network events to the available signatures is relatively low cost [72].

Because both techniques have their strengths and weaknesses, traditional intrusion detection systems use systems that implement both anomaly-based intrusion detection and misuse intrusion detection models. This allows such systems to utilize the fast evaluation of the misuse intrusion detection system, but still recognize abnormal system behavior.

Intrusion Detection in Wireless Sensor Networks

Typically a wireless sensor network uses cryptography to secure itself against unauthorized external nodes gaining entry into the network. But cryptography can only protect the network against the external nodes and does little to thwart malicious nodes that already possess one or more keys. Brutch and Ko classify intrusion detection systems (IDS) into two categories: *host based* and *network based*. They further classify intrusion detection schemes into those that are signature based, anomaly based, and specification based [9].

Simply put, a host-based IDS system operates on operating systems audit trails, system call audit trails, logs, and so on. A network-based IDS, on the other hand, operates entirely on packets that have been captured from the network [9]. A signature-based IDS simply monitors the network for specific predetermined signatures that are indicative of an intrusion. In an anomaly-based scheme, a standard behavior is defined and any deviation from that behavior triggers the intrusion detection system. Finally, a specification-based scheme defines a set of constraints that are indicative of a program's or protocol's correct operation [9].

Brutch and Ko describe a series of attacks against several aspects of a wireless sensor network and also introduce three architectures for intrusion detection in wireless sensor networks. The first is termed the stand-alone architecture. In this case, as its name implies, each node functions as an independent intrusion detection system and is responsible for detecting attacks directed toward itself. Nodes do not cooperate in any way [9].

The second architecture is the distributed and cooperative architecture. In this case, an intrusion detection agent still resides on each node (as in the case of the stand-alone architecture) and nodes are still responsible for detecting attacks against themselves (local attacks), but also cooperate to share information in order to detect global intrusion attempts [9].

The third technique proposed by Brutch and Ko is called the hierarchical architecture. These architectures are suitable for multilayered wireless sensor networks. In this case, Brutch and Ko describe a multilayered network as one in which the network is divided into clusters with cluster-head nodes responsible for routing within the cluster. The multilayered network is used primarily for event correlation.

Albers et al. describe an intrusion detection architecture based on the implementation of a local intrusion detection system (LIDS) at each node [2]. In order to extend each node's "vision" of the network, Albers suggests that the LIDS existing within the network should collaborate with one another. All LIDS within the network will exchange two types of data, security data and intrusion alerts. The security data are simply used to exchange information with other network hosts. The intrusion alerts, however, are used to inform other LIDS of a locally detected intrusion [2].

A pictorial representation of the LIDS architecture is depicted in Figure 16.2. MIB (management information base) variables are accessed through SNMP running on the mobile host, where the LIDS components are depicted within the block labeled LIDS. The local MIB is designed to interface with the SNMP agent to provide MIB variable collection from the local LIDS agent or mobile agents. The mobile agents are responsible for both the collection and processing of data from remote hosts, specifically SNMP requests. The agents are capable of migration between individual hosts and are capable of transferring data back to their home LIDS. The local LIDS agent is responsible for detecting and responding to local intrusions as well as responding to events generated by remote nodes [2].

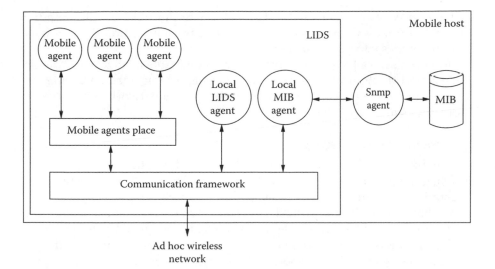

FIGURE 16.2
The LIDS architecture from [2].

Albers et al. propose to use SNMP auditing as the audit source for each LIDS. Rather than simply sending the SNMP messages over an unreliable UDP connection, it is suggested that mobile agents will be responsible for message transporting. In order to detect an intrusion, Albers suggests using either misuse or anomaly detection. When a LIDS detects an intrusion, it should communicate this intrusion to other LIDS on the network. Possible responses include forcing the potential intruder to reauthenticate, or to simply ignore the suspicious node when performing cooperative actions [2]. Although this approach cannot be applied to wireless sensor network directly, it is an interesting idea that explores the local information only, which is the key to any intrusion detection techniques in a sensor network [22]. In summary, we envision that the intrusion detection in wireless sensors remains an open problem, and more study is needed. Taking the predeployment information, such as sensing data distribution, into consideration is a possible direction.

16.5.9 Secure Data Aggregation

As wireless sensor networks continue to grow in size, so does the amount of data that the sensor networks are capable of sensing. However, due to the computational constraints placed on individual sensors, a single sensor is typically responsible for only a small part of the overall data. Because of this, a query of the wireless sensor network is likely to return a great deal of raw data, much of which is not of interest to the individual performing the query.

Thus, it is advantageous for the raw data to first be processed so that more meaningful data can be gleaned from the network. This is typically done using a series of aggregators. An aggregator is responsible for collecting the raw data from a subset of nodes and processing/aggregating the raw data from the nodes into more usable data.

However, such a technique is particularly vulnerable to attacks as a single node is used to aggregate multiple data. Because of this, secure information aggregation techniques are needed in wireless sensor networks where one or more nodes may be malicious.

Introduction to Data Aggregation and Its Utility

Before discussing the security aspects of secure information aggregation, we first begin with an overview of several information aggregating techniques. Clustering techniques are discussed in [22]. They develop a localized algorithm that uses the directed diffusion technique to achieve a global perspective using only local nodes. In their algorithm, nodes are assigned levels, with level 0 being the lowest level. When a node transmits a message, the number of hops that the message travels is proportional to the node's level. A node can be promoted and demoted. Using this technique, higher-level nodes are able to communicate across clusters, while their lower-level siblings cannot. This effectively enables localized cluster computation while the higher-level nodes can coordinate their cluster's local information to achieve a global solution [22].

If an aggregation node is itself compromised, then all of the data being delivered from the sensor network to the base station may be forged. To detect this, Ye et al. describe a statistical en-route filtering mechanism [91]. It utilizes multiple MACs along the path from the aggregator to the base station. Any packet that fails any of the MAC tests will be disregarded.

A more recent technique called TAG is proposed in [54]. In this case, the authors propose an SQL-like language that is used for generating queries over the sensor network. The TAG approach is one of general-purpose aggregation. That is, it has not been designed with an application-specific intent. Its operation is fairly simple, the base station defines a query using the SQL-like language designed for use in TAG. The sensors then route data back to the base station according to a routing tree. At each point in the tree, data is aggregated according to the routing tree and according to the particular aggregation function that is defined in the initial query [54].

More recently Shrivastava et al. propose a summary structure that is able to support fairly complex aggregate functions, such as median and range queries [75]. It is important to note that typical aggregate functions are capable of performing min/max, sum, and average. The more complex aggregates, such as finding the most frequent data values, are typically not supported. They note that the added aggregate functions are not exact. However, they prove strict guarantees on the approximation quality of the queries [75].

Wagner analyzes the resilience of all aggregation techniques in [82], and argues that current aggregation schemes were designed without security in mind and that there are easy attacks against them. Wagner proposes a mathematical framework for formally evaluating the security for aggregation, allowing them to quantify the robustness of an aggregation operator against malicious data. This seminal work opens the door to secure data aggregation in sensor networks; however, the one-level homogeneous aggregation model is too simple to represent real sensor network deployments. Extending the model to a more realistic model, e.g., multilevel and heterogeneous, is an interesting direction.

Secure Data Aggregation Techniques

As was shown above, the idea of information aggregation has been studied in reasonable depth. The problem with the standard information aggregation techniques, however, is that they assume that all nodes are trustworthy. Of course, this is not the case and secure data aggregation techniques will be necessary in many wireless sensor networks.

Przydatek et al. describe a secure information aggregation technique (SIA) [68]. They note that sensor networks and data aggregation techniques are vulnerable to a variety of attacks including denial-of-service attacks as described in Section 16.4.2. However, [68] focus their efforts on defending specifically against a type of attack called the stealthy attack. In a stealthy attack, the attacker seeks to provide incorrect aggregation results to the user without the user knowing that the results are incorrect. Therefore, the goal of [68] is to ensure that if a user accepts an aggregate value as correct, then there is a high probability that the value is close to the true aggregation value [68]. In the event that the aggregate value has been tampered with, the user should reject the incorrect results with high probability.

The approach that [68] provide is termed the aggregate-commit-prove technique. As the name would suggest, the technique is composed of three phases. In the first stage, aggregate, the aggregator collects data from the sensors and computes the aggregation result according to a specific aggregate function. Each sensor should share a key with the aggregator. This allows the aggregator to verify that the sensor reading is authentic. However, it is possible that a sensor has been compromised and possesses the key, or that the sensor is simply malfunctioning. The aggregate phase does not prevent such malfunctioning.

In the second phase, the commit phase, the aggregator is responsible for committing to the collected data. This commitment ensures that the aggregator actually uses the data collected from the sensors. One way to perform this commitment is to use a Merkle hash-tree construction [56]. Using this technique the aggregator computes a hash of each input value and the internal nodes are computed as the hash of their children concatenated. The commitment is the root value. The hashing is used to ensure that the aggregator cannot change any input values after having hashed them.

In the final phase, the aggregator is charged with proving the results to the user. The aggregator first communicates the aggregation result and the commitment. The aggregator then uses an interactive proof to prove the correctness of the results. This generally requires two steps. In the first, the user/home server checks to ensure that the committed data is a good representation of the data values in the sensor network. In the second step, the user/home server decides whether the aggregator is lying. This can be done by checking whether the aggregation result is close to the committed result [68]. The interactive proof differs depending on the aggregation function that is being used.

Hu and Evans propose a secure aggregation technique that uses the μTESLA protocol for security [33]. In this case, the nodes organize into a tree-based hierarchy where the internal nodes act as aggregators. Recall that the μTESLA protocol achieves asymmetry through delayed disclosure of symmetric keys. Therefore, a child's parent will be unable to immediately verify the authenticity of the child's data as the key used to generate the MAC will not have been revealed. This technique, however, does not guarantee that nodes and aggregators are providing correct values. To address this problem, the base station is responsible for distributing temporary keys to the network as well as the base station's current μTESLA key, used for validating MACs. Using the μTESLA key, nodes verify their children's MAC and are responsible for ensuring that the MACs are consistent.

To this end, we argue that secure aggregation techniques play an important role in adopting wireless sensor networks, because of the large amount of raw data and the necessity of the localized in-network processing, and much more investigation is needed.

16.5.10 Defending Against Physical Attacks

Physical attacks, as we argued in the beginning of the chapter, pose a great threat to wireless sensor networks, because of their unattended feature and limited resources. Sensor nodes may be equipped with physical hardware to enhance protection against various attacks. For example, to protect against tampering with the sensors, one defense involves tamper-proofing the node's physical package [88]. References [3, 4, 43] focus on building tamper-resistant hardware in order to make the actual data and memory contents on the sensor chip inaccessible to attack. Another way is to employ special software and hardware outside the sensor to detect physical tampering.

As the price of the hardware itself gets cheaper, tamper-resistant hardware may become more appropriate in a variety of sensor network deployments. One possible approach to protect the sensors from physical attacks is self-termination. The basic idea is that the sensor kills itself, including destroying all data and keys, when it senses a possible attack. This is particularly feasible in the large-scale wireless sensor network, which has enough redundancy of information, and the cost of a sensor is much cheaper than the cost of being broken (attacked). The key to this approach is detecting the physical attack. A simple solution is periodically conducting neighborhood checking in static deployment. For mobile sensor networks, this is still an open problem.

In [3, 4, 43], the authors describe techniques for extracting protected software and data from smartcard processors. This includes manual microprobing, laser cutting, focused ion-beam manipulation, glitch attacks, and power analysis, most of which are also possible physical attacks on the sensor.

Based on an analysis of these attacks, Andersen et al. give examples of low-cost protection countermeasures that make such attacks considerably more difficult, including [4]:

- **Randomized Clock Signal** Inserting random-time delays between any observable reaction and critical operations that might be subject to an attack.

- **Randomized Multithreading** Designing a multithread processor architecture that schedules the processor by hardware between two or more threads of execution randomly at a per-instruction level.

- **Robust Low-Frequency Sensor** Building an intrinsic self-test into the detector. Any attempt to tamper with the sensor should result in the malfunction of the entire processor.

- **Destruction of Test Circuitry** Destroying or disabling the special test circuitry that is for the test engineers, closing the door to micro-probing attackers.

- **Restricted Program Counter** Avoid providing a program counter that can run over the entire address space.

- **Top-Layer Sensor Meshes** Introducing additional metal layers that form a sensor mesh above the actual circuit and that do not carry any critical signals to be effective annoyances to microprobing attackers.

For the deployment of components outside the sensor, various approaches have been proposed to protect the sensor, and are summarized in [17]. Sastry et al. [71] introduce the concept of secure location verification and propose a secure localization scheme, the ECHO protocol, to make sure the location claims are legitimate. In their work, the security rests on physical properties of sound and RF signal propagation. An adversary cannot cheat and claim a shorter distance by starting the ultrasound response early, because it will not have the nonce. Hu et al. [34] introduce directional antennas to defend against wormhole attacks. In [85] the authors study the modeling and defense of sensor networks against *Search-Based Physical Attacks*. They define a search-based physical attack model, where the attacker walks through the sensor network using signal-detecting equipment to locate active sensors, and then destroys them. In a prior work, they have identified and modeled blind physical attacks [84]. The defense algorithm is executed by individual sensors in two phases: in the first phase, sensors detect the attacker and send out attack notification messages to other sensors; in the second phase, the recipient sensors of the notification message schedule their states to switch. A mechanism named SWATT to verify whether the memory of a sensor node has been changed [74] is proposed by Seshadri et al.

16.5.11 Trust Management

Trust is an old but important issue in any networked environment, whether social networking or computer networking. Trust can solve some problems beyond the power of the traditional cryptographic security. For example, judging the quality of the sensor nodes and the quality of their services, and providing the corresponding access control, e.g., does the data aggregator perform the aggregation correctly? Does the forwarder send out the packet in a timely fashion? These questions are important, but difficult, if not impossible, to answer using existing security mechanisms. We argue that trust management is the key to build trusted, dependable wireless sensor network applications. The trust issue is emerging as sensor networks thrive. However, it is not easy to build a good trust model within a sensor network given the resource limits. Furthermore, in order to keep the sensor nodes independent, we should not assume there is a trust among sensors in advance.

According to the small world principle in the context of social networks and peer-to-peer computing [60], one can employ a pathfinder to find paths from a source node to a designated target node efficiently. Based on this observation, Zhu et al. [92] provide a practical approach to compute trust in wireless networks by viewing individual mobile devices as a node of a delegation graph G and mapping a delegation path from the source node S to the target node T into an edge in the correspondent transitive closure of the graph G, from which the trust value is computed. In this approach, an undirected transitive signature scheme is used within the authenticated transitive graphs.

In [90], a trust evaluation-based security solution is proposed to provide effective security decisions on data protection, secure routing, and other network activities. Logical and computational trust analysis and evaluation are deployed among network nodes. Each node's evaluation of trust on other nodes is based on serious study and inference from trust factors such as experience statistics, data value, intrusion detection results, and references to other nodes, as well as a node owner's preference and policy. Ren et al. describe a technique to establish sufficient trust relationships in ad hoc networks with minimum local storage capacity requirements on the mobile nodes [70]. The authors propose a probabilistic solution based on a distributed trust model. A secret dealer is introduced only in the system bootstrapping phase to complement the assumption in trust initialization. With the help of the secret dealer, much shorter and more robust trust chains are able to be constructed with high probability. A fully self-organized trust establishment approach is then adopted to conform to the dynamic membership changes. But the shortcoming of this approach for the common sensor network is that it is not reasonable to introduce a dealer in a totally decentralized ad hoc environment.

The approaches described above are proposed in the context of ad hoc network. For the wireless sensor network, they cannot be employed directly because of the capacity of the sensor. Some researchers specifically focus on the sensor networks that have been proposed recently. Ganeriwal and Srivastava propose a reputation-based framework for high-integrity sensor networks [23]. Within this framework the authors employ a beta reputation system for reputation representation, updates, and integration.

Tanachaiwiwat et al. [80] propose a mechanism of location-centric isolation of misbehavior and trust routing in sensor networks. In their trust model, the trustworthiness value is derived from the capacity of the cryptography, availability, and packet forwarding. If the trust value is below a specific trust threshold, then this location is considered insecure and is avoided when forwarding packets.

Liang and Shi focus on trust model developing and the analysis of rating aggregation algorithms in the open untrusted environment [48, 49, 50]. Their findings and observations can be applied to wireless sensor networks directly, although the work is performed in the context of peer-to-peer settings. They propose a personalized trust model called PET in [50], which supports the customization of trustworthiness from the view of individual sensors. Regarding how to aggregate the ratings from referrals, they recently analyzed the effect of ratings on the trust inference in a comprehensive way [48]. They find that the rating is not always helpful given the limitations of other factors. In the open environment with high dynamics the rating performance degrades and can produce negative effects. They observe that the storage space for saving self-knowledge is a potential bottleneck to the effect of ratings. Their recent simulation results show that it is better to treat the ratings from different evaluators equally given the dynamics of the open environment, and simply averaging ratings is appropriate considering the simplicity of the algorithm design and the low cost in running the system. They argue that the most important issue for building a trust model is adjusting parameters according to environment changes. These suggestions are quite useful for building trust models in the wireless sensor network given their simplicity and cost savings.

16.6 Conclusions

In this chapter we have described the four main aspects of wireless sensor network security: obstacles, requirements, attacks, and defenses. Within each of those categories we have also subcategorized the major topics including routing, trust, denial of service, and so on. Our aim is to provide both a general overview of the rather broad area of wireless sensor network security, and give the main citations such that further review of the relevant literature can be completed by the interested researcher.

As wireless sensor networks continue to grow and become more common, we expect that further expectations of security will be required of these wireless sensor network applications. In particular, the addition of public key cryptography and the addition of public key-based key management described in Section 16.5.1 will likely make strong security a more realistic expectation in the future. We also expect that the current and future work in privacy and trust will make wireless sensor networks a more attractive option in a variety of new arenas.

Bibliography

1. I. F. Akyildiz, W. Su, Y. Sankarasubramaniam, and E. Cayirci. A survey on sensor networks. *IEEE Communications Magazine*, 40(8):102–114, August 2002.
2. P. Albers and O. Camp. Security in ad hoc networks: A general intrusion detection architecture enhancing trust-based approaches. In *First International Workshop on Wireless Information Systems, 4th International Conference on Enterprise Information Systems*, 2002.
3. R. Anderson and M. Kuhn. Tamper resistance — a cautionary note. In *The Second USENIX Workshop on Electronic Commerce Proceedings*, Oakland, CA, 1996.
4. R. Anderson and M. Kuhn. Low-cost attacks on tamper-resistant devices. In *IWSP: International Workshop on Security Protocols, LNCS*, 1997.
5. T. Aura, P. Nikander, and J. Leiwo. Dos-resistant authentication with client puzzles. In *Revised Papers from the 8th International Workshop on Security Protocols*, pages 170–177. Springer-Verlag, 2001.
6. A. R. Beresford and F. Stajano. Location privacy in pervasive computing. *IEEE Pervasive Computing*, 2(1):46–55, 2003.
7. P. Bose, P. Morin, I. Stojmenović, and J. Urrutia. Routing with guaranteed delivery in ad hoc wireless networks. *Wirel. Netw.*, 7(6):609–616, 2001.
8. D. Braginsky and D. Estrin. Rumor routing algorithm for sensor networks. In *WSNA '02: Proceedings of the 1st ACM International Workshop on Wireless Sensor Networks and Applications*, pages 22–31, New York, 2002. ACM Press.
9. P. Brutch and C. Ko. Challenges in intrusion detection for wireless ad-hoc networks. In *2003 Symposium on Applications and the Internet Workshops (SAINT'03 Workshops)*, 2003.
10. D. W. Carman, P. S. Krus, and B. J. Matt. Constraints and approaches for distributed sensor network security. Technical Report 00-010, NAI Labs, Network Associates, Inc., Glenwood, MD, 2000.
11. H. Chan and A. Perrig. Security and privacy in sensor networks. *IEEE Computer Magazine*, pages 103–105, 2003.
12. H. Chan and A. Perrig. Pike: Peer intermediaries for key establishment in sensor networks. In *IEEE Infocom 2005*, 2005.
13. H. Chan, A. Perrig, and D. Song. Random key predistribution schemes for sensor networks. In *Proceedings of the 2003 IEEE Symposium on Security and Privacy*, page 197. IEEE Computer Society, 2003.
14. http://www.xbow.com/wireless_home.aspx, 2006.
15. J. Deng, R. Han, and S. Mishra. INSENS: Intrusion-tolerant routing in wireless sensor networks. In *Technical Report CU-CS-939-02, Department of Computer Science, University of Colorado*, 2002.
16. J. Deng, R. Han, and S. Mishra. Countermeasures against traffic analysis in wireless sensor networks. Technical Report CU-CS-987-04, University of Colorado at Boulder, 2004.
17. J. Deng, R. Han, and S. Mishra. *Security, Privacy, and Fault Tolerance in Wireless Sensor Networks*. Artech House, August 2005.
18. J. Douceur. The Sybil attack. In *Proc. of the 1st International Workshop on Peer-to-Peer Systems (IPTPS'02)*, February 2002.
19. W. Du, J. Deng, Y. S. Han, and P. K. Varshney. A pairwise key pre-distribution scheme for wireless sensor networks. In *CCS '03: Proceedings of the 10th ACM*

Conference on Computer and Communications Security, pages 42–51, New York, 2003. ACM Press.

20. S. Duri, M. Gruteser, X. Liu, P. Moskowitz, R. Perez, M. Singh, and J. Tang. Framework for security and privacy in automotive telematics. In *2nd ACM International Worksphop on Mobile Commerce*, 2000.

21. L. Eschenauer and V. D. Gligor. A key-management scheme for distributed sensor networks. In *Proceedings of the 9th ACM Conference on Computer and Communications Security*, pages 41–47. ACM Press, 2002.

22. D. Estrin, R. Govindan, J. S. Heidemann, and S. Kumar. Next century challenges: Scalable coordination in sensor networks. In *Mobile Computing and Networking*, pages 263–270, 1999.

23. S. Ganeriwal and M. Srivastava. Reputation-based framework for high integrity sensor networks. In *Proceedings of the 2nd ACM Workshop on Security of Ad Hoc and Sensor Networks*, Washington, DC, 2004.

24. S. Ganeriwal, S. Čapkun, C.-C. Han, and M. B. Srivastava. Secure time synchronization service for sensor networks. In *WiSe '05: Proceedings of the 4th ACM Workshop on Wireless Security*, pages 97–106, New York, 2005. ACM Press.

25. G. Gaubatz, J.P. Kaps, and B. Sunar. Public key cryptography in sensor networks — revisited. In *1st European Workshop on Security in Ad-Hoc and Sensor Networks (ESAS 2004)*, 2004.

26. M. Gruteser and D. Grunwald. Anonymous usage of location-based services through spatial and temporal cloaking. In *Proceedings of the First International Conference on Mobile Systems, Applications, and Services (MobiSys)*. USENIX, 2003.

27. M. Gruteser and D. Grunwald. A methodological assessment of location privacy risks in wireless hotspot networks. In *First International Conference on Security in Pervasive Computing*, 2003.

28. M. Gruteser, G. Schelle, A. Jain, R. Han, and D. Grunwald. Privacy-aware location sensor networks. In *9th USENIX Workshop on Hot Topics in Operating Systems (HotOS IX)*, 2003.

29. N. Gura, A. Patel, A. Wander, H. Eberle, and S. Shantz. Comparing elliptic curve cryptography and rsa on 8-bit cpus. In *2004 Workshop on Cryptographic Hardware and Embedded Systems*, August 2004.

30. C. Hartung, J. Balasalle, and R. Han. Node compromise in sensor networks: The need for secure systems. Technical Report CU-CS-988-04, Department of Computer Science, University of Colorado at Boulder, 2004.

31. U. Hengartner and P. Steenkiste. Protecting access to people location information. In *Proceedings of First International Conference on Security in Pervasive Computing (to appear)*, LNCS. Springer, March 2003.

32. J. Hill, R. Szewczyk, A. Woo, S. Hollar, D. E. Culler, and K. Pister. System architecture directions for networked sensors. In *Architectural Support for Programming Languages and Operating Systems*, pages 93–104, 2000.

33. L. Hu and D. Evans. Secure aggregation for wireless networks. In *SAINT-W '03: Proceedings of the 2003 Symposium on Applications and the Internet Workshops (SAINT'03 Workshops)*, page 384. IEEE Computer Society, 2003.

34. L. Hu and D. Evans. Using directional antennas to prevent wormhole attacks. In *11th Annual Network and Distributed System Security Symposium*, February 2004.

35. Y. Hu, A. Perrig, and D. B. Johnson. Packet leashes: A defense against wormhole attacks in wireless networks. In *INFOCOM 2003. Twenty-Second Annual*

Joint Conference of the IEEE Computer and Communications Societies, volume 3, pages 1976–1986, 2003.

36. Q. Huang, J. Cukier, H. Kobayashi, B. Liu, and J. Zhang. Fast authenticated key establishment protocols for self-organizing sensor networks. In *Proceedings of the 2nd ACM International Conference on Wireless Sensor Networks and Applications*, pages 141–150. ACM Press, 2003.

37. J. Hwang and Y. Kim. Revisiting random key pre-distribution schemes for wireless sensor networks. In *Proceedings of the 2nd ACM Workshop on Security of Ad Hoc and Sensor Networks (SASN '04)*, pages 43–52, New York, 2004. ACM Press.

38. C. Intanagonwiwat, R. Govindan, and D. Estrin. Directed diffusion: A scalable and robust communication paradigm for sensor networks. In *Mobile Computing and Networking*, pages 56–67, 2000.

39. C. Karlof, N. Sastry, and D. Wagner. Tinysec: A link layer security architecture for wireless sensor networks. In *Second ACM Conference on Embedded Networked Sensor Systems (SensSys 2004)*, pages 162–175, November 2004.

40. C. Karlof and D. Wagner. Secure routing in wireless sensor networks: Attacks and countermeasures. *Elsevier's AdHoc Networks Journal, Special Issue on Sensor Network Applications and Protocols*, 1(2–3):293–315, September 2003.

41. B. Karp and H. T. Kung. GPSR: Greedy perimeter stateless routing for wireless networks. In *Proceedings of the 6th Annual International Conference on Mobile Computing and Networking*, pages 243–254. ACM Press, 2000.

42. T. Kaya, G. Lin, G. Noubir, and A. Yilmaz. Secure multicast groups on ad hoc networks. In *Proceedings of the 1st ACM Workshop on Security of Ad Hoc and Sensor Networks (SASN '03)*, pages 94–102. ACM Press, 2003.

43. O. Kömerling and M. G. Kuhn. Design principles for tamper-resistant smartcard processors. In *The USENIX Workshop on Smartcard Technology Proceedings*, Chicago, May 1999.

44. Y. W. Law, J. Doumen, and P. Hartel. Survey and benchmark of block ciphers for wireless sensor networks. *ACM Trans. Sen. Netw.*, 2(1):65–93, 2006.

45. L. Lazos and R. Poovendran. Secure broadcast in energy-aware wireless sensor networks. In *IEEE International Symposium on Advances in Wireless Communications (ISWC'02)*, 2002.

46. L. Lazos and R. Poovendran. Energy-aware secure multicast communication in ad-hoc networks using geographic location information. In *Proceedings of IEEE International Conference on Acoustics Speech and Signal Processing*, 2003.

47. L. Lazos and R. Poovendran. Serloc: Robust localization for wireless sensor networks. *ACM Trans. Sen. Netw.*, 1(1):73–100, 2005.

48. Z. Liang and W. Shi. Analysis of recommendations on trust inference in the open environment. Technical Report MIST-TR-2005-002, Department of Computer Science, Wayne State University, February 2005.

49. Z. Liang and W. Shi. Enforcing cooperative resource sharing in untrusted peer-to-peer environment. *ACM Journal of Mobile Networks and Applications (MONET)*, 10(6):771–783, 2005.

50. Z. Liang and W. Shi. PET: A PErsonalized Trust model with reputation and risk evaluation for P2P resource sharing. In *Proceedings of the HICSS-38*, Waikoloa Village, Hawaii, January 2005.

51. D. Liu and P. Ning. Efficient distribution of key chain commitments for broadcast authentication in distributed sensor networks. In *Proceedings of the 10th Annual Network and Distributed System Security Symposium*, pages 263–276, 2003.

52. D. Liu and P. Ning. Multilevel μTESLA: Broadcast authentication for distributed sensor networks. *Trans. on Embedded Computing Sys.*, 3(4):800–836, 2004.

53. D. Liu, P. Ning, and R. Li. Establishing pairwise keys in distributed sensor networks. *ACM Trans. Inf. Syst. Secur.*, 8(1):41–77, 2005.

54. S. Madden, M. J. Franklin, J. M. Hellerstein, and W. Hong. Tag: A tiny aggregation service for ad-hoc sensor networks. *SIGOPS Oper. Syst. Rev.*, 36(SI):131–146, 2002.

55. D. J. Malan, M. Welsh, and M. D. Smith. A public key infrastructure for key distribution in tinyos based on elliptic curve cryptography. In *First Annual IEEE Communications Society Conference on Sensor and Ad Hoc Communications and Networks, 2004. IEEE SECON*, 2004.

56. R. C. Merkle. Protocols for public key cryptosystems. In *Proceedings of the IEEE Symposium on Research in Security and Privacy*, April 1980.

57. D. Molnar and D. Wagner. Privacy and security in library rfid: Issues, practices, and architectures. In *ACM CCS*, 2004.

58. G. Myles, A. Friday, and N. Davies. Preserving privacy in environments with location-based applications. *IEEE Pervasive Computing*, 2(1):56–64, 2003.

59. J. Newsome, E. Shi, D. Song, and A. Perrig. The Sybil attack in sensor networks: Analysis and defenses. In *Proceedings of the Third International Symposium on Information Processing in Sensor Networks*, pages 259–268. ACM Press, 2004.

60. A. Oram. *Peer-to-Peer: Harnessing the Power of Disruptive Technologies*. O'Reilly & Associates, March 2001.

61. C. Ozturk, Y. Zhang, and W. Trappe. Source-location privacy in energy-constrained sensor network routing. In *Proceedings of the 2nd ACM Workshop on Security of Ad hoc and Sensor Networks*, 2004.

62. P. Papadimitratos and Z. J. Haas. Secure routing for mobile ad hoc networks. In *Proceedings of the SCS Communication Networks and Distributed System Modeling and Simulation Conference (CNDS 2002)*, 2002.

63. B. Parno, A. Perrig, and V. Gligor. Distributed detection of node replication attacks in sensor networks. In *Proceedings of IEEE Symposium on Security and Privacy*, May 2005.

64. A. Perrig, J. Stankovic, and D. Wagner. Security in wireless sensor networks. *Commun. ACM*, 47(6):53–57, 2004.

65. A. Perrig, R. Szewczyk, J. D. Tygar, V. Wen, and D. E. Culler. Spins: Security protocols for sensor networks. *Wireless Networking*, 8(5):521–534, 2002.

66. R. Di Pietro, L. V. Mancini, Y. W. Law, S. Etalle, and P. Havinga. LKHW: A directed diffusion-based secure multicast scheme for wireless sensor networks. In *First International Workshop on Wireless Security and Privacy (WiSPr'03)*, 2003.

67. N. B. Priyantha, A. Chakraborty, and H. Balakrishnan. The cricket location-support system. In *Proc. of the Sixth Annual ACM International Conference on Mobile Computing and Networking (MOBICOM)*, August 2000.

68. B. Przydatek, D. Song, and A. Perrig. Sia: Secure information aggregation in sensor networks, 2003.

69. S. Rafaeli and D. Hutchison. A survey of key management for secure group communication. *ACM Comput. Surv.*, 35(3):309–329, 2003.

70. K. Ren, T. Li, Z. Wan, F. Bao, R. H. Deng, and K. Kim. Highly reliable trust establishment scheme in ad hoc networks. *Computer Networks: The International Journal of Computer and Telecommunications Networking*, 45:687–699, August 2004.

71. N. Sastry, U. Shankar, and D. Wagner. Secure verification of location claims. In *ACM Workshop on Wireless Security*, September 2003.

72. I. Sato, Y. Okazaki, and S. Goto. An improved intrusion detection method based on process profiling. *IPSJ Journal*, 43(11):3316–3326, 2002.

73. B. Schneier. *Applied Cryptography*. Second Edition, John Wiley & Sons, 1996.

74. A. Seshadri, A. Perrig, L. van Doorn, and P. Khosla. Swatt: Software-based attestation for embedded devices. In *Proceedings of the IEEE Symposium on Security and Privacy*, May 2004.

75. N. Shrivastava, C. Buragohain, D. Agrawal, and S. Suri. Medians and beyond: New aggregation techniques for sensor networks. In *SenSys '04: Proceedings of the 2nd International Conference on Embedded Networked Sensor Systems*, pages 239–249. ACM Press, 2004.

76. A. Smailagic, D. P. Siewiorek, J. Anhalt, Y. Wang, and D. Kogan. Location sensing and privacy in a context aware computing environment. In *Pervasive Computing*, 2001.

77. E. Snekkenes. Concepts for personal location privacy policies. In *Proceedings of the 3rd ACM Conference on Electronic Commerce*, pages 48–57. ACM Press, 2001.

78. J. A. Stankovic et al. Real-time communication and coordination in embedded sensor networks. *Proceedings of the IEEE*, 91(7):1002–1022, July 2003.

79. S. Tanachaiwiwat, P. Dave, R. Bhindwale, and A. Helmy. Poster abstract secure locations: Routing on trust and isolating compromised sensors in location-aware sensor networks. In *Proceedings of the 1st International Conference on Embedded Networked Sensor Systems*, pages 324–325. ACM Press, 2003.

80. S. Tanachaiwiwat, P. Dave, R. Bhindwale, and A. Helmy. Location-centric isolation of misbehavior and trust routing in energy-constrained sensor networks, April 2004.

81. S. Čapkun and J.-P. Hubaux. Secure positioning in wireless networks. *IEEE Journal on Selected Areas in Communications*, 24(2):221–232, 2006.

82. D. Wagner. Resilient aggregation in sensor networks. In *Proceedings of the 2nd ACM workshop on Security of Ad hoc and Sensor Networks (SASN '04)*, pages 78–87, New York, 2004. ACM Press.

83. W. Wang and B. Bhargava. Visualization of wormholes in sensor networks. In *WiSe '04: Proceedings of the 2004 ACM Workshop on Wireless Security*, pages 51–60, New York, 2004. ACM Press.

84. X. Wang, W. Gu, S. Chellappan, K.T Schoseck, and Dong Xuan. Lifetime optimization of sensor networks under physical attacks. In *Proc. of IEEE International Conference on Communications*, May 2005.

85. X. Wang, W. Gu, S. Chellappan, Dong Xuan, and Ten H. Laii. Search-based physical attacks in sensor networks: Modeling and defense. Technical Report, Dept. of Computer Science and Engineering, Ohio State University, February 2005.

86. X. Wang, W. Gu, K. Schosek, S. Chellappan, and D. Xuan. Sensor network configuration under physical attacks. Technical Report (OSU-CISRC-7/ 04-TR45), Dept. of Computer Science and Engineering, Ohio State University, July 2004.

87. R. Watro, D. Kong, S. Cuti, C. Gardiner, C. Lynn, and P. Kruus. Tinypk: securing sensor networks with public key technology. In *Proceedings of the 2nd ACM Workshop on Security of Ad hoc and Sensor Networks (SASN '04)*, pages 59–64, New York, 2004. ACM Press.

88. A. D. Wood and J. A. Stankovic. Denial of service in sensor networks. *Computer*, 35(10):54–62, 2002.

89. Y. Xi, L. Schwiebert, and W. Shi. Preserving privacy in monitoring-based wireless sensor networks. In *Proceedings of the 2nd International Workshop on Security in Systems and Networks (SSN '06)*. IEEE Computer Society, 2006.

90. Z. Yan, P. Zhang, and T. Virtanen. Trust evaluation-based security solution in ad hoc networks. In *NordSec 2003, Proceedings of the Seventh Nordic Workshop on Secure IT Systems*, 2003.

91. F. Ye, H. Luo, S. Lu, and L. Zhang. Statistical en-route detection and filtering of injected false data in sensor networks. In *IEEE INFOCOM 2004*, 2004.

92. H. Zhu, F. Bao, R. H. Deng, and K. Kim. Computing of trust in wireless networks. In *Proceedings of 60th IEEE Vehicular Technology Conference*, Los Angeles, September 2004.

93. S. Zhu, S. Setia, and S. Jajodia. Leap: Efficient security mechanisms for large-scale distributed sensor networks. In *CCS '03: Proceedings of the 10th ACM Conference on Computer and Communications Security*, pages 62–72, New York, NY, USA, 2003. ACM Press.

94. http://www.zigbee.org/, 2005.

Index

A

Access control, 6–7
  cryptographic, 8
  discretionary, 6
  mandatory, 6
  originator controlled, 7
  policies, 15–16
  role-based, 6–7
  rule-based, 7
  secure distributed file system, 7–11
Active Bat, 391
Ad hoc routing protocols, 162
Adapter, 58, 59
Advanced Encryption Standard, 94
Akenti system, 310
ANODR. *See* Anonymous-On-Demand
    Routing (ANODR)
Anomaly detection, 120–121
Anonymity, 96
  conventional concept of, 163
  maintaining, 219
  mechanisms, 391–392
  mobility and, 163–170
  techniques for supporting, 18
  unlinkability and, 163
  unobservability and, 163
  venue, 161, 163–166
Anonymous global trapdoor,
    175, 176
Anonymous virtual circuit, 175
Anonymous-On-Demand Routing
    (ANODR), 175–178
  analysis, 178
  anonymous data forwarding, 177
  anonymous route maintenance, 177
  route discovery, 175–177
    anonymous global trapdoor, 175
    trapdoored boomerang onion,
      175, 176
Asymmetric cryptography, 90, 97, 104, 133,
    138–139, 233, 373
Attacks, 73–79
  collision, 118
  countermeasures, 74, 75–76, 77–78,
    78–79

denial-of-service, 73–74, 116, 120
  dictionary, 100
  exhaustion, 118–119
  HELLO flood, 119
  network authentication-based, 76–77
  on data link layer, 118–119
    collision attack, 118
    exhaustion attack, 118–119
    unfairness attack, 118
  on net layer, 119–120
  opponent, 335–338
    broker masquerading as consumer,
      338
    broker masquerading as owner,
      337–338
    consumer masquerading as broker,
      337
    consumer masquerading as owner,
      337
    owner masquerading as broker, 335
    owner masquerading as consumer,
      335–337
  path-based denial of service, 75–76
  physical node, 78–79
  replay, 100
  selective forwarding, 119
  selfishness-based denial of service, 76
  sinkhole, 119
  Sybil, 119
  unfairness, 118
  wormhole, 119
Authentication, 116, 227–228, 271, 275,
    374–375
  ad hoc networks, 105–108
  basis, 89
  bootstrapping phase, 90
  data aggregation, 142
    algorithm SDFC, 148–149
    communication overhead, 152–154
    computational overhead, 151–152
    modified, 154
    simulation, 151
    system, 146–148
  defined, 87
  EAP framework, 102–103
  grid computing, 220–223, 233